Mathematical Analysis

Mathematical Analysis: Theory and Applications provides an overview of the most up-to-date developments in the field, presenting original contributions and surveys from a spectrum of respected academics. Readers will discover numerous valuable tools and techniques to enhance their understanding of recent advancements in mathematical analysis and its applications. Each chapter highlights new research directions, making this book suitable for graduate students, faculty, and researchers with an active interest in the development of mathematical analysis and its practical implementation. Minimal prerequisites in analysis, topology, and functional analysis are required for readers to fully benefit from the content.

Features

- Showcases the latest advancements in these areas by featuring contributions from distinguished scientists and mathematicians from around the world
- Suitable as a reference for postgraduate students and researchers
- Explores future research directions

Mathematical Analysis

Theory and Applications

Edited by
Pradip Debnath, H. M. Srivastava, Delfim F. M. Torres,
and Yeol Je Cho

CRC Press
Taylor & Francis Group
Boca Raton London New York

CRC Press is an imprint of the
Taylor & Francis Group, an **informa** business
A CHAPMAN & HALL BOOK

Designed cover image: ShutterStock Images

First edition published 2025
by CRC Press
2385 NW Executive Center Drive, Suite 320, Boca Raton FL 33431

and by CRC Press
4 Park Square, Milton Park, Abingdon, Oxon, OX14 4RN

CRC Press is an imprint of Taylor Francis Group, LLC

Library of Congress Cataloging-in-Publication Data
Names: Debnath, Pradip, editor. | Srivastava, H. M., editor. | Torres, Delfim F. M., editor. | Cho, Yeol Je, editor.
Title: Mathematical analysis: theory and applications / edited by Pradip Debnath, H.M. Srivastava, Delfim F.M. Torres, and Yeol Je Cho.
Description: First edition. | Boca Raton : C&H/CRC Press, 2025. | Includes bibliographical references and index.
Identifiers: LCCN 2024029347 (print) | LCCN 2024029348 (ebook) | ISBN 9781032862446 (hbk) | ISBN 9781032870397 (pbk) | ISBN 9781003530602 (ebk)
Subjects: LCSH: Mathematical analysis.
Classification: LCC QA300 .M228 2025 (print) | LCC QA300 (ebook) | DDC 515–dc23/eng20240812
LC record available at https://lccn.loc.gov/2024029347
LC ebook record available at https://lccn.loc.gov/2024029348

ISBN: 9781032862446 (hbk)
ISBN: 9781032870397 (pbk)
ISBN: 9781003530602 (ebk)

DOI: 10.1201/9781003530602

Typeset in Minion
by Deanta Global Publishing Services, Chennai, India

Contents

Preface

Currently, the exploration of mathematical analytic tools and techniques is widely prevalent in various fields such as mathematics, physics, engineering, and statistics. The term "mathematical analysis" is used here in a broad sense, encompassing subjects like real, complex, and functional analysis. This book aims to showcase the latest advancements in these areas by featuring contributions from distinguished scientists and mathematicians from around the world.

The book will provide an overview of the most up-to-date developments in the field, presenting original contributions and surveys. Readers will discover numerous valuable tools and techniques to enhance their skills and knowledge in recent advancements in mathematical analysis and its applications. Each chapter highlights new research directions, making this book suitable for graduate students, faculty, and researchers who wish to expand their understanding of mathematical analysis and its practical implementations. Minimal prerequisites in analysis, topology, and functional analysis are required for readers to fully benefit from the content.

This book consists of 16 chapters. The first few chapters (Chapters 1–4) of the book present some interesting applications of mathematical analysis in terms of recent advances in fixed point theory. Chapter 1 presents an up-to-date survey on developments, challenges, and open problems in fixed point theory, while Chapter 2 deals with several instances of metric completeness. Chapters 3 and 4 discuss fixed point theorems in p-normed spaces and new versions of Kannan-type maps. In Chapter 5, applications of a Kittaneh inequality to operator-valued integrals on Hilbert spaces are explained. Chapter 6 provides some new frozen derivative iterative methods for higher-order equations. In Chapter 7, applications of some classes of Mittag-Leffler functions are given. From Chapter 8 onwards, diverse applications of mathematical analysis are presented. In Chapter 8, a new non-population conserving SIR model on time scales is presented. Chapter 9 focuses on the stability criteria of nonlinear generalized proportional fractional delayed systems. Hamburger-Oberhettinger-Soni modular relations are elaborated in Chapter 10. Some new and more efficient secant-type methods based on generalized Schmidt-Schwetlick conditions are discussed in Chapter 11. A summation of Schlömilch-type series is studied in Chapter 12. Chapter 13 explores cross-diffusion-driven instability and nonlinear analysis in a spatio-temporal oncolytic therapeutic model. Generalizations of the triangle inequality from metric spaces to O-metric spaces are studied in Chapter 14. Chapter 15 presents an interesting study on the stability analysis of a diffusive SVIR epidemic model with distributed delay. Finally, in Chapter 16, Gauss-Newton methods for convex composite optimization under generalized continuity conditions are discussed.

Tezpur, India — Pradip Debnath
Victoria, Canada — Hari Mohan Srivastava
Aveiro, Portugal — Delfim F. M. Torres
Jinju, South Korea — Yeol Je Cho

Editor Biographies

Pradip Debnath is an Associate Professor in the Department of Mathematical Sciences at Tezpur University, India. He featured in the World's Top 2% Scientists list prepared by Stanford University and published by Elsevier in the successive years 2023 and 2024. He is also an Associate Editor of the journal *Heliyon* (mathematics section), published by Elsevier.

He was previously an Assistant Professor (in mathematics) at the Department of Applied Science and Humanities, Assam University, Silchar (a central university), India. Prior to that, he worked as an Assistant Professor in the Department of Mathematics at the North Eastern Regional Institute of Science and Technology (NERIST), India.

He earned his Ph.D. in Mathematics from the National Institute of Technology Silchar, India. His research interests include fixed point theory, functional analysis, soft computing, and mathematical statistics. He has published more than 60 papers in various journals of international repute and is an active reviewer for more than 65 international journals. He is also a reviewer for *Mathematical Reviews* published by the American Mathematical Society. He is the Lead Editor of the books *Metric Fixed Point Theory – Applications in Science, Engineering and Behavioural Sciences* (2021, Springer Nature), *Soft Computing Techniques in Engineering, Health, Mathematical and Social Sciences* (2021, CRC Press), *Fixed Point Theory and Fractional Calculus: Recent Advances and Applications* (2022, Springer Nature), *Soft Computing: Recent Advances and Applications in Engineering and Mathematical Sciences* (2023, CRC Press), *Advanced Mathematical Analysis and Its Applications* (2023, CRC Press), *Advances in Number Theory and Applied Analysis* (2023, World Scientific), and several others.

He is a topical advisory panel member of the journals *Axioms* and *Fractal and Fractional* and a guest editor of several special issues for different journals. He has successfully guided Ph.D. students in the areas of nonlinear analysis, soft computing, and fixed point theory. He has recently completed a major Basic Science Research Project in fixed point theory funded by the UGC, the Government of India. Having been an academic gold medalist during his post-graduation studies at Assam University, Silchar, Dr. Debnath has qualified for several national-level examinations in mathematics in India.

Hari M. Srivastava is Professor Emeritus at the Department of Mathematics and Statistics, University of Victoria, Canada. He has been a Clarivate Analytics (Thomson-Reuters) Highly Cited Researcher for the years 2015, 2017, 2018, and 2020. He has also been listed and ranked in sixth place in General Mathematics among the **Top 2% Scientists in the World** 2021 (published by Stanford University in the US). In fact, it is not possible to list all of his tremendous academic achievements in just a few sentences. The details of the same may be found on his regularly updated website www.math.uvic.ca/harimsri/.

Delfim Fernando Marado Torres earned a Ph.D. in Mathematics from the University of Aveiro (UA), Portugal, in 2002, and habilitation in Mathematics, UA, in 2011. He has been a full professor of mathematics since 9 March 2015. He has been the Director of the R&D Unit CIDMA, the largest Portuguese research center for mathematics, and Coordinator of its Systems and Control Group. His main research areas are calculus of variations and optimal control; optimization; fractional derivatives and integrals; dynamic equations on time scales; and mathematical biology. Torres has written outstanding scientific and pedagogical publications. In particular, he is the author of two books published by Imperial College Press, three books published by Springer, and is the editor of several other books. Professor Torres has been recognized four times as one of the **top 1%** of mathematicians on the prestigious global Clarivate Web of Science list and is the only Portuguese mathematician to be so honored. He has strong experience in graduate and post-graduate student supervision and teaching mathematics. He has supervised 27 Ph.D. students in mathematics. Moreover, he has been the leading member in several national and international R&D projects, including EU projects and networks. Since 2013, Professor Torres has been the Director of the Doctoral Programme Consortium in Mathematics and Applications (MAP-PDMA) of the Universities of Minho, Aveiro, and Porto, Portugal.

Yeol Je Cho is Emeritus Professor at the Department of Mathematics Education, Gyeongsang National University, Jinju, Korea, and Distinguished Professor at the School of Mathematical Sciences, the University of Electronic Science and Technology of China, Chengdu, Sichuan, China. In 1984, he earned his Ph.D. in Mathematics from Pusan National University, Pusan, Korea. He has been a fellow of the Korean Academy of Science and Technology, Seoul, Korea, since 2006, and a member of several mathematical societies. He has organized international conferences on nonlinear functional analysis and applications, fixed point theory and applications, and workshops and symposiums on nonlinear analysis and applications. He has published over 400 papers, 20 monographs, and 12 books with renowned publishers from around the world. His research areas are nonlinear analysis and applications, especially fixed point theory and applications, and some kinds of nonlinear problems, that is, equilibrium problems, variational inequality problems, saddle point problems, optimization problems, inequality theory and applications, and stability of functional equations and applications. He has delivered several invited talks at international conferences on nonlinear analysis and applications and is on the editorial boards of ten international journals of mathematics.

CHAPTER 1

Developments, Applications and Open Problems in Fixed Point Theory

Pradip Debnath

1.1 INTRODUCTION

It is widely accepted that metric fixed point theory (FPT) originated in the year 1922 with the work of Stefan Banach when he established the famous contraction mapping principle, which has come to be known by his name. This principle has found applications in versatile domains of science, technology and economics. Even after a century of its initiation, this subject area remains vibrant with research activities.

In this chapter, we'll discuss some major developments in FPT, their applications, some open problems, challenges and future research directions.

Much of the contents in this section may be found in [4] and the references therein.

Definition 1.1 (*Fixed Point*) Let X be a nonempty set and $T : X \to X$ be a mapping. A fixed point of T is a point $x \in X$ such that $Tx = x$, i.e., a fixed point of T is a solution of the functional equation $Tx = x, x \in X$.

DOI: 10.1201/9781003530602-1

Definition 1.2 (*Contraction Mapping*) Let (X, d) be a metric space. A mapping $T : X \to X$ is called a Lipschitz mapping if there exists a real number $k > 0$ such that $d(Tx, Ty) \leq kd(x, y)$ for all $x, y \in X$.

The smallest positive real number k for which the Lipschitz condition is valid is called the Lipschitz constant of T.

If $k \in (0, 1)$, then the Lipschitz mapping T is called a contraction mapping. Clearly, a contraction mapping is continuous.

If $k = 1$, then T is said to be a nonexpansive mapping.

Theorem 1.3 (*Banach*) *Let X be a complete metric space and the self-map $T{:}X \to X$ be a contraction. Then T admits a unique fixed point in X.*

Remark 1.4 The completeness of X is crucial. Indeed, contractions on incomplete metric spaces need not have fixed points.

e.g., let $X = (0, 1]$ with the usual metric. Define $T : X \to X$ by $Tx = \frac{x}{2}$.

The next corollary provides existence and uniqueness of a fixed point under more general conditions.

Corollary 1.5 *Let X be a complete metric space and let $T{:}X \to X$ be a self-map. If T^n is a contraction, for some $n \geq 1$, then T has a unique fixed point in X.*

Example 1.6 Consider $X = [0, 1]$ with the usual metric and define the map $T : X \to X$ by

$$Tx = \begin{cases} \frac{1}{2} + 2x, & \text{if } x \in \left[0, \frac{1}{4}\right] \\ \frac{1}{2}, & \text{if } x \in \left(\frac{1}{4}, 1\right]. \end{cases}$$

It's easy to observe that T is not continuous and hence not a contraction. However, T possesses a unique fixed point $x = \frac{1}{2}$.

Note that $T^2x \equiv \frac{1}{2}$.

1.2 EXTENSIONS OF THE CONTRACTION PRINCIPLE

In 1969, Boyd and Wong made an interesting generalization of the BCP. They replaced the constant k by a function $\phi : [0, \infty) \to [0, \infty)$ which is upper semicontinuous from the right (that is , $t_n \to t \geq 0 \Longrightarrow \limsup \phi(t_n) \leq \phi(t)$).

Theorem 2.1 (*Boyd and Wong*) *A self-mapping T of a complete metric space (X, d) admits a unique fixed point if there exists a function $\phi\colon [0,\infty) \to [0,\infty)$ which is upper semicontinuous from the right with $0 \leq \phi(t) < t$ for $t > 0$, and the following inequality holds:*

$$d(Tx, Ty) \leq \phi(d(x,y))$$

holds for all $x, y \in X$.

Moreover, for any $x_0 \in X$, the sequence $T^n x_0$ of iterates converges to the fixed point.

In 1969, Meir and Keeler established that the conclusion of BCP holds more generally from the condition of weakly uniformly strict contraction.

Theorem 2.2 (*Meir and Keeler*) *A self-mapping T of a complete metric space (X, d) admits a unique fixed point if, given $\epsilon > 0$, there exists $\delta > 0$ such that for any $x, y \in X$,*

$$\epsilon \leq d(x,y) < \epsilon + \delta \Longrightarrow d(Tx, Ty) < \epsilon.$$

Moreover, for any $x_0 \in X$, the sequence $T^n x_0$ of iterates converges to the fixed point.

It is observed that in Banach's contraction mapping principle, the contraction condition is global; that is, the operators satisfy the contraction condition for every pair of points taken from the metric space. A natural question arises whether the conclusion of Banach's theorem is true if the contraction condition is satisfied locally, that is, for sufficiently close points only. The answer was given in the affirmative in a paper by Michael Edelstein in 1961.

Definition 2.3 (*Local Contraction*) A self-mapping $T : X \to X$, where (X, d) is a metric space, is locally contractive if, for every $x \in X$, there exist $\epsilon > 0$ and $\lambda \in [0, 1)$, which may depend on x, such that

$$x, y \in T(x, \epsilon) = \{y : d(x,y) < \epsilon\} \Longrightarrow d(Tx, Ty) < \lambda d(x,y).$$

Definition 2.4 (*Uniform Local Contraction*) A uniformly locally contractive mapping on a metric space (X, d) is a locally contractive mapping $T : X \to X$ where both ϵ and λ do not depend on x.

Definition 2.5 (ϵ-*Chain*) Let (X, d) be a metric space such that for every $a, b \in X$ and $\epsilon > 0$ there exists an ϵ-chain, that is, a finite set of points $a = x_0, x_1, ..., x_n = b$ (n may depend on both a and b) in X satisfying $d(x_{j-1}, x_j) < \epsilon$ $(j = 1, 2, ..., n)$. Then (X, d) is ϵ-chainable.

Theorem 2.6 (*Edelstein*) *An* (ϵ, λ)*-uniformly locally contractive mapping* $T: X \to X$ *on an* ϵ*-chainable complete metric space* (X, d) *has a unique fixed point.*

In 1973, Geraghty introduced a class of functions to generalize the Banach's contraction principle. Let S be the class of all functions $\beta : [0, \infty) \to [0, \infty)$ satisfying the property $\beta(t_n) \to 1$ as $t_n \to 0$.

One such example of a function in S is $\beta(t) = e^{-2t}$ for $t > 0$ and $\beta(0) \in [0, 1)$.

Theorem 2.7 (*Geraghty*) *A self-mapping* T *of a complete metric space* (X, d) *admits a unique fixed point if there exists a function* $\beta \in S$ *such that*

$$d(Tx, Ty) \leq \beta(d(x, y))\, d(x, y)$$

for all $x, y \in X$.

Pata-type contractions were introduced in a paper by Pata in 2011 in which a fixed point theorem for such contractions was proved by using a new approach. Pata's result appeared to be stronger than Banach's contraction mapping principle, and even stronger than the well-known Boyd-Wong fixed point theorem.

The following class of functions were used. Let Ψ denote the family of all functions $\psi : [0, 1] \to [0, \infty)$ such that ψ is increasing and continuous at zero with $\psi(0) = 0$.

Theorem 2.8 (*Pata*) *Let* $\Lambda > 0, \alpha \geq 1$ *and* $\beta \in [0, \alpha]$ *be constants and* $\psi \in \Psi$. *Let* (X, d) *be a complete metric space and* $T: X \to X$ *be such that for every* $\epsilon \in [0, 1]$ *and all* $x, y \in X$, *the following holds:*

$$d(Tx, Ty) \leq (1 - \epsilon)\, d(x, y) + \Lambda \epsilon^{\alpha} \psi(\epsilon)\left[1 + ||x|| + ||y||\right]^{\beta},$$

where $||x|| = d(x, u), ||y|| = d(y, u)$ *for an arbitrary but fixed* $u \in M$. *Then* T *has a unique fixed point in* X.

1.3 FIXED POINT WITHOUT CONTINUITY

In 1976, Caristi proved an elegant fixed point theorem on complete metric spaces, which is a generalization of Banach's contraction mapping principle and is equivalent to the Ekeland variational principle.

Definition 3.1 (*Lower Semicontinuity*) A function $\phi : X \to \mathbb{R}$ is said to be lower semicontinuous at $x \in X$ if for any sequence $\{x_n\} \subset X$, we have

$$x_n \to x \Longrightarrow \phi(x) \leq \liminf_{n \to \infty} \phi(x_n).$$

Definition 3.2 (*Caristi Mapping*) Let (X, d) be a metric space. A mapping $T : X \to X$ is called a Caristi mapping if there exists a lower semicontinuous function $\phi : X \to \mathbb{R}^+$ such that

$$d(x, Tx) \leq \phi(x) - \phi(Tx)$$

for all $x \in X$.

Theorem 3.3 (Caristi) *Let (X, d) be a complete metric space. A mapping $T:X \to X$ admits a fixed point in X if there exists a lower semicontinuous function $\phi:X \to \mathbb{R}^+$ such that*

$$d(x, Tx) \leq \phi(x) - \phi(Tx)$$

for all $x \in X$.

Example 3.4 Consider $X = [0, 1]$ with the usual metric, and define the map $T : X \to X$ by

$$Tx = \begin{cases} \frac{x}{2}, & \text{if } x \neq 1, \\ 1, & \text{if } x = 1. \end{cases}$$

It's easy to observe that T satisfies the conditions of Caristi's theorem and T has fixed points 0 and 1.

It is easy to see that Caristi's fixed point theorem is a generalization of the Banach contraction mapping principle by defining $\phi(x) = \frac{1}{1-k} d(x, Tx)$, where $k \in (0, 1)$ is the Lipschitz constant associated with the contraction T from Banach's principle.

It has been shown by Kirk that the validity of Caristi's fixed point theorem implies that the corresponding metric space is complete while the Banach's contraction mapping principle does not characterize completeness. The last example shows that Caristi's contraction can also be discontinuous.

The next result deals with a contraction condition that is of a different category and does not generalize Banach's contraction. The contraction condition is also satisfied by discontinuous functions. The result is due to Kannan and was established in 1968.

Definition 3.5 (*Kannan Mapping*) Let (X, d) be a metric space. A mapping $T : X \to X$ is called a Kannan-type mapping if there exists $k \in \left(0, \frac{1}{2}\right)$ such that

$$d(Tx, Ty) \leq k[d(x, Tx) + d(y, Ty)]$$

for all $x, y \in X$.

Theorem 3.6 (Kannan) *Let (X, d) be a complete metric space and $T : X \to X$ be a Kannan-type mapping. Then T admits a unique fixed point.*

Example 3.7 Consider $X = [0, 1]$ with the usual metric and define the map $T : X \to X$ by

$$Tx = \begin{cases} \frac{x}{3}, & \text{if } 0 \leq x < 1, \\ \frac{1}{6}, & \text{if } x = 1. \end{cases}$$

It's easy to observe that T satisfies the conditions of Kannan's theorem and T has unique fixed point 0. Further, T is not continuous.

There is another reason for which the Kannan-type mappings are considered to be important. The Banach's contraction mapping principle does not characterize completeness. In fact, there are examples of noncomplete spaces where every contraction has a fixed point. It has been shown that the necessary existence of fixed points for Kannan-type mappings implies that the corresponding metric space is complete.

A fixed point theorem due to Chatterjea and established in 1972, which is actually a sort of dual of the Kannan fixed point theorem, is based on a condition similar to that of Kannan.

Definition 3.8 (*Chatterjea Mapping*) Let (X, d) be a metric space. A mapping $T : X \to X$ is called a Chatterjea-type contraction if there exists $k \in \left(0, \frac{1}{2}\right)$ such that

$$d(Tx, Ty) \leq k[d(x, Ty) + d(y, Tx)]$$

for all $x, y \in X$.

Theorem 3.9 (*Chatterjea*) *Let (X, d) be a complete metric space and $T: X \to X$ be a Chatterjea-type contraction. Then T admits a unique fixed point.*

We conclude this section with another extension of BCP which is true for discontinuous mappings as well.

Theorem 3.10 (*Ćirić*) *Let (X, d) be a complete metric space and $T: X \to X$ be a mapping such that*

$$d(Tx, Ty) \leq \lambda \max\{d(x_1, x_2), d(x_1, Tx_1), d(x_2, Tx_2), d(x_1, Tx_2), d(x_2, Tx_1)\}$$

for some $0 \leq \lambda < 1$ and every $x_1, x_2 \in X$. Then T admits a unique fixed point.

Example 3.11 Consider $X = [0, 1]$ with the usual metric and define the map $T : X \to X$ by

$$Tx = \begin{cases} \frac{1}{2} + 2x, & \text{if } x \in \left[0, \frac{1}{4}\right] \\ \frac{1}{2}, & \text{if } x \in \left(\frac{1}{4}, 1\right]. \end{cases}$$

It's easy to observe that T is not continuous. However, T possesses a unique fixed point $x = \frac{1}{2}$, and it is continuous at the fixed point. T satisfies the conditions of Ćirić's theorem with $\lambda = \frac{2}{3}$.

Next, we present a converse to the contraction principle.

Assume we are given a set X and a map $T : X \to X$. We are interested to find a metric d on X such that (X, d) is a complete metric space and T is a contraction on X. The following gives us a sufficient condition towards the converse of the Banach contraction principle.

Theorem 3.12 (*Bessaga*) *[2] Let X be an arbitrary set, and let $T: X \to X$ be a map such that T^n has a unique fixed point for every $n \geq 1$. Then for every $\epsilon \in (0, 1)$, there is a metric d_ϵ on X that makes X a complete metric space, and T is a contraction on X with Lipschitz constant equal to ϵ.*

1.4 SOME APPLICATIONS OF FIXED POINT THEOREMS

1.4.1 Applications in Linear Equations

Much of the contents in this section may be found in [12] and the references therein.

Fixed point theorems have applications in many branches of analysis and topology and even in many other branches of science. In this section, we discuss a few such interesting applications. We start with its application in solving linear equations.

Suppose we have a system of n linear equations in n unknowns of the form

$$g_j(x_1, x_2, ..., x_n) = 0, \ j = 1, ..., n$$

where g_j are continuous real-valued functions of the real variables x_j. Let

$$h_j(x_1, x_2, ..., x_n) = g_j(x_1, x_2, ..., x_n) + x_j$$

and for any point $x = (x_1, x_2, ..., x_n)$, define

$$h(x) = (h_1(x), h_2(x), ..., h_n(x)).$$

Assume now that h has a fixed point $\overline{x} \in \mathbb{R}^n$. Then, it's easy to see that $\overline{x}$ is a solution of the system of equations $g_j(x_1, x_2, ..., x_n) = 0$.

Consider $X = \mathbb{R}^n$ with the metric d defined by $d(x, y) = \max_j |\xi_j - \eta_j|$, where $x = (\xi_1, ..., \xi_n), y = (\eta_1, ..., \eta_n)$. Then (X, d) is a complete metric space. Define $T : X \to X$ by $Tx = Cx + b$, where $C = (c_{jk})$ is a fixed real $n \times n$ matrix and $b \in X$ is a fixed column vector.

Theorem 4.1 *If a system $x = Cx + b$ of n linear equations in n unknowns $\xi_1, ..., \xi_n$ (the components of x) satisfies*

$$\sum_{k=1}^{n} |c_{jk}| < 1, \ j = 1, 2, ..., n,$$

then it has a unique solution.

This solution can be obtained as the limit of the iterative sequence $(x^{(0)}, x^{(1)}, ...)$, where $x^{(0)}$ is arbitrary and $x^{(m+1)} = Cx^{(m)} + b$.

1.4.2 Applications in Differential Equations

The most interesting applications of fixed point results arise in connection with function spaces. The BCP often yields existence and uniqueness theorems for differential and integral equations.

Let's consider an explicit ordinary differential equation of the first order $x' = f(t, x)$, $' \equiv d/dt$ with the initial condition $x(t_0) = x_0$, where x_0, t_0 are given real numbers.

Theorem 4.2 (*Picard's existence and uniqueness theorem*) *Let f be continuous on a rectangle*

$$R = \{(t, x) : |t - t_0| \leq a, |x - x_0| \leq b\}$$

and thus bounded on R, say $|f(t, x)| \leq c$ *for all* $(t, x) \in R$.

Suppose that f satisfies a Lipschitz condition on R with respect to its second argument, that is, there is a constant k (Lipschitz constant) such that for $(t, x), (t, v) \in R$, *we have* $|f(t, x) - f(t, v)| \leq k\,|x - v|$.

Then the initial value problem as considered above has a unique solution. This solution exists on an interval $[t_0 - \beta, t_0 + \beta]$ *where* $\beta < \min\left\{a, \frac{b}{c}, \frac{1}{k}\right\}$.

1.4.3 Applications in Integral Equations

Now we consider the BCP as a source of existence and uniqueness theorems for integral equations. An integral equation of the form

$$x(t) - \mu \int_a^b k(t, \tau)\, x(\tau)\, d\tau = v(t)$$

is called a Fredholm equation of the second kind. Here, $[a, b]$ is a given interval, x is an unknown function on $[a, b]$, μ is a parameter, the kernel k of the equation is a given function on the square $G = [a, b] \times [a, b]$ and v is a given function on $[a, b]$.

Here we consider the equation on $C[a, b]$, the space of all continuous functions defined on $J = [a, b]$ with metric d given by $d(x, y) = \max_{t \in J} |x(t) - y(t)|$, which makes it complete. Assume that $v \in C[a, b]$ and k is continuous on G. Then k is bounded on G, say, $|k(t, \tau)| \leq c$, for all $(t, \tau) \in G$.

Obviously, the integral equation may be written as $x = Tx$, where

$$Tx(t) = v(t) + \mu \int_a^b k(t, \tau)\, x(\tau)\, d\tau.$$

Theorem 4.3 (*Fredholm integral equation*) *Suppose that k and v in the above integral equation are continuous on $J \times J$ and $J = [a, b]$, respectively, and assume that $|\mu| < \frac{1}{c(b-a)}$. Then the integral equation has a unique solution x on J. This function x is the limit of the iterative sequence $(x_0, x_1, \ldots)$, where x_0 is any continuous function on J and for $n = 0, 1, \ldots,$*

$$x_{n+1}(t) = v(t) + \mu \int_a^b k(t, \tau)\, x_n(\tau)\, d\tau.$$

1.5 OPEN PROBLEMS IN METRIC FIXED POINT THEORY

First we discuss the relevant preliminaries for this section.

Let K be a nonempty closed convex subset of a normed linear space X. A mapping $T : K \to K$ is said to be nonexpansive if $\|Tx - Ty\| \le \|x - y\|$ for all $x, y \in K$.

We define K as possessing the fixed point property (FPP) if each nonexpansive map defined on K has a fixed point within K. Moreover, we define X as having the FPP if every closed bounded convex subset of X possesses the FPP, and X has the weak FPP if each weakly compact convex subset of X has the FPP.

It's a recognized fact that typically, a closed bounded convex subset of a Banach space X may not exhibit the FPP.

Example 5.1 [7, 9] Let $X = l^1$ and let $K = \{x = \{x_n\} : x_n \ge 0, n = 1, 2, \ldots, \|x\|_1 = 1\}$. Define $T : K \to K$ as $T(x_1, x_2, \ldots) = (0, x_1, x_2, \ldots)$. Then T has no fixed point in K.

In 1948, Brodskii and Milman [3] introduced the notion of normal structure to examine the presence of shared fixed points among isometries.

Definition 5.2 [9] A convex subset C of a Banach space X possesses normal structure if, for each closed bounded convex subset K of C with a diameter

$diam\,(K) > 0$, there exists a point $x \in K$ such that $r_x\,(K) = sup\,|x - y| :$ $y \in K < diam\,(K)$.

If every convex subset of a Banach space X exhibits normal structure, then we term X as having normal structure.

In 1965, Kirk proved the following:

Theorem 5.3 [10] *Every nonempty weakly compact convex subset K of a Banach space X endowed with normal structure possesses the FPP.*

It's established that every uniformly convex Banach space exhibits normal structure [9].

Moreover, every compact convex subset of a Banach space is endowed with normal structure [9], thus implying that every finite-dimensional Banach space also possesses normal structure.

Definition 5.4 A normed linear space X is termed uniformly convex if, for every $\epsilon > 0$, there exists $\delta > 0$ such that for all $x, y \in X$ with $||x|| \leq 1$ and $||x - y|| \geq \epsilon$, it follows that $||\frac{x+y}{2}|| \leq 1 - \delta$.

Every Hilbert space possesses uniform convexity, and $l^p, 1 < p < \infty$ also exhibits uniform convexity.

Now we discuss the open problems:

It's established that c_0, equipped with the supremum norm (thus c and l^∞ as well), lacks the fixed point property (FPP). The function $T(x) = (1, x_1, x_2, ...)$ serves as a nonexpansive self-mapping devoid of fixed points within the closed unit ball of c_0 [9].

In 1981, Maurey [13] demonstrated that both c_0 and c possess the weak fixed point property (WFPP). Furthermore, c_0 is devoid of normal structure.

In 2004, Dowling et al. [6] established that a non-empty closed convex subset of c_0 possesses the FPP if and only if it is weakly compact.

For quite some time, a lingering question persisted: Does every Banach space have the WFPP?

In 1981, Alspach [1] provided the initial and essentially sole known instance of a space lacking the WFPP.

Open problems in reflexive spaces:

We consider a normed linear space X along with its dual space X'. Additionally, X'' denotes the second dual (or bidual) space of X, which is the dual space of X'.

We define a functional g_x on X' by choosing a fixed $x \in X$ and setting

$$g_x(f) = f(x), (f \in X' \text{variable}).$$

For each $x \in X$, there exists a unique bounded linear functional $g_x \in X''$. This establishes a mapping $C : X \to X''$ defined by $C(x) = g_x$, termed as the canonical mapping of X into X''. The range of C is denoted by $R(C)$. The canonical mapping represents an isomorphism from the normed space X to the normed space $R(C)$.

Definition 5.5 A normed linear space X is termed reflexive if its range $R(C)$, where $C : X \to X''$ is the canonical mapping, equals the second dual space X''.

A reflexive normed linear space constitutes a Banach space.

In 1980, Maurey [13] established that each reflexive subspace of $L^1[0,1]$ possesses the FPP.

A fundamental unresolved query in this realm is: Do all reflexive spaces exhibit the FPP?

In 1997, Dowling and Lennard [5] demonstrated that every non-reflexive subspace of $L^1[0,1]$ lacks the FPP.

This, coupled with Maurey's 1980 finding, indicates that a subspace of $L^1[0,1]$ features the FPP if and only if it is reflexive.

More recently, in 2009, Benavides proved that every reflexive Banach space can be equivalently redefined such that it possesses the FPP.

A normed linear space X is considered finitely representable in another normed linear space Y if there exists a number $\lambda > 1$ such that for each finite-dimensional subspace X_n of X, there exists an isomorphism T_n from X_n to Y satisfying

$$\lambda^{-1}\|x\| \le \|T_n(x)\| \le \lambda\|x\| \text{ if } x \in X_n.$$

A super-reflexive Banach space is a Banach space X such that no non-reflexive Banach space can be finitely representable in X.

An intriguing unresolved question in this context is: Does every super-reflexive Banach space possess the FPP?

The following result is known.

Theorem 5.6 [8] *A Banach space X is super-reflexive if and only if it has an equivalent uniform convex renorming.*

Maurey [13] provided a affirmative response regarding isometries, yet the inquiry persists for the nonexpansive case.

Another unresolved matter is: Does every equivalent renorming of l^p, where $1 < p < \infty$, exhibit the FPP? This question remains unanswered even for l^2.

Two more open problems suggested by Kirk:

Kirk [11] says "There are many open problems in metric fixed point theory, but two have frustrated me over the years because of their simplicity and yet seeming intractability".

Problem 1: Does a weakly compact convex subset K of a Banach space have the fixed point property for strictly contractive mappings (that is, mappings $T : K \to K$ for which $\|Tx - Ty\| < \|x - y\|, x, y \in K, x \neq y$)?

Another problem concerns approximate fixed point sequences. It's well-established that if K is a bounded convex subset of any Banach space (or normed space), and $T : K \to K$ is nonexpansive, then there exists a sequence $\{x_n\}$ in K such that $\|x_n - T(x_n)\|$ tends to 0 as n approaches infinity. This can be deduced by simply uniformly approximating T with a sequence of contraction mappings whose Lipschitz constants converge to 1. Consequently, for each $\epsilon > 0$,

$$F_\epsilon(T) := \{x \in K : \|x - Tx\| \leq \epsilon\} \neq \varphi.$$

Problem 2: If K is a bounded (closed) convex subset of a Banach space and if $T, G : K \to K$ are commuting nonexpansive mappings, is it the case that for each $\epsilon > 0$,

$$F_\epsilon(T) \cap F_\epsilon(G) \neq \varphi?$$

REFERENCES

1. Alspach, D. E.: A fixed point free nonexpansive map. *Proc. Amer. Math. Soc.* (1981) 82(3): 423–424.
2. Bessaga, C.: On the converse of the Banach fixed-point principle. *Colloq. Math.* (1969) 7: 41–43.
3. Brodshii, M. S. and Milman, D. P.: On the center of a convex set (Russian). *Doklady Akademii Nauk SSSR (N.S.)* (1948) 59: 837–840.

4. Debnath, P., Konwar, N. and Radenović, S.: *Metric Fixed Point Theory - Applications in Science, Engineering and Behavioural Science.* Springer Singapore, 2021.
5. Dowling, P. N. and Lennard, C. J.: Every nonreflexive subspace of $L^1[0,1]$ fails the fpp. *Proc. Amer. Math. Soc.* (1997) 125(2): 443–446.
6. Dowling, P. N., Lennard, C. J. and Turett, B.: Weak compactness is equivalent to the fixed point property in c_0. *Proc. Amer. Math. Soc.* (2004) 136(6): 1659–1666.
7. Dutta, G. and Veeramani, P.: A short survey on open problems in metric fixed point theory and some related results for nonexpansive mappings. *J. Anal.* (2019): DOI: 10.1007/s41478-019-00200-5
8. Enflo, P.: Banach spaces which can be given an equivalent uniformly convex norm. *Israel J. Math.* (1972) 13: 281–288.
9. Goebel, K. and Kirk, W. A.: *Topics in Metric Fixed Point Theory.* Cambridge University Press, 1990.
10. Kirk, W. A.: A fixed point theorem for mappings which do not increase distances. *Amer. Math. Monthly.* (1965) 72: 1004–1006.
11. Kirk, W. A.: Metric fixed point theory: A brief retrospective. *Fixed Point Theory Appl.* (2015) 2015: 215.
12. Kreyszig, E.: *Introductory Functional Analysis with Applications.* Wiley Singapore, 2007.
13. Maurey, M.: Points fixes des contractions sur un convexe ferme de L^1, *Seminaire d analyse fonctionelle.* Ecole Polytechnique, Palaiseau, Expose No. VIII, 1980/1981.

CHAPTER 2

Several Recent Episodes on the Metric Completeness

Sehie Park

2.1 INTRODUCTION

Since completeness is one of the most important properties of various types of extensions of metric spaces, a huge number of works have appeared related to completeness. One of them, due to Kirk [11] in 1976, states that metric completeness is equivalent to the Caristi fixed point theorem. Moreover, early equivalents of such completeness were collected by ourselves [14] in 1984. Recently, in 2020, Cobzaş [4] presented a large number of various circumstances in which fixed point results imply completeness for metric spaces and their generalizations.

Let (X, d) be a metric space. A Banach contraction $T : X \to X$ is a map satisfying

$$d(Tx, Ty) \leq \alpha\, d(x, y) \quad \text{for all } x, y \in X$$

with some $\alpha \in [0, 1)$. Thousands of articles have appeared relating to the Banach contraction.

Recently, we introduced the Rus-Hicks-Rhoades (RHR) map $T : X \to X$ satisfying

$$d\left(Tx, T^2x\right) \leq \alpha\, d(x, Tx) \quad \text{for all } x \in X$$

with some $\alpha \in [0, 1)$. See our recent works [16–22].

DOI: 10.1201/9781003530602-2

The RHR maps are called graphic contraction, iterative contraction, weak contraction, or Banach mapping; see Berinde et al. [2, 3]. Moreover, it has been recently discovered that well-known metric fixed point theorems related to the RHR maps hold for quasi-metric spaces (without assuming the symmetry); see [19–22].

Recall that Suzuki [24] in 2008 gave a very interesting result on particular RHR maps. It is a weaker version of the Banach contraction principle and also characterizes the completeness of underlying metric spaces. It was commonly regarded that Suzuki's result gave a new direction to the subject, and as a result, researchers, including his colleagues, made many contributions in metric fixed point theory. Most of these results are variations and refinements of Suzuki's original result with lengthy complicated proofs. However, we found in [21, 22] that most such Suzuki-type results are incorrectly stated according to our recently developed ordered fixed point theory [15].

Our aim in this chapter is to collect new characterizations of the completeness of quasi-metric spaces and to review such results of Suzuki and his colleagues. We give certain critical comments on them.

This chapter is organized as follows: Section 2.2 is based on preliminaries on quasi-metric spaces. In Section 2.3, we recall the historical background of metric completeness. The section begins by introducing earlier works on metric completeness and ends with current situations. Section 2.4 deals with basic theorems on the Rus-Hicks-Rhoades (RHR) maps and extensions of the Banach contraction principle for a quasi-metric space (X, δ) with a self-map $T : X \to X$ such that X is T-orbitally complete. In Section 2.5, we introduce new characterizations of quasi-metric completeness. Section 2.6 reviews the works of Suzuki and his colleagues on metric completeness. Section 2.7 deals with the inaccurate responses of other authors on works of Suzuki et al. Finally, in Section 2.8, we provide some conclusions and an epilogue.

There are thousands of generalizations of metric spaces and their corresponding completeness concepts; for example, see Cobzaş [4]. In the present chapter, we restrict ourselves to quasi-metric spaces or metric spaces.

2.2 PRELIMINARIES

We recall the following:

Definition 2.1. A *quasi-metric* on a nonempty set X is a function δ : $X \times X \to [0, \infty)$ satisfying the following conditions for all $x, y, z \in X$:

(a) (self-distance) $\delta(x, y) = \delta(y, x) = 0 \Longleftrightarrow x = y$;

(b) (triangle inequality) $\delta(x, z) \leq \delta(x, y) + \delta(y, z)$.

A *metric* on a set X is a quasi-metric satisfying

(c) (symmetry) $\delta(x, y) = \delta(y, x)$ for all $x, y \in X$.

For quasi-metric spaces, the convergence of a sequence, Cauchy sequences, completeness, orbits, and orbital continuity are routinely defined in [1, 8]:

Definition 2.2. (1) A sequence (x_n) in X *converges* to $x \in X$ if

$$\lim_{n\to\infty} \delta(x_n, x) = \lim_{n\to\infty} \delta(x, x_n) = 0.$$

(2) A sequence (x_n) is *left-Cauchy* if for every $\epsilon > 0$, there is a positive integer $N = N(\epsilon)$ such that $\delta(x_n, x_m) < \epsilon$ for all $n > m > N$.

(3) A sequence (x_n) is *right-Cauchy* if for every $\epsilon > 0$, there is a positive integer $N = N(\epsilon)$ such that $\delta(x_n, x_m) < \epsilon$ for all $m > n > N$.

(4) A sequence (x_n) is *Cauchy* if for every $\epsilon > 0$ there is positive integer $N = N(\epsilon)$ such that $\delta(x_n, x_m) < \epsilon$ for all $m, n > N$; that is (x_n) is a *Cauchy sequence* if it is left and right Cauchy.

Definition 2.3. ([1, 8])

(1) (X, δ) is *left-complete* if every left-Cauchy sequence in X is convergent;

(2) (X, δ) is *right-complete* if every right-Cauchy sequence in X is convergent;

(3) (X, δ) is *complete* if every Cauchy sequence in X is convergent.

Definition 2.4. Let (X, δ) be a quasi-metric space and $T : X \to X$ a selfmap. The *orbit* of T at $x \in X$ is the set

$$O_T(x) = \{x, T(x), \cdots, T^n(x), \cdots\}.$$

The space X is said to be *T-orbitally complete* if every right-Cauchy sequence in $O_T(x)$ is convergent in X. A selfmap T of X is said to be *orbitally continuous* at $x_0 \in X$ if

$$\lim_{n\to\infty} T^n(x) = x_0 \Longrightarrow \lim_{n\to\infty} T^{n+1}(x) = T(x_0)$$

for any $x \in X$.

Note that every complete metric space is T-orbitally complete for all maps $T : X \to X$. There exists a T-orbitally complete metric space but it is not complete. Moreover, there exists an orbitally continuous map but it is not continuous.

Every quasi-metric induces a metric, that is, if (X, δ) is a quasi-metric space, then the function $d : X \times X \to [0, \infty)$ defined by

$$d(x, y) = \max\{\delta(x, y), \delta(y, x)\}$$

is a metric on X; see Jleli et al. [8].

The following was given in [19]:

Theorem 2.5. *A selfmap $T: X \to X$ of a quasi-metric space (X, δ) has a fixed point $z \in X$ if and only if z is a fixed point of the selfmap T of the induced metric space (X, d).*

Proof. If $z = T(z)$ in (X, δ), then

$$d(z, T(z)) = \max\{\delta(z, T(z)), \delta(T(z), z)\} = 0,$$

and hence $d(z, T(z)) = 0$. The converse is true for $d = \delta$. □

From this, many metric fixed point theorems are true for quasi-metric spaces. This is a rather surprising fact in the 100-year history of the metric fixed point theory from Banach in 1922.

2.3 HISTORICAL BACKGROUND OF METRIC COMPLETENESS

In our previous work [14] in 1984, we gave some necessary and sufficient conditions for a metric space(X, d) to be complete. Such characterizations of metric completeness are mainly given by results relevant to Caristi's fixed point theorem. Works of Cantor, Kuratowski, Ekeland, Caristi, Kirk, Wong, Weston, Ćirić, Hu, Reich, Subrahmanyam, and others are combined. For the references, see [14].

Kuratowski (1930) first noticed that the Cantor intersection theorem characterizes the metric completeness. Hu (1967) showed that a metric space is complete if and only if any Banach contraction on closed subsets thereof has a fixed point. On the other hand, Kirk (1976) showed that Caristi's theorem characterizes the metric completeness. Later, motivated by Wong's proof (1976) of Caristi's theorem, Weston (1977) showed that a metric space X is complete if and only if X satisfies a condition of Ekeland (1972,1974), that is, for each lower semicontinuous function $h : X \to (-\infty, \infty)$ bounded from below on X, there is a point p in X such that $h(p) - h(x) < d(p,x)$ for every point $x \in X$. Reich (1971) and Subrahmanyam (1975) also obtained characterizations of the metric completeness using Kannan's result (1968) similar to the Banach contraction principle, which is known to be a consequence of Caristi's theorem. On the other hand, Kolodner (1967) and Boyd-Wong (1968) noticed that the Banach contraction principle follows from the Cantor intersection theorem. More later, Sullivan (1981) and Tasković (1984) also obtained similar characterizations by using Ekeland's principle and general contractive type fixed point results, respectively.

In our previous work [14], we combined those results and stated our characterizations of the metric completeness as Theorem of [14]. The first response to the article was: "Who dare use this kind of things to check the completeness of a metric space?"

A few years later, the author and Billy E. Rhoades published the study in [23]. Its Abstract says: "Several authors have characterized completeness of a metric space by using a fixed point theorem. The two theorems of this paper encompass some previous results as well as future theorems of this type."

We introduced the two theorems in [23]:

Let $\mathfrak{B}$ be a class of selfmaps of closed subsets of a metric space X such that if any $g \in \mathfrak{B}$ has a fixed point then X is complete. Examples of $\mathfrak{B}$ are the classes of the Banach contractions (Hu (1967)) and the Kannan type contractions.

Let $\mathfrak{A}$ be a class of selfmaps of closed subsets of X containing $\mathfrak{B}$ such that completeness of X implies the existence of a fixed point for any map in $\mathfrak{A}$. Examples of $\mathfrak{A}$ containing the preceding examples of $\mathfrak{B}$ are classes of maps satisfying the conditions of Meir-Keeler (1969), Hegedüs-Szilágyi (1980), Caristi (1976), Tasković (1978, 1984), and Hikida (1984).

The following is the main result in [23]:

Theorem 3.1. *X is complete if and only if any map in $\mathfrak{A}$ has a fixed point.*

Let $\mathfrak{B}'$ be a class of selfmaps defined on X such that every map in $\mathfrak{B}'$ satisfies a certain condition Q, then X is complete. An example of $\mathfrak{B}'$ is the map satisfying an equivalent formulation of Caristi's theorem as in Weston (1977).

Let $\mathfrak{A}'$ be a class of maps defined on X containing $\mathfrak{B}'$ such that completeness of X implies that every map in $\mathfrak{A}'$ satisfies a condition P, where P implies Q. An example of $\mathfrak{A}'$ is the maps satisfying Ekeland's variational principle as in Sullivan (1981).

Theorem 3.2. *X is complete if and only if every map in $\mathfrak{A}'$ satisfies the condition P.*

In MR835839 (87m:54125), the reviewer J. Matkowski stated: "There are many papers in which the completeness of a metric space is characterized by using a fixed point theorem. In the present paper, the authors prove two very simple and general theorems that 'encompass some previous as well as future theorems of this type.'"

In 2020, S. Cobzaş [4] published an article entitled "Fixed points and completeness in metric and generalized metric spaces" with the following:

> Abstract. The famous Banach contraction principle holds in complete metric spaces, but completeness is not a necessary condition: there are incomplete metric spaces on which every contraction has a fixed point. The aim of this paper is to present various circumstances in which fixed point results imply completeness. For metric spaces, this is the case of Ekeland's variational principle and of its equivalent, Caristi's fixed point theorem. Other fixed point results having this property will also be presented in metric spaces, in quasi-metric spaces, and in partial metric spaces. A discussion on topology and order and on fixed points in ordered structures and their completeness properties is included as well.

At the end of his paper, Cobzaş quoted Matkowski's review of the paper [23] under the title "Appendix. A Pessimistic Conclusion" and stated as follows:

"Under these circumstances, it seems that the best we can hope to do in this domain is to prove some particular cases of these very general results."

2.4 BASIC FIXED POINT THEOREMS

In this section, we introduce extensions of the Banach contraction principle. The first one is the Rus-Hicks-Rhoades (RHR) theorem for a quasi-metric space (X, δ) with a self-map $T : X \to X$ such that X is T-orbitally complete.

In our previous work [22], we obtained the following RHR theorem:

Theorem H(γ1). *Let (X, δ) be a quasi-metric space, $0 < \alpha < 1$, and $f{:}X \to X$ be a map satisfying*

$$\delta\left(f(x), f^2(x)\right) \leq \alpha\, \delta\,(x, f(x)) \quad \textit{for all} \quad x \in X \backslash \{f(x)\}.$$

Then f has a fixed point $v \in X$ if and only if X is f-orbitally complete.

Recall that the Banach contraction principle was extended to multimaps by Nadler [12] in 1969. A more general form of Nadler's theorem was established by Covitz-Nadler [5] in 1970 for metric spaces.

Let (X, δ) be a quasi-metric space and Cl(X) denote the family of all nonempty closed subsets of X (not necessarily bounded). For $A, B \in \mathrm{Cl}\,(X)$, set

$$H(A,B) = \begin{cases} \max\{sup\{\delta\,(a,B) : a \in A\}, & \text{if the maximum exists} \\ \quad sup\{\delta\,(b,A) : b \in B\}\} & \\ \infty & \text{if otherwise.} \end{cases} \tag{1}$$

where $\delta\,(a, B) = inf\{\delta\,(a, b) : b \in B\}$.

Such a map H is called generalized Hausdorff quasimetric induced by δ. Notice that H is a quasimetric on Cl(X). A point $p \in X$ is called a fixed point of $T : X \to \mathrm{Cl}\,(X)$ if $p \in T(p)$. A function $f : X \to \mathbb{R}$ is said to be T-orbitally lower semi-continuous if $\{x_n\}$ is a sequence in $O(T, x_0)$ and $x_n \to \zeta$ implies $f(\zeta) \leq \lim_n inff(x_n)$.

Moreover, we have the following from Theorem H in [15, 18, 22]:

Theorem H(δ1). *Let (X, δ) be a complete quasi-metric space, and $0 < \alpha < 1$. Let $T{:}X \to \mathrm{Cl}\,(X)$ be a multimap such that, for any $x \in X \backslash T(x)$, there exists $y \in X \backslash \{x\}$ satisfying*

$$H(T(x), T(y)) \leq \alpha\, \delta\,(x, y).$$

Then T has a fixed point $v \in X$, that is, $v \in T(v)$.

The following is given in [19], which can be called the Rus-Hicks-Rhoades (RHR) contraction principle:

Theorem P. *Let (X, δ) be a quasi-metric space and let $T: X \to X$ be an RHR map; that is,*

$$\delta\left(T(x), T^2(x)\right) \leq \alpha\, \delta(x, T(x)) \text{ for every } x \in X,$$

where $0 < \alpha < 1$.

(i) *If X is T-orbitally complete, then, for each $x \in X$, there exists a point $x_0 \in X$ such that*

$$\lim_{n \to \infty} T^n(x) = x_0$$

and

$$\delta(T^n(x), x_0) \leq \frac{\alpha^n}{1-\alpha} \delta(x, T(x)), \text{ for each } n = 1, 2, \cdots.$$

(ii) *x_0 is a fixed point of T, and, equivalently,*

(iii) *$T: X \to X$ is orbitally continuous at $x_0 \in X$.*

Theorem P was originated from Hicks-Rhoades [7]. The first example of Theorem P was given by Kannan in 1969.

In our previous work [21], we showed that most results on RHR maps due to Suzuki and his colleagues are simple consequences of the three theorems in this section.

2.5 COMPLETENESS OF QUASI-METRIC SPACES

Recently, as a basis of ordered fixed point theory [15, 18], we obtained the 2023 Metatheorem and Theorem H including Nadler's fixed point theorem [12] in 1969 and its extended version by Covitz-Nadler [5] in 1970.

From Theorem H(γ1) and Metatheorem, we have the following new version:

Theorem H ([18, 22]). *Let (X, δ) be a quasi-metric space and $0 < r < 1$. Then the following statements are equivalent:*

(0) (X, δ) *is complete.*

(α) *For a multimap* $T: X \to \mathrm{Cl}(X)$, *there exists an element* $v \in X$ *such that* $H(Tv, Tw) > r\,\delta(v, w)$ *for any* $w \in X \backslash \{v\}$.

(β) *If* $\mathfrak{F}$ *is a family of maps* $f: X \to X$ *such that, for any* $x \in X \backslash \{fx\}$, *there exists a* $y \in X \backslash \{x\}$ *satisfying* $\delta(fx, fy) \leq r\,\delta(x, y)$, *then* $\mathfrak{F}$ *has a common fixed element* $v \in X$, *that is,* $v = fv$ *for all* $f \in \mathfrak{F}$.

(γ) *If* $\mathfrak{F}$ *is a family of maps* $f: X \to X$ *satisfying* $\delta\left(fx, f^2x\right) \leq r\,d(x, fx)$ *for all* $x \in X \backslash \{fx\}$, *then* $\mathfrak{F}$ *has a common fixed element* $v \in A$, *that is,* $v = fv$ *for all* $f \in \mathfrak{F}$.

(δ) *Let* $\mathfrak{F}$ *be a family of multimaps* $T: X \to \mathrm{Cl}(X)$ *such that, for any* $x \in X \backslash Tx$, *there exists* $y \in X \backslash \{x\}$ *satisfying* $H(Tx, Ty) \leq r\,\delta(x, y)$. *Then* $\mathfrak{F}$ *has a common fixed element* $v \in X$, *that is,* $v \in Tv$ *for all* $T \in \mathfrak{F}$.

(ϵ) *If* $\mathfrak{F}$ *is a family of multimaps* $T: X \to \mathrm{Cl}(X)$ *satisfying* $H(Tx, Ty) \leq r\,\delta(x, y)$ *for all* $x \in X$ *and any* $y \in Tx \backslash \{x\}$, *then* $\mathfrak{F}$ *has a common stationary element* $v \in X$, *that is,* $\{v\} = Tv$ *for all* $T \in \mathfrak{F}$.

(η) *If* Y *is a subset of* X *such that for each* $x \in X \backslash Y$ *there exists a* $z \in X \backslash \{x\}$ *satisfying* $H(Tx, Tz) \leq r\,\delta(x, z)$ *for a* $T: X \to \mathrm{Cl}(X)$, *then there exists a* $v \in X \cap Y = Y$.

Proof. The equivalency (α)-(η) follows from our Metatheorem in [15, 18]. When $\mathfrak{F}$ is a singleton, (β)-(ϵ) are denoted by (β1)-(ϵ1), respectively, They are also logically equivalent to (α)-(η) by our Metatheorem. Note that (γ1) follows from Theorem H(γ1). The equivalency of (0) and (γ1) is given in [21, 22]. Then Theorem H holds. □

Remark 4.1. (1) The completeness in (0) can be replaced by *f*-orbitally or *T*-orbitally completeness according to the corresponding situation.

(2) (β1) properly extends the Banach contraction principle, which does not characterize the metric completeness.

(3) (γ1) is the Rus-Hicks-Rhoades theorem and equivalent to (0).

(4) Further, (δ1) and (ϵ1) extend the well-known theorems of Nadler [12] and Covitz-Nadler [5] on multi-valued contraction.

We have a single-valued version of Theorem H(α) as follows:

Theorem H(α1). *Let (X, δ) be a quasi-metric space, $f: X \to X$ a map, and $0 < r < 1$. Then X is f-orbitally complete if and only if there exists an element $v \in X$ such that $\delta(fv, fw) > r\delta(v, w)$ for any $w \in X \setminus \{v\}$.*

This is also equivalent to all items in Theorem H. In some sense, this shows that the Banach contraction principle does not characterize the metric completeness. But so does the RHR theorem or Theorem H(γ1).

2.6 THEOREMS DUE TO SUZUKI ET AL.

From 2008, Suzuki found fixed point theorems for certain RHR type maps with very sophisticated proofs. He and his colleagues continued to publish more related papers within a few years. Their papers became very popular and many followers published scores of papers of the same nature.

In this section, we collect some theorems related to metric completeness due to Suzuki and his colleagues.

2.6.1 Suzuki [24] in 2008

In Section 4 of [24], Suzuki discussed the metric completeness:

Theorem 4. *Let (X, d) be a metric space and define a nonincreasing function θ from $[0, 1)$ onto $(1/2, 1]$ by*

$$\theta(r) = \begin{cases} 1 & \text{if } 0 \le r \le \left(\sqrt{5}-1\right)/2, \\ (1-r)r^{-2} & \text{if } \left(\sqrt{5}-1\right)/2 \le r \le 2^{-1/2}, \\ (1+r)^{-1} & \text{if } 2^{-1/2} \le r < 1. \end{cases} \tag{2}$$

For $r \in [0, 1)$ and $\eta \in (0, \theta(r)]$, let $A_{r,\eta}$ be the family of mappings T on X satisfying the following:

(a) *For $x, y \in X$,*

$$\eta\, d(x, Tx) \le d(x, y) \text{ implies } d(Tx, Ty) \le r\, d(x, y).$$

Let $B_{r,\eta}$ be the family of mappings T on X satisfying (a) *and the following:*

(b) *$T(X)$ is countably infinite.*

(c) *Every subset of $T(X)$ is closed.*

Then the following are equivalent:

(i) *X is complete.*

(ii) *Every mapping* $T \in A_{r,\theta(r)}$ *has a fixed point for all* $r \in [0, 1)$.

(iii) *There exist* $r \in (0, 1)$ *and* $\eta \in (0, \theta(r)]$ *such that every mapping* $T \in B_{r,\eta}$ *has a fixed point.*

We prove this theorem for quasi-metric spaces in view of Theorem H(γ1) as follows:

New proof. Note that

$$B_{r,\eta} \subset A_{r,\eta} \subset \{\text{RHR maps on } X\}.$$

If X is complete, by Theorem H(γ1), (i) ⇒ (ii) ⇒ (iii).

In order to show (iii) ⇒ (i), assume (iii) and that X is not complete, that is, there exists a Cauchy sequence $\{u_n\}$ which does not converge. Suzuki defined a map $T : X \to X$ without having fixed points and satisfying (a)–(c). By (iii), T has a fixed point which yields a contradiction. Hence we obtain that X is complete.

This completes the proof. □

As a direct consequence of Theorem 4, Suzuki obtained the following.

Corollary 1. *For a metric space* (X, d), *the following are equivalent:*

(i) *X is complete.*

(ii) *There exists* $r \in (0, 1)$ *such that every mapping T on X satisfying the following has a fixed point:*

• $\frac{1}{10000} d(x, Tx) \leq d(x, y)$ *implies* $d(Tx, Ty) \leq r\, d(x, y)$ *for all* $x, y \in X$.

Comment. For a complete metric space, Theorem 4(ii) and (iii) hold. However (ii) and (iii) seem to be inconvenient to check whether X is complete or not.

2.6.2 Suzuki [25] in 2008

In order to characterize the completeness of underlying metric spaces, Suzuki introduced a weaker notion of contractions and proved the following theorem:

Theorem 1 [25]. *Define a nonincreasing function* θ *from* $[0, 1)$ *onto* $(1/2, 1]$ *as in* [21].

Then for a metric space (X, d), *the following are equivalent:*

(i) *X is complete.*

(ii) *There exists* $r \in (0,1)$ *such that every mapping T on X satisfying the following has a fixed point:*

- $\theta(r)\, d(x, Tx) \le d(x, y)$ implies $d(Tx, Ty) \le r d(x, y)$ *for all* $x, y \in X$.

Theorem 1 is meaningful because contractions do not characterize the metric completeness while Caristi and Kannan mappings do.

2.6.3 Kikkawa and Suzuki [9] in 2008

The authors prove three fixed point theorems for generalized contractions with constants in complete metric spaces, which are generalizations of fixed point theorems due to Suzuki in the same year.

Kikkawa-Suzuki [9] stated that the following is in Suzuki [24] in 2008:

Theorem 1. (Suzuki) *For a metric space* (X, d), *the following are equivalent:*

(i) *X is complete.*

(ii) *Every mapping T on X satisfying the following has a fixed point:*
- *There exists* $r \in [0,1)$ *such that* $\theta(r)\, d(x, Tx) \le d(x, y)$ implies $d(Tx, Ty) \le r d(x, y)$ *for all* $x, y \in X$.

(iii) *There exists* $r \in (0,1)$ *such that every mapping T on X satisfying the following has a fixed point:*

- $\frac{1}{10000} d(x, Tx) \le d(x, y)$ implies $d(Tx, Ty) \le r d(x, y)$ *for all* $x, y \in X$.

They stated: "The authors are very attracted by $\theta(r)$ because $\theta(r)$ does not seem to be natural. We know $\theta(r)$ is the best constant because of the existence of counterexamples. To find an intuitive reason is another motivation. However, we have not found such a reason yet. On the contrary, we have to raise one problem concerning $\theta(r)$."

Comment. Theorem 1 can be extended as follows:

(i) $\Longleftrightarrow$ Theorem H(γ1) for metric spaces $\Longleftrightarrow$ (ii) $\Longleftrightarrow$ (iii).

Therefore, we have the following :

Theorem. *A quasi-metric space X is complete if and only if any RHR selfmap of X has a fixed point.*

Another proof. In Theorem 3.2 in Section 2.3 of the present paper, let

$\mathfrak{B}'$ be the class of maps T in Theorem 1(ii);

$\mathfrak{A}'$ be the class of RHR maps T; and

both of P and Q mean "having a fixed point."

Then our new theorem follows from Theorem 3.2. □

Consequently, the RHR maps characterize the metric completeness, but the Banach contractions does not as is well-known.

2.6.4 Kikkawa and Suzuki [10] in 2008

The following is a Kannan version of the Suzuki theorem [24]:

Theorem 2.2. [10] *Let T be a mapping on complete metric space (X, d) and let ϕ be a non-increasing function from $[0, 1)$ into $(1/2, 1]$ defined by*

$$\phi(r) = \begin{cases} 1, & \text{if } 0 \le r \le \frac{1}{\sqrt{2}}, \\ \frac{1}{1+r}, & \text{if } \frac{1}{\sqrt{2}} \le r < 1. \end{cases} \tag{3}$$

Let $\alpha \in [0, 1/2)$ and $r = \alpha/(1-\alpha) \in [0, 1)$. Suppose that

$$\phi(r)\, d(x, Tx) \le d(x, y) \text{ implies } d(Tx, Ty) \le \alpha d(x, Tx) + \alpha d(y, Ty)$$

for all $x, y \in X$. Then, T has a unique fixed point z, and $\lim_n T^n x = z$ holds for every $x \in X$.

This gives an example of RHR maps and Theorem P can be applied.

They said the following theorem shows that $\phi(r)$ is the best constant for every r:

Theorem 2.4. *Define a function ϕ as in Theorem 2.2. For every $\alpha \in [0, 1/2)$, putting $r = \alpha/(1-\alpha)$, there exist a complete metric space (X, d) and a mapping T on X such that T has no fixed points and*

$$\phi(r)\, d(x, Tx) < d(x, y) \text{ implies } d(Tx, Ty) \le \alpha\, d(x, Tx) + \alpha\, d(y, Ty)$$

for all $x, y \in X$.

New proof. Note that $\theta(r)\, d(x, Tx) < d(x, y)$ means T can not have a fixed point $x = y = Tx$. □

In this paper, it is noted that $d(Tx, Ty) \le r\,d(x,y)$ in the original Suzuki type map in [24] can be replaced by one of the following in Theorem 3.1:

$d(Tx, Ty) \le r \max\{d(x, Tx), d(y, Ty)\}$,

$d(Tx, Ty) \le \max\{\alpha d(x, Tx), \beta d(y, Ty)\}, \alpha, \beta \in [0, 1)$.

Similarly, Theorem 3.3 is modified by applying such conditions.

In fact, the authors gave unnecessarily lengthy proofs for their theorems.

2.6.5 Enjouji, Nakanishi, and Suzuki [6] in 2009

In order to observe the condition of Kannan mappings, the authors prove a generalization of Kannan's fixed point theorem. Their theorem involves constants and they obtain the best constants to ensure a fixed point. They consider "$\alpha d(x, Tx) + \beta d(y, Ty)$" instead of "$\alpha d(x, Tx) + \alpha d(y, Ty)$."

Let $\Delta = \{(\alpha, \beta) : \alpha \ge 0, \beta \ge 0, \alpha + \beta < 1\}$. Define a nonincreasing function $\psi : \Delta \to (1/2, 1]$. Let T be a map on a complete metric space such that there exists $\alpha, \beta \in \Delta$ satisfying

$$\psi(\alpha, \beta)\, d(x, Tx) \le d(x, y) \ \text{ implies } \ d(Tx, Ty) \le \alpha d(x, Tx) + \beta d(y, Ty)$$

for all $x, y \in X$.

Note that T is an RHR map and that the condition $\psi(\alpha, \beta)\, d(x, Tx) < d(x, y)$ in Theorem 4.1 implies nonexistence of a fixed point of T.

2.6.6 Nakanishi and Suzuki [13] in 2010

In this paper, in order to observe the condition of Kannan maps more deeply, they prove a generalization of Kannan's fixed point theorem.

Moreover, they assumed something like $\theta(\alpha)\, d(x, Tx) < d(x, y)$, which is false for $x = y = Tx$. Hence, from the beginning, such T can not have a fixed point.

Further, for $(\alpha, \beta) \in \Delta = [0, 1)^2$ and a function $\phi : \Delta \to (1/2, 1]$, let T be a selfmap of a complete metric space (X, d) satisfying

$$\phi(\alpha, \beta)\, d(x, Tx) \le d(x, y) \ \text{ implies } \ d(Tx, Ty) \le \max\{\alpha d(x, Tx), \beta d(y, Ty)\},$$

for all $x, y \in X$.

Note that T is an RHR map and that the condition $\phi(\alpha, \beta)\, d(x, Tx) < d(x, y)$ does not hold for $x = y = Tx$.

2.7 OTHER AUTHORS' COMMENTS RELATED TO SUZUKI ET AL.

A large number of followers of Suzuki's works published inaccurate comments on completeness and others. We list them in this section by indicating the year published and the initials of their authors.

[2011DD] "We obtain multi-valued mapping generalizations of two recent theorems of Kikkawa and Suzuki in 2008 and the main theorem of Enjouji, Nakanishi, and Suzuki in 2009."

[2011DL] "In this article we obtain a Suzuki-type generalization of a fixed point theorem for generalized multivalued mappings of Ćirić in 1972. The obtained results extend furthermore the recently developed Kikkawa-Suzuki-type contractions. Applications to certain functional equations arising in dynamic programming are also considered."

[2011P] "The remarkable generalization of the classical Banach contraction theorem, due to Suzuki in 2008, has lead to some important contributions in metric fixed point theory (see, for instance, eight papers are listed)."

[2012A] "Inspired by the work of Suzuki in 2008, we prove a fixed point theorem for contractive mappings that generalizes a theorem of Geraghty in 1973, and characterizes metric completeness."

[2012GR] "Suzuki has shown that metric completeness can be characterized by a family of functions satisfying a generalized Banach's contraction principle."

[2014HS] "In 2008, in order to characterize the completeness of underlying metric spaces, Suzuki introduced a weaker notion of contraction. ... As an application of our results we deduce Suzuki type results for GF-contractions."

[2015MAK] "The Banach contraction ⋯ cannot describe the metric completeness. To overcome this issue, Suzuki in 2008 established a fixed point theorem which generalized the Banach contraction theorem and characterized the metric completeness. In 2008, Kikkawa and Suzuki extended the main theorem of Suzuki for the case of multivalued mappings."

[2016ND] "Suzuki also introduced a new type of mapping. This mapping not only generalizes the Banach Contraction Mapping Principle but also characterizes the completeness of the underlying metric space. Kikkawa and Suzuki presented Kannan version of Suzuki theorem. In this paper we state and prove a fixed point theorem that uses Suzuki mapping and Kannan type of contraction."

[2018J] "Suzuki proved a fixed point theorem which is a generalization of Banach's fixed point theorem and characterizes the metric completeness. ... Kikkawa and Suzuki proved a generalization of Kannan's fixed point theorem. ... Enjouji et al. proved a generalization of Theorem of Kikkawa and Suzuki."

[2022CJJ] "In this paper, we establish some fixed point theorems for single valued and multi-valued mappings on a complete metric space. Suzuki's and some other fixed point theorems are generalized by taking a more general contractive condition for single valued mappings. It is also proved that our result characterizes the completeness of the metric space. Further, taking generalized contractive condition, a fixed point theorem is also established for multi-valued mappings."

[2022KA] "Suzuki presented a fixed point theorem that generalized BCP and characterized metric completeness as well. Recently, Ali et al. obtained completeness characterizations of b-metric spaces via the fixed point of Suzuki type contractions."

[2022KAG] "Another important generalization of the Banach contraction principle was obtained by Suzuki in 2008. Suzuki generalized the principle for the class of weak contractive mappings in complete metric spaces and established a new version of the Banach contraction principle. ... The Banach contraction principle does not characterize the metric completeness, however the result of Suzuki characterizes the metric completeness."

[2022R] "In this note we show the somewhat surprising fact that the proof of the 'if part' of the distinguished characterizations of metric completeness due to Kirk, and Suzuki and Takahashi, respectively, can be deduced in a straightforward manner from Hu's theorem that a metric space is complete if and only if any Banach contraction on bounded and closed subsets thereof has a fixed point. We also take advantage of this approach to easily deduce a characterization of metric completeness via fixed point theorems for α-ψ-contractive mappings."

[2022R′] "Suzuki published in 2008 his renowned article in which he presented a necessary and sufficient condition for the metric completeness by utilizing an appealing generalization of the Banach contraction principle."

[2023LKPS] "Suzuki's generalization of the BCP introduced a new class of contractive maps that satisfy contraction conditions only for specific elements of the underlying space. … Numerous mathematicians have contributed to the generalization of Nadler's theorem, with Kikkawa and Suzuki achieving significant progress in the study of generalized multi-valued maps."

[2023O] "In this paper, we prove a common fixed point theorem for Suzuki type contractions on complete partial metric spaces. We also state some corollaries related to Suzuki type common fixed point theorem. We also give an example where we apply our main theorem on complete partial metric spaces. Finally, to show usability of our results, we give its an application showing existence and uniqueness of a common solution for a class of functional equations in dynamic programming."

[2023PK] "In 2008, Suzuki introduced a new class of contraction mappings where the contraction condition to be hold only on certain elements of the underlying space. He presented a remarkable generalization of the BCP which also characterizes completeness of the metric space."

[2024RT] From Abstract: "In an outstanding article published in 2008, Suzuki obtained a nice generalization of the Banach contraction principle from which derived a characterization of metric completeness. Although Suzuki's theorem has been successfully generalized and extended in several directions and contexts, we here show by means of a simple example that the problem of achieving, in an obvious way, its full extension to the framework of w-distances does not have an emphatic response. Motivated by this fact we introduce the concept of presymmetric w-distance on metric spaces, we give some properties and examples of this new structure and show that it provides a reasonable setting to obtain a real and hardly forced w-distance generalization of Suzuki's theorem. This is realized in our main result, which is a fixed point theorem that involves presymmetric w-distances and certain contractions of Suzuki-type."

From Introduction: "In his already classical article, Suzuki presented an elegant generalization of Banach's contraction principle that he used to characterize complete metric spaces. … The last part of the paper is devoted to obtain necessary and sufficient conditions for a metric space to be complete which is made by combining our fixed point results with both Suzuki's characterization and Suzuki-Takahashi's characterization."

2.8 EPILOGUE

In our previous works [19, 20] and the present one, we showed many metric fixed point theorems hold for quasi-metric spaces from the beginning.

Even for the Banach contraction principle, certain traditional monographs or text-books on fixed point theory or general topology stated for metric spaces only. All of them stated the principle for metric spaces only, but their proofs do not use the symmetry of a metric. Of course, they do not mention the concept of quasi-metric spaces.

Since our study in [15] in 2022, we have published nearly two dozen articles on our 2023 Metatheorem and related topics. These could add up many new results to ordered fixed point theory in [15]. Moreover, our Metatheorem has numerous applications; for recent examples, see [18–22].

In the present article, we corrected only some of inaccurate statements in metric fixed point theory. However there are many results related to the Rus-Hicks-Rhoades theorem due to scores of other authors. Since the theorem and its extensions properly include the corresponding ones of the Banach contraction, the future study on metric fixed point theory would be concentrated to extend the theorem and its new applications.

There are thousands of artificial generalizations of metric spaces. Most of them assume the symmetry which might be eliminated as in the present article.

REFERENCES

1. H. Aydi, M. Jellali, E. Karapinar, On fixed point results for α-implicit contractions in quasi-metric spaces and consequences, *Nonlinear Anal. Model. Control*, 21(1) (2016) 40–56.
2. V. Berinde, On the approximation of fixed points of weak contractive mappings, *Carpathian J. Math.* 19(1) (2003) 7– 22.
3. V. Berinde, A. Petrusel, I. Rus, Remarks on the terminology of the mappings in fixed point iterative methods in metric spaces, *Fixed Point Theory*, 24(2) (2023) 525–540. DOI: 10.24193/fpt-ro.2023.2.05
4. S. Cobzaş, Fixed points and completeness in metric and generalized metric spaces *Jour. Math. Sci.* 250(3) (2020) 475–535. DOI 10.1007/s10958-020-05027-1

5. H. Covitz, S.B. Nadler, Jr. Multi-valued contraction mappings in generalized metric spaces, *Israel J. Math.* 8 (1970) 5–11.
6. Y. Enjouji, M. Nakanishi, T. Suzuki, A generalization of Kannan's fixed point theorem, *Fixed Point Theory Appl.* 1 (2009) 192872, 10pp.
7. T.L. Hicks, B.E. Rhoades, A Banach type fixed point theorem, *Math. Japon.* 24 (1979) 327–330.
8. M. Jleli, B. Samet, Remarks on G-metric spaces and fixed point theorems, *Fixed Point Theory Appl.* 2012 (2012) 210.
9. M. Kikkawa, T. Suzuki, Three fixed point theorems for generalized contractions with constants in complete metric spaces, *Nonlinear Analysis* 69 (2008) 2942–2949.
10. M. Kikkawa, T. Suzuki, Some similarity between contractions and Kannan mappings, *Fixed Point Theory Appl.* (2008) 649749, 8. DOI: 10.1155/2007/49749
11. W.A. Kirk, Caristi's fixed point theorem and metric convexity, *Colloq. Math.* 36(1) (1976) 81–86.
12. S.B. Nadler, Jr. Multi-valued contraction mappings, *Pacific J. Math.* 30 (1969) 475-488.
13. M. Nakanishi1, T. Suzuki, An observation on Kannan mappings, *Cent. Eur. J. Math.* 8(1) (2010) 170–178. DOI: 10.2478/s11533-009-0065-9
14. S. Park, Characterizations of metric completeness, *Colloq. Math.* 49(1) (1984) 21–26,
15. S. Park, Foundations of ordered fixed point theory, *J. Nat. Acad. Sci., Rok, Nat. Sci. Ser.* 61(2) (2022) 1–51.
16. S. Park, Relatives of a theorem of Rus-Hicks-Rhoades, *Letters Nonlinear Anal. Appl.* 1 (2023) 57–63. ISSN 2958-874X
17. S. Park, Almost all about Rus-Hicks-Rhoades maps in quasi-metric spaces, *Adv. Th. Nonlinear Anal. Appl.* 7(2) (2023) 455–471. DOI: 0.31197/atnaa.1185449
18. S. Park, History of the metatheorem in ordered fixed point theory, *J. Nat. Acad. Sci., Rok, Nat. Sci. Ser.* 62(2) (2023) 373–410.
19. S. Park, All metric fixed point theorems hold for quasi-metric spaces, *Results in Nonlinear Analysis* 6(4) (2023) 116–127. https://doi.org/10.31838/rna/2023.06.04.012 (Dec. 1, 2023). A revised and corrected version appears in Research Gate.
20. S. Park, *The use of quasi-metric in the metric fixed point theory*, Manuscript.

21. S. Park, Comments on the Suzuki type fixed point theorems, *Adv. Theory Nonlinear Anal. Appl.* 7(3) (2023) 67–78.
22. S. Park, The realm of the Rus-Hicks-Rhoades maps in the metric fixed point theory, *J. Nat. Acad. Sci., Rok, Nat. Sci. Ser.* 63(1) (2024) 1–50.
23. S. Park, B.E. Rhoades, Comments on characterizations for metric completeness, *Math. Japon.* 31(1) (1986) 95–97.
24. T. Suzuki, A generalized Banach contraction principle that generalizes metric completeness, *Proc. Amer. Math. Soc.* 136(5) (2008) 1861–1869.
25. T. Suzuki, Fixed point theorems and convergence theorems for some generalized nonexpansive mappings, *Jour. Math. Anal. Appl.* 340 (2008) 1088–1095.

CHAPTER 3

Fixed Point Theorems in p-Normed Spaces

George Xianzhi Yuan and Jian-Zhong Xiao

3.1 INTRODUCTION

It is known that the class of p-seminorm spaces for $p \in (0, 1]$ is an important generalization of usual normed spaces with rich topological and geometrical structures, and related study has received a lot of attention (see Balachandran [2], Bayoumi [3], Chang et al. [9], Ennassik and Taoudi [11], Ennassik et al. [10], Granas and Dugundji [13], Jarchow [14], Kalton [15, 16], Kalton et al. [17], Park [21], Xiao and Zhu [25], Yuan [26], Zeidler [27], and many others). However, to the best of our knowledge, the corresponding basic tools and associated results in the category of nonlinear functional analysis have not been well developed. The goal of this paper is to establish fixed point theorems for s-convex subsets in p-normed spaces by applying the existence of homeomorphisms for s-convex subsets in p-normed spaces in a way different by comparing with those used in the existing literature (e.g., see Bayoumi [3] and related references), where $s, p \in (0, 1]$ by the fact that the corresponding results established in this paper reduce to the classical fixed point theorems in normed spaces when $s = p = 1$. Thus, these results not only provide fundamental tools for the study of Schauder's conjecture and related nonlinear analysis under the setting of p-vector or locally p-convex spaces for either single-valued or set-valued mappings but can be used to establish fixed point theorems for continuous mappings in p-normed spaces which unify or improve corresponding results in the existing literature (see Chang et al. [9], Yuan [26], or Zeidler [27] and related references for more details).

DOI: 10.1201/9781003530602-3

This paper consists of three sections. Section 3.1 provides a brief introduction; Section 3.2 describes general concepts for the p-convex subsets in p-vector spaces for $p \in (0, 1]$, and related basic facts; and in Section 3.3, a fixed point theorem for continuous mappings is proved by the existence of homeomorphisms for s-convex subsets in p-normed spaces, where $s, p \in (0, 1]$.

For the convenience of our discussion, throughout this paper, all p-convex vector spaces are assumed to be Hausdorff, and p satisfying the condition for $p \in (0, 1]$ unless specified. We denote by $\mathbb{N}$ the set of all positive integers, i.e., $\mathbb{N} = \{1, 2, \cdots, \}$; and for a given set X, the 2^X denotes the family of all subsets of X.

For a vector space X over a field K (which is either a real field or a complex field), the span of a set A of vectors (not necessarily finite) is defined to be the intersection W of all subspaces of X that contain A. W is referred to as the subspace spanned by W, or by the vectors in A. Conversely, A is called a spanning set of W, and we say that A spans W, and it is also denoted by $span\,(A)$ for W.

In what follows we denote the interior, the closure, and the boundary of a given subset A in a $(p-)$ normed space X by $A^0, \overline{A}$ and $\partial\,(A)$, respectively. By $B\,(x, r)$ we mean the open ball of X with center $x \in X$ and radius $r > 0$. If there is $r > 0$ such that $B\,(x, r) \cap span\,(A) \subset A$, then x is said to be an interior point with respect to its (i.e., $span\,(A)$) relative topology and is denoted by $x \in ri\,(A)$.

Here we remark that although most notation, facts, and basic results for p-vector spaces have been used by Yuan [26], they are included here only for the purpose of reader's convenience.

3.2 SOME BASIC RESULTS OF *P*-VECTOR SPACES

We now recall some notion and definitions on p-convexity, p-vector and locally p-convex spaces, and some related fundamental facts (see Jarchow [14], Kalton [15], Bayoumi [3], or Ennassik and Taoudi [11]) that will be used in this paper.

Definition 2.1. Let $p \in (0, 1]$. A set A in a vector space X is said to be p-convex if for any $x, y \in A$, we have $sx + ty \in A$, whenever $0 \leq s, t \leq 1$ with $s^p + t^p = 1$; the set A is said to be absolutely p-convex if for any $x, y \in A$, we have $sx + ty \in A$, whenever $|s|^p + |t|^p \leq 1$. In the case $p = 1$, the concept of the (absolutely) 1-convexity is simply the usually (absolutely) convexity defined in vector spaces.

Definition 2.2. Let $p \in (0, 1]$. If A is a subset of a topological vector space X, then the p-convex hull of A and its closed p-convex hull are denoted by $co_p(A)$, and $\overline{co}_p(A)$, respectively, which is the smallest p-convex set containing A, and the smallest closed p-convex set containing A, respectively.

Definition 2.3. Let $p \in (0, 1]$ and A be p-convex and $x_1, \cdots, x_n \in A$, and $t_i \geq 0, \sum_1^n t_i^p = 1$. Then $\sum_1^n t_i x_i$ is called a p-convex combination of $\{x_i\}$ for $i = 1, 2, \cdots, n$. If $\sum_1^n |t_i|^p \leq 1$, then $\sum_1^n t_i x_i$ is called an absolutely p-convex combination. It is easy to see that $\sum_1^n t_i x_i \in A$ for a p-convex set A.

Definition 2.4. A subset A of a vector space X is called balanced (or circled) if $\lambda A \subset A$ holds for all scalars λ satisfying $|\lambda| \leq 1$. We say that A is absorbing if for each $x \in X$, there is a real number $\rho_x > 0$ such that $\lambda x \in A$ for all $\lambda > 0$ with $|\lambda| \leq \rho_x$.

Definition 2.5. Let X is a vector space and $\mathbb{R}^+$ is a non-negative part of a real line $\mathbb{R}$. Then a mapping $P : X \longrightarrow \mathbb{R}^+$ is said to be a p-seminorm if it satisfies the following requirements for $p \in (0, 1]$:

(i) $P(x) \geq 0$ for all $x \in X$;
(ii) $P(\lambda x) = |\lambda|^p P(x)$ for all $x \in X$ and $\lambda \in R$;
(iii) $P(x + y) \leq P(x) + P(y)$ for all $x, y \in X$.

A p-seminorm P is called an p-norm if $x = 0$ whenever $P(x) = 0$. A topological vector space with a specific p-norm is called a p-normed space. Of course if $p = 1$, X is is a usual normed space. By Lemma 3.2.5 of Balachandra [2], the following proposition gives a necessary and sufficient condition for a p-seminorm to be continuous.

Proposition 2.1. *Let X be a topological vector space, P a p-seminorm on X, and $V := \{x \in X : P(x) < 1\}$. Then P is continuous if and only if $0 \in int(V)$, where $int(V)$ is the interior of V.*

Now given a p-seminorm P, the p-seminorm topology determined by P (in short, the p-topology) is the class of unions of open balls $B(x, \epsilon) := \{y \in X : P(y - x) < \epsilon\}$ for $x \in X$ and $\epsilon > 0$.

We also need following notion for the so-called p-gauge (see Balachandra [2]).

Definition 2.6. Let A be an absorbing subset of a vector space X. For $x \in X$ and $p \in (0,1]$, set $P_A = inf\left\{\alpha > 0 : x \in \alpha^{\frac{1}{p}} A\right\}$, then the non-negative real-valued function P_A is called the p-gauge (gauge if $p = 1$). The p-gauge of A is also known as the Minkowski p-functional.

By Proposition 4.1.10 of Balachandra [2], we have the following proposition.

Proposition 2.2. *Let A be an absorbing subset of X and $p \in (0,1]$. Then p-gauge P_A has the following properties:*

(i) $P_A(0) = 0$;

(ii) $P_A(\lambda x) = |\lambda|^p P_A(x)$ if $\lambda \geq 0$;

(iii) $P_A(\lambda x) = |\lambda|^p P_A(x)$ for all $\lambda \in R$ provided A is balanced (circled);

(iv) $P_A(x + y) \leq P_A(x) + P_A(y)$ for all $x, y \in A$ provided A is p-convex.

In particular, P_A is a p-seminorm if A is absolutely p-convex (and also absorbing).

Remark 2.1. It is worthwhile noting that a zero-neighborhood in a topological vector space being an absolutely 0-neighborhood is also absorbing (by Lemma 2.1.16 of Balachandran [2], or Proposition 2.2.3 of Jarchow [14]), this leads us to have the following definition for a topological vector space E being a topological p-vector space (in short, p-vector space) for $p \in (0,1]$ by using the concept of the Minkowski p-functional as given below.

Definition 2.7. A topological vector space X is said to be a topological p-vector space (in short, p-vector space) if the base of the origin in X is generated by a family of Minkowski p-functionals (p-gauges) (see the Definition 2.6 above), where $p \in (0,1]$.

By incorporating Proposition 2.2, it seems for us that the following one is in a nature way to have the definition for locally p-convex spaces, where $p \in (0,1]$.

Definition 2.8. A topological (p-) vector space X is said to be locally p-convex if the origin in X has a fundamental set of absolutely p-convex

0-neighborhoods. This topology can be determined by p-seminorms which are defined in the obvious way (see pp.52 of Bayoumi [3], Jarchow [14]. When $p = 1$, a locally p-convex space X is reduced to be a usual locally convex space.

By Proposition 4.1.12 of Balachandra [2], we also have the following proposition.

Proposition 2.3. *Let A be a subset of a vector space X, which is absolutely p-convex for $p \in (0, 1]$ and absorbing. Then, we have that*

(i) The p-gauge P_A is a p-seminorm such that if $B_1 := \{x \in X : P_A(x) < 1\}$ and $\overline{B_1} = \{x \in X : P_A(x) \leq 1\}$. Then $B_1 \subset A \subset \overline{B_1}$; in particular, $kerP_A \subset A$, where $kerP_A := \{x \in X : P_A(x) = 0\}$.

(ii) $A = B_1$ or $\overline{B_1}$ according as A is open or closed in the P_A-topology.

Remark 2.2. Let X be a topological vector space and let U be an open absolutely p-convex neighborhood of the origin, and let ϵ be any given positive number. If $y \in \epsilon^{\frac{1}{p}} U$, then $y = \epsilon^{\frac{1}{p}} u$ for some $u \in U$ and $P_U(y) = P_U\left(\epsilon^{\frac{1}{p}} u\right) = \epsilon P_U(u) \leq \epsilon$ (as $u \in U$ implies that $P_U(u) \leq 1$). Thus, P_U is continuous at *zero*, and therefore, P_U is continuous everywhere. Moreover, we have $U = \{x \in X : P_U(x) < 1\}$.

The following Lemma 2.1 is a very important and useful result which allows use to make the approximation for convex subsets containing zero in topological vector spaces by p-convex subsets in locally p-convex spaces (see Lemma 2.1 of Ennassik and Taoudi [11] and Remark 2.1 of Qiu and Rolewicz [22]; thus we omit their proof).

Lemma 2.1. *Let A be a subset of a vector space X, then*

(i) *If A is r-convex, with $r \in (0, 1)$, then $\alpha x \in A$ for any $x \in A$ and any $\alpha \in (0, 1]$;*

(ii) *If A is convex and $0 \in A$, then A is s-convex for any $s \in (0, 1]$; and*

(iii) *If A is r-convex for some $r \in (0, 1)$, then A is s-convex for any $s \in (0, r]$.*

Remark 2.3. We like to point out that the results (i) and (iii) of Lemma 2.1 do not hold for $p = 1$. Indeed, any singleton $\{x\} \subset X$ is convex in topological vector spaces; but if $x \neq 0$, then it is not p-convex for any $p \in (0, 1)$.

We also need the following results which are Lemma 2.10 and Lemma 2.11 of Xiao and Zhu [25], respectively.

Lemma 2.2. *Let* $(X, \|\cdot\|_p)$ *be a complete separable p-normed space* $(0 < p \leq 1)$ *and D a bounded closed subset of X. Let* $T : D \to X$ *be a continuous bounded operator with* $0 \in ri(\overline{co_s}T(D))$. *Then T has a continuous extension* $S : X \to X$ *such that* $S(X) \subset \overline{co_s}(T(D)$.

Proof. It is Lemma 2.10 of Xiao and Zhu [25]. The proof is complete. □

Lemma 2.3. *Let* C_s *be the closed cuboid in* l^p *defined by*

$$C_s := \{x = (\alpha_1, \cdots, \alpha_n, \cdots) \in l^p : |\alpha_n|^s \leq 1/n^2\}$$

and $T : C_s \to C_s$ *a continuous operator, where* $0 < p \leq 1$, $0 < s \leq p$. *Then* C_s *is s-convex and compact, and there exists* $z \in C_s$ *such that* $T(z) = z$.

Proof. It is Lemma 2.11 of Xiao and Zhu [25]. The proof is complete. □

3.3 FIXED POINT THEOREMS IN *P*-NORMED SPACE

The goal of this section is to establish fixed point theorems for s-convex subsets in p-normed spaces by applying the existence of homeomorphisms for s-convex subsets in p-normed spaces, with a different proof by comparing those in the existing literature, where $s, p \in (0, 1]$. We note that the corresponding results established in this paper reduce to the classical fixed point theorems in normed spaces when $s = p = 1$.

First by Proposition 2 of Shapiro [23], for a space l^p, where $0 < p \leq 1$, we have the following result which is often called the universal property.

Lemma 3.1 *Every complete, separable p-normed space* $(0 < p \leq 1)$ *is a continuous linear image of* l^p.

By applying Lemma 3.1 and the space decomposition method skill, we have the following result which is Corollary (of Theorem 4.3.1) by Wang

[24] (see also the closely related Corollary 2.11 on pp.26 by Kalton et al. [17]), and thus we omit its proof here.

Lemma 3.2. *Let $(X, \|\cdot\|_p)$ be a complete and separable p-normed space $(0 < p \leq 1)$. Then there exists a closed subspace l_X^p of l^p and a linear operator $T : l^p \to X$ such that $T : l_X^p \to X$ is a homeomorphism.*

Now we have the following result with its proof in a way different from original one given by Xiao and Zhu [25] below.

Theorem 3.3. *Let $(X, \|\cdot\|_p)$ be a complete p-normed space and D a s-convex compact subset of X, where $0 < p \leq 1$, $0 < s \leq p$. Then there exists a linear operator $F : D \to l^p$ such that $F(D) \subset C_s$ and $F : D \to F(D)$ is a homeomorphism, where*

$$C_s := \{x = (\alpha_1, \cdots, \alpha_n, \cdots) \in l^p : |\alpha_n|^s \leq 1/n^2\}.$$

Proof. Let $Y = \overline{\text{span}\, D}$. Since D is compact, thus Y is separable. As Y is a closed subspace of X, Y is also complete. By Lemma 3.2, there exists a closed subspace l_Y^p of l^p and a linear operator $T : l^p \to Y$ such that $T : l_Y^p \to Y$ is a homeomorphism. Let $F = T^{-1}$. Then $F : D \to l^p$ is a linear operator such that $F : D \to F(D)$ is a homeomorphism. In addition, it is easily to verify that $F(D) \subset C_s$. This completes the proof. □

As an application of Theorem 3.3, we have the following fixed point theorem in complete p-normed spaces for $s, p \in (0, 1]$.

Theorem 3.4. *Let $(X, \|\cdot\|_p)$ be a complete p-normed space and C a compact s-convex subset of X, where $s, p \in (0, 1]$. If $T : C \to C$ is continuous, then there exists $z \in C$ such that $Tz = z$.*

Proof. We prove this result by following two steps below.

The first step: we prove the conclusion for $p \in (0, 1]$ and $0 < s \leq p \leq 1$. By following Remark 3.1 below, without loss of the generality, we may assume that original point $0 \in ri(C)$ (where $ri(C)$ is the relatively topological interior of C in X). Now by Theorem 3.3, there exists a linear homeomorphism $F : C \to F(C)$ such that $F(C) \subset C_s$, where C_s is an s-convex compact subset of l^p defined by Lemma 2.3. Then $F(C)$ is an s-convex compact and $0 \in ri(F(C))$. Since T is continuous, the mapping $FTF^{-1} : F(C) \to F(C)$

is also continuous. By Lemma 2.2, the mapping FTF^{-1} has a continuous extension $S : C_s \to C_s$ such that $S(C_s) \subset \overline{co_s}F(C)$. By Lemma 2.3, which is a fixed point theorem for a compact s-convex subset C_s in l^p spaces (where, $0 < s \leq p \leq 1$), there is $u \in C_s$ such that $u = S(u) \in F(C)$, and so $u = S(u) = FTF^{-1}(u)$. Let $z = F^{-1}(u)$. It follows that $z \in C$ and $T(z) = z$, which is a fixed point of the mapping T.

The second step: we prove the case $0 < p < s < 1$ and $0 < p < s = 1$. For the case $0 < p < s < 1$, by Lemma 2.1, C is also p-convex. Applying the conclusion given in the first step above, it follows that there exists $x \in C$ such that $T(x) = x$.

Now the only case left to prove is for $(0 < p <)s = 1$, which means C is convex. We choose an arbitrary $x_0 \in C$ and let $C_0 := \{x - x_0 : x \in C\}$. Then it is clear that C_0 is a compact convex subset of X which contains the zero element. By Lemma 2.1 again, we conclude that C_0 is s-convex for any $s \in (0, 1)$. Now, define the mapping $S : C_0 \to C_0$ by $S(x - x_0) = T(x) - x_0$ for each $x - x_0 \in C_0$. Clearly, S is continuous, and applying the result of the first step to S, we conclude that there exists $x \in C$ such that $S(x - x_0) = x - x_0$, it implies that $x = T(x)$, which is a fixed point of T. This completes the proof. □

Remark 3.1. In the proof of Theorem 3.4, the condition $0 \in ri(C)$ is used. Indeed, if $0 \notin ri(C)$, we can take $x_0 \in ri(C)$ and let $D := C - \overline{co_s}\{x_0\}$. Then we have $0 \in ri(D)$, and D is compact and s-convex. In addition, the translation operator $T_0 : D \to D$ (of T) is also continuous (e.g., see Lemma 2.3(a) of Xiao and Zhu [25]). This implies that T_0 has a fixed point, thus T has a fixed point, also by Lemma 2.3(b) of Xiao and Zhu [25].

In order to prove Theorem 3.4 also holds for a p-normed space which may not be complete for $p \in (0, 1]$, we need the following result.

Lemma 3.5. *Let $(X, \|\cdot\|_p)$ be a p-normed space, where $p \in (0, 1]$. Then there is a complete p-normed space $\hat{X}$ and a linear isometry mapping i from X onto the subspace $W := i(X)$ which is the dense in $\hat{X}$.*

Proof. This is Theorem 2.2 of Ennassik and Taoudi [11]. Its proof completely follows the way for the proof of Theorem 2.3.2 by Kreyszig [18]. This completes the proof. □

As applications of Theorem 3.4 and Lemma 3.5, we have the following general fixed point theorem in p-normed spaces which may not be complete.

Theorem 3.6. *Let* $(X, \|\cdot\|_p)$ *be a* p*-normed space and* C *a compact* s*-convex subset of* X*, where* $s, p \in (0, 1]$*. If* $T : C \to C$ *is continuous, then there exists* $z \in C$ *such that* $Tz = z$*.*

Proof. Let $\hat{X}$ be the completion of X. By Lemma 3.5, there exists a linear isometric embedding $i : X \to \hat{X}$ with $i(X)$ dense in X. We define $\hat{T} : i(C) \to i(C)$ by $\hat{T}(i(x)) = i(T(x))$ for each $x \in C$. Then this mapping is easily checked to be well defined, and it is continuous since i is a linear isometry and T is continuous on C. Furthermore, the set $i(C)$ is compact, being the image of a compact set under a continuous mapping. It is also s-convex as it is the image of an s-convex set under a linear mapping.

Now by Theorem 3.4, there exists $x \in C$ such that $\hat{T}(i(x)) = i(x)$. Thus, $i(T(x)) = i(x)$ so $T(x) = x$, which means T has a fixed point in C. This completes the proof. □.

Remark 3.2. By applying Theorem 3.3, we establish general fixed point theorems for s-convex subsets in p-normed spaces, which is Theorem 3.6 above, where $s, p \in (0, 1]$. Here we like to share with readers that the proof for the existence of homeomorphisms for s-convex subsets in p-normed spaces is new, with different idea by comparing those given by Xiao and Zhu [25] and references therein. As each p-normed space includes normed spaces as a special class (with $p = 1$), Theorem 3.6 unifies corresponding results in the existing literature. For more details, see Agarwal et al. [1], Ben-El-Mechaiekh [4], Browder [5], Cauty [6–8], Chang et al. [9], Ennassik et al. [10], Fan [12], Granas and Dugundji [13], Mauldin [19], Park [20, 21], Xiao and Zhu [25], Zeidler [27], and references therein. In addition, for the fixed point theorem in p-vector and locally p-convex spaces, we refer to Ennassik et al. [10], Ennassik and Taoudi [11], and Yuan [26]'s recent works for more details.

Conflict of interest: The authors declares that no conflict of interests.

Ethical approval: This article does not contain any studies with human participants or animals performed by the authors.

Informed consent: For this type of study informed consent was not required.

REFERENCES

1. Agarwal, R.P., Meehan, M. and O'Regan, D. 2001. *Fixed Point Theory and Applications. Cambridge Tracts in Mathematics*, vol. 141. Cambridge University Press, Cambridge, 2001.
2. Balachandran, V.K. 2000. *Topological Algebras*, vol. 185. Elsevier, Amsterdam, 2000.
3. Bayoumi, A. 2003. *Foundations of Complex Analysis in Non Locally Convex Spaces. Function Theory Without Convexity Condition, North-Holland Mathematics Studies*, vol. 193. Elsevier Science B.V., Amsterdam.
4. Ben-El-Mechaiekh, H. 2014. Approximation and selection methods for set-valued maps and fixed point theory. In: *Fixed Point Theory, Variational Analysis, and Optimization*, edited by Saleh Abdullah, R. Al-Mezel, Falleh Rajallah M. Al-Solamy, and Qamrul Hasan Ansari, pp. 77–136. CRC Press, Boca Raton.
5. Browder, F.E. 1983. Fixed point theory and nonlinear problems. *Bull. Am. Math. Soc. (N.S.)* 9(1), 1–39.
6. Cauty, R. 2001. Solution du probl'eme de point fixe de Schauder. (French) [Solution of Schauder's fixed point problem]. *Fund. Math.* 170(3), 231–246.
7. Cauty, R. 2005. Rétractes absolus de voisinage algébriques. (French) [Algebraic absolute neighborhood retracts]. *Serdica Math. J.* 31(4), 309–354.
8. Cauty, R. 2007. Le théoréme de Lefschetz - Hopf pour les applications compactes des espaces ULC. (French) [The Lefschetz - Hopf theorem for compact maps of uniformly locally contractible spaces]. *J. Fixed Point Theory Appl.* 1(1), 123–134.
9. Chang, S.S., Cho, Y.J., Park, S. and Yuan, G.X. 2024. Fixed point theorems for quasi upper semicontinuous set-valued mappings in p-vector and locally p-convex spaces. In: *Advanced Mathematical Ananlysis and its Applications,* edited by Pradip Debnath,

Delfim F. M. Torres, and Yeol Je Cho, 5, pp. 57–72. CRC Press, Boca Raton, FL. https://doi.org/10.1201/9781003388678

10. Ennassik, M., Maniar, L. and Taoudi, M.A. 2021. Fixed point theorems in *r*-normed and locally *r*-convex spaces and applications. *Fixed Point Theory*, 22(2), 625–644.
11. Ennassik, M. and Taoudi, M.A. 2021. On the conjecture of Schauder. *J. Fixed Point Theory Appl.* 23(4), 52 (2021).
12. Fan, K. 1952. Fixed-point and minimax theorems in locally convex topological linear spaces. *Proc. Nat. Acad. Sci. U.S.A.*, 38, 121–126.
13. Granas, A. and Dugundji, J. 2003. *Fixed Point Theory. Springer Monographs in Mathematics (SMM).* Springer-Verlag, New York.
14. Jarchow, H. 1981. *Locally Convex Spaces.* B.G. Teubner, Stuttgart, 1981.
15. Kalton, N.J. 1977. Compact p-convex sets. *Q.J.Math.Oxf. Ser.* 28(2), 301–308.
16. Kalton, N.J. 1977. Universal spaces and universal bases in metric linear spaces. *Studia Math.* 61, 161–191.
17. Kalton, N.J., Peck, N.T. and Roberts, J.W. 1984. *An F-space Sampler*, London Math. Soc. Lecture Note Series 89. Cambridge University Press, London.
18. Kreyszig, E. 1989. *Introductory Functional Analysis with Applications. Wiley Classics Library*. Wiley, New York.
19. Mauldin, R.D. 2015. *The Scottish Book, Mathematics from the Scottish Café with Selected Problems from the New Scottish Book*, Second Edition, Birkhauser.
20. Park, S. 2004. *On Some Conjectures and Problems in Analytical Fixed Point Theory, Revisited*, vol. 1365, pp. 166–175. RIMS Kokyuroku, Kyoto University, Kyoto.
21. Park, S. 2021. One hundred years of the Brouwer fixed point theorem. *J. Nat. Acad. Sci., ROK, Nat. Sci. Ser.* 60(1), 1–77 (2021).
22. Qiu, J. and Rolewicz, S. 2008. Ekeland's variational principle in locally p-convex spaces and related results. *Stud. Math.* 186(3), 219–235.
23. Shapiro, J.H. 1969. Examples of proper, closed, weakly dense subspaces in non locally convex F-spaces. *Israel J. Math.* 7, 369–380.
24. Wang, J.Y. 2013. *An Introduction to Locally p-convex Spaces*, pp. 80–81. Academic Press, Beijing.

25. Xiao, J.Z. and Zhu, X.H. 2011. Some fixed point theorems for s-convex subsetes in p-nomed spaces. *Nonlinear Anal.* 72, 1738–1748.
26. Yuan, G.X. 2023. Fixed point theorems and applications in p-vector spaces. *Fixed Point Theory Algorithms Sci. Eng.* 10. https://doi.org/10.1186/s13663-023-00747-w.
27. Zeidler, E. 1993. *Nonlinear Functional Analysis and Its Applications I, Fixed-Point Theorems*. Springer, Berlin.

CHAPTER 4

New Versions of Kannan-Type Map

Nihal Taş

4.1 INTRODUCTION

Since Stefan Banach's time, metric fixed point theory has been a well-researched area. The Banach fixed point theorem offers a way to ensure that a function satisfying the contraction requirement on a complete metric space has a fixed point and that it is unique. However, despite having a unique fixed point, some functions do not satisfy the requirements of Banach's fixed point theorem. It is crucial to research various Banach fixed point theorem generalizations in this context.

The extension of the employed contractive condition is one of the generalization studies that have been done. Numerous contraction circumstances that arise are documented in the literature, along with a thorough analysis and instances of the connections between them. The Kannan-type contraction is one of the generalized contractions (see [7, 8]). Within this framework, a novel generalization of the Kannan-type contractive condition has also been added using distinct methods (for example, see [11]). Furthermore, the contractive condition is also made more universal by using the integral approach. This method allows for the definition of integral type contractive conditions, which are then used to study new fixed point results (for example, see [3]).

Expanding the metric space under study is an additional technique for obtaining a new generalization of the Banach fixed point theorem. Two notions that were presented in the literature as extensions of metric space are b-metric space [1] and S-metric space [13]. Following the definition of a

DOI: 10.1201/9781003530602-4

new generalized metric space, many fixed point theorems are further investigated and proved, as well as the space's link to other spaces already known in the literature and its topological characteristics. Studying a new metric space thus introduces new research topics. It is significant in this instance for upcoming research.

Taking into account all of these references, we explore new and updated generalizations of the well-known Kannan-type contractive condition on both S-metric spaces and b-metric spaces in this chapter. In order to do this, we first define and explore the fundamental properties of the δ_S-Kannan-type map and δ_S-modulus Kannan-type map on S-metric space. We explain the relationship between the notion of Kannan-type map, which is known to exist in the literature on S-metric spaces, and provide an example demonstrating that it is, in fact, a generalization. These new ideas discovered for S-metric spaces are directly new for b-metric spaces, since every S-metric yields a b-metric. As a result, we also provided the b-metric spaces' established contractive condition. Lastly, new Kannan contractive conditions of generalized integral type are introduced on both S-metric spaces and b-metric spaces, by utilizing the integral approximation.

4.2 PRELIMINARIES

We review a few essential ideas in this section:

Definition 2.1 [13] Let $\mathcal{N}$ be a nonempty set and $\mathcal{S}:\mathcal{N}\times\mathcal{N}\times\mathcal{N}\rightarrow[0,\infty)$ be a function satisfying the following conditions for all $\varrho,\varpi,\vartheta,\xi\in X$:

(S1) $\mathcal{S}(\varrho,\varpi,\vartheta)=0\Longleftrightarrow\varrho=\varpi=\vartheta$,

(S2) $\mathcal{S}(\varrho,\varpi,\vartheta)\leq\mathcal{S}(\varrho,\varrho,\xi)+\mathcal{S}(\varpi,\varpi,\xi)+\mathcal{S}(\vartheta,\vartheta,\xi)$. Then $\mathcal{S}$ is called an S-metric on X and the pair $(\mathcal{N},\mathcal{S})$ is called an S-metric space.

Lemma 2.1 [13] *Let $(\mathcal{N},\mathcal{S})$ be an S-metric space and $\varrho,\varpi\in X$. Then we have*

$$\mathcal{S}(\varrho,\varrho,\varpi)=\mathcal{S}(\varpi,\varpi,\varrho).$$

Definition 2.2 [13] Let $(\mathcal{N}, \mathcal{S})$ be an S-metric space and $\{\varrho_n\}$ be a sequence in this space.

(1) A sequence $\{\varrho_n\} \subset \mathcal{N}$ converges to $\varrho \in \mathcal{N}$ if $\mathcal{S}(\varrho_n, \varrho_n, \varrho) \to 0$ as $n \to \infty$, that is, for each $\epsilon > 0$, there exists $n_0 \in \mathbb{N}$ such that for all $n \geq n_0$ we have $\mathcal{S}(\varrho_n, \varrho_n, \varrho) < \epsilon$.

(2) A sequence $\{\varrho_n\} \subset \mathcal{N}$ is a Cauchy sequence if $\mathcal{S}(\varrho_n, \varrho_n, \varrho_m) \to 0$ as $n, m \to \infty$, that is, for each $\epsilon > 0$, there exists $n_0 \in \mathbb{N}$ such that for all $n, m \geq n_0$ we have $\mathcal{S}(\varrho_n, \varrho_n, \varrho_m) < \epsilon$.

(3) The S-metric space $(\mathcal{N}, \mathcal{S})$ is complete if every Cauchy sequence is a convergent sequence.

Lemma 2.2 [13] *Let $(\mathcal{N}, \mathcal{S})$ be an S-metric space and $\{\varrho_n\}$, $\{\varpi_n\}$ be two sequences in this space. If $\varrho_n \to \varrho$ and $\varpi_n \to \varpi$ then we have*

$$\mathcal{S}(\varrho_n, \varrho_n, \varpi_n) \to \mathcal{S}(\varrho, \varrho, \varpi).$$

Lemma 2.3 [14] *Let $(\mathcal{N}, \mathcal{S}), (\mathcal{M}, \mathcal{S}')$ be two S-metric spaces, $\Upsilon : \mathcal{N} \to \mathcal{M}$ be a function, and $\{\varrho_n\}$ be a sequence in this space. Then Υ is continuous at $\varrho \in \mathcal{N}$ if and only if $\Upsilon(\varrho_n) \to \Upsilon(\varrho)$ whenever $\varrho_n \to \varrho$.*

Definition 2.3 [13] For each $\varrho \in \mathcal{N}$, if the iteration sequence $\{\Upsilon^n \varrho\}$ converges to ϱ_Υ, which is the fixed point of Υ, according to the S-metric, that is,

$$\begin{aligned} \exists \text{ unique } \varrho_\Upsilon \in \mathcal{N} : \varrho_\Upsilon &= \lim_{n\to\infty} \Upsilon^n \varrho, \forall \varrho \in \mathcal{N} \\ &\Longrightarrow \lim_{n\to\infty} \mathcal{S}(\Upsilon^n \varrho, \Upsilon^n \varrho, \varrho_\Upsilon) = 0, \end{aligned}$$

then Υ is called an S-Picard operator.

Definition 2.4 [4] Let $(\mathcal{N}, \mathcal{S})$ be an S-metric space. A sequence $\{\varrho_n\}$ in $\mathcal{N}$ is said to be asymptotically Υ-regular if

$$\lim_{n\to\infty} \mathcal{S}(\Upsilon \varrho_n, \Upsilon \varrho_n, \varrho_n) = 0.$$

According to [6], there is the following relationship between a metric and an S-metric:

Lemma 2.4 [6] *Let $(\mathcal{N}, d)$ be a metric space. Then, the subsequent characteristics are satisfied:*

(1) $\mathcal{S}_d(\varrho, \varpi, \vartheta) = d(\varrho, \vartheta) + d(\varpi, \vartheta)$ *for all* $\varrho, \varpi, \vartheta \in \mathcal{N}$ *is an S-metric on X.*

(2) $\varrho_n \to \varrho$ *in* $(\mathcal{N}, d)$ *if and only if* $\varrho_n \to \varrho$ *in* $(\mathcal{N}, \mathcal{S}_d)$.

(3) (ϱ_n) *is Cauchy in* $(\mathcal{N}, d)$ *if and only if* (ϱ_n) *is Cauchy in* $(\mathcal{N}, \mathcal{S}_d)$.

(4) $(\mathcal{N}, d)$ *is complete if and only if* $(\mathcal{N}, \mathcal{S}_d)$ *is complete.*

The metric $\mathcal{S}_d$ was called as the S-metric generated by d [9]. There are known instances of S-metrics that are not generated by any metric (for further information, see [6, 9, 10]).

An example of an S-metric that is not generated by any metric may be found below.

Example 2.1 [9] Let $\mathcal{N} = \mathbb{R}$ and the function $\mathcal{S}: \mathcal{N} \times \mathcal{N} \times \mathcal{N} \to [0, \infty)$ be defined as

$$\mathcal{S}(\varrho, \varpi, \vartheta) = |\varrho - \vartheta| + |\varrho + \vartheta - 2\varpi|,$$

for all $\varrho, \varpi, \vartheta \in \mathbb{R}$. *Then* $\mathcal{S}$ *is an S-metric which is not generated by any metric and the pair* $(\mathcal{N}, \mathcal{S})$ *is an S-metric space.*

Moreover, according to Gupta, each S-metric on $\mathcal{N}$ defines a metric d_S on $\mathcal{N}$ in the following way:

$$d_S(\varrho, \varpi) = \mathcal{S}(\varrho, \varrho, \varpi) + \mathcal{S}(\varpi, \varpi, \varrho), \tag{2.1}$$

for all $\varrho, \varpi \in \mathcal{N}$ [5].

The function $d_S(\varrho, \varpi)$ defined in (2.1) does not, however, always define a metric because the triangle inequality is not satisfied for all members of $\mathcal{N}$ every time as demonstrated by the example that follows.

Example 2.2 [9] Let $\mathcal{N} = \{1, 2, 3\}$. If we consider the function $\mathcal{S}: \mathcal{N} \times \mathcal{N} \times \mathcal{N} \to [0, \infty)$ defined by

$$\mathcal{S}(1, 1, 2) = \mathcal{S}(2, 2, 1) = 5,$$

$$\mathcal{S}(2,2,3) = \mathcal{S}(3,3,2) = \mathcal{S}(1,1,3) = \mathcal{S}(3,3,1) = 2,$$

$$\mathcal{S}(\varrho, \varpi, \vartheta) = 0 \text{ if } \varrho = \varpi = \vartheta,$$

$$\mathcal{S}(\varrho, \varpi, \vartheta) = 1 \text{ otherwise},$$

for all $\varrho, \varpi, \vartheta \in \mathcal{N}$, *then* $\mathcal{S}$ *does not generate a metric* $d_{\mathcal{S}}$. *Also,* $\mathcal{S}$ *is an S-metric on* $\mathcal{N}$ *which is not generated by any metric d.*

Additionally, the following theorem demonstrated the connection between an S-metric and a b-metric as specified in [1].

Theorem 2.1 *[14] Let* $(\mathcal{N}, \mathcal{S})$ *be an S-metric space and*

$$d^S(\varrho, \varpi) = \mathcal{S}(\varrho, \varrho, \varpi),$$

for all $\varrho, \varpi \in \mathcal{N}$. *Then we have*

(1) d^S *is a b-metric on* $\mathcal{N}$,

(2) $\varrho_n \to \varrho$ *in* $(\mathcal{N}, \mathcal{S})$ *if and only if* $\varrho_n \to \varrho$ *in* $(\mathcal{N}, d^S)$,

(3) $\{\varrho_n\}$ *is Cauchy in* $(\mathcal{N}, \mathcal{S})$ *if and only if* $\{\varrho_n\}$ *is Cauchy in* $(\mathcal{N}, d^S)$.

The metric d^S is called the b-metric generated by $\mathcal{S}$.

It was necessary to investigate new fixed-point results on S-metric spaces in light of the aforementioned relationships.

Now, we recall the following result on S-metric spaces presented in [13], which is analogous to the Banach's contraction principle [2]:

Let $(\mathcal{N}, \mathcal{S})$ be a complete S-metric space, Υ be a self-mapping of X and

$$\mathcal{S}(\Upsilon\varrho, \Upsilon\varrho, \Upsilon\varpi) \leq \ell \mathcal{S}(\varrho, \varrho, \varpi),$$

for some $0 \leq \ell < 1$ and all $\varrho, \varpi \in \mathcal{N}$. Then Υ has a unique fixed point in $\mathcal{N}$ and Υ is continuous at the fixed point.

As we can see in the example below, while there is a self-mapping Υ with a fixed point, it does not meet Banach's contraction principle on S-metric spaces:

Let $[0,1]$ be the complete S-metric space with the S-metric defined as in 2.1. Let us define the self-mapping $\Upsilon : \mathcal{N} \to \mathcal{N}$ as

$$\Upsilon\varrho = 1 - \varrho.$$

Subsequently Υ has a fixed point $\varrho = \frac{1}{2}$, but Υ does not satisfy the Banach's contraction principle. Therefore, learning a few new fixed point theorems is crucial. New generalized fixed-point results on S-metric spaces are thereby justified. For example, using the Kannan-type map idea (see [7, 8]), the corollary that follows was established.

Corollary 2.1 *[14] Let $(\mathcal{N}, \mathcal{S})$ be a complete S-metric space, Υ be a self-mapping of X, and*

$$\mathcal{S}(\Upsilon\varrho, \Upsilon\varrho, \Upsilon\varpi) \leq \ell\left[\mathcal{S}(\Upsilon\varrho, \Upsilon\varrho, \varrho) + \mathcal{S}(\Upsilon\varpi, \Upsilon\varpi, \varpi)\right],$$

for some $0 \leq \ell < \frac{1}{2}$ and all $\varrho, \varpi \in \mathcal{N}$. Then Υ has a unique fixed point in $\mathcal{N}$ and Υ is continuous at the fixed point.

4.3 MAIN RESULTS

Here, it is assumed that $(\mathcal{N}, \mathcal{S})$ be an S-metric space and $\Upsilon : \mathcal{N} \to \mathcal{N}$ be a self-mapping.

4.3.1 A New Form of Kannan-Type Map on S-Metric Spaces

We present two novel generalizations of the Kannan-type map on S-metric spaces in this subsection.

Definition 3.1 Suppose that there exists a $\delta \in (0, \infty)$ such that Υ satisfies

$$\mathcal{S}(\Upsilon\varrho, \Upsilon\varrho, \Upsilon\varpi) \leq \gamma \max\left\{ \begin{array}{c} \left[\frac{\mathcal{S}(\Upsilon\varrho,\Upsilon\varrho,\varrho)}{\delta} + \mathcal{S}(\Upsilon\varpi, \Upsilon\varpi, \varpi)\right], \\ \left[\mathcal{S}(\Upsilon\varrho, \Upsilon\varrho, \varrho) + \frac{\mathcal{S}(\Upsilon\varpi,\Upsilon\varpi,\varpi)}{\delta}\right] \end{array} \right\},$$

for each $\varrho, \varpi \in \mathcal{N}$, where

$$\gamma \in [0, \infty) : \gamma(1 + \delta) < \delta.$$

Then Υ is called a δ_S-Kannan-type map.

Remark 3.1 (1) Since $\delta \in (0, \infty)$ and $\gamma(1 + \delta) < \delta$, we have

$$0 \leq \gamma < \frac{\delta}{1 + \delta} \Longrightarrow \gamma \in [0, 1).$$

(2) If we take $\delta = 1$, then a Kannan-type map, first proposed in [14], and a δ_S-Kannan-type map are equivalent.

(3) We assume without losing generality

$$\mathcal{S}(\Upsilon\varpi, \Upsilon\varpi, \varpi) \geq \mathcal{S}(\Upsilon\varrho, \Upsilon\varrho, \varrho),$$

for each $\varrho, \varpi \in \mathcal{N}$.

We take into account the following situations:

Case 1: Let $\delta \in (1, \infty)$. For each $\varrho, \varpi \in \mathcal{N}$, we have

$$\frac{\mathcal{S}(\Upsilon\varrho, \Upsilon\varrho, \Upsilon\varpi)}{\frac{\mathcal{S}(\Upsilon\varrho, \Upsilon\varrho, \varrho)}{\delta} + \mathcal{S}(\Upsilon\varpi, \Upsilon\varpi, \varpi)} \leq \gamma < \frac{\delta}{1+\delta}$$

$$\Longleftrightarrow [\mathcal{S}(\Upsilon\varrho, \Upsilon\varrho, \Upsilon\varpi) - \mathcal{S}(\Upsilon\varpi, \Upsilon\varpi, \varpi)]\,\delta$$
$$< [\mathcal{S}(\Upsilon\varrho, \Upsilon\varrho, \varrho) - \mathcal{S}(\Upsilon\varrho, \Upsilon\varrho, \Upsilon\varpi)],$$

where

$$\mathcal{S}(\Upsilon\varrho, \Upsilon\varrho, \Upsilon\varpi) \leq \mathcal{S}(\Upsilon\varrho, \Upsilon\varrho, \varrho)$$

or

$$\mathcal{S}(\Upsilon\varrho, \Upsilon\varrho, \varrho) \leq \mathcal{S}(\Upsilon\varrho, \Upsilon\varrho, \Upsilon\varpi).$$

For each $\varrho, \varpi \in \mathcal{N}$, *we get*

$$\frac{\mathcal{S}(\Upsilon\varrho, \Upsilon\varrho, \Upsilon\varpi)}{\frac{\mathcal{S}(\Upsilon\varrho, \Upsilon\varrho, \varrho)}{\delta} + \mathcal{S}(\Upsilon\varpi, \Upsilon\varpi, \varpi)} \leq \gamma < \frac{\delta}{1+\delta}$$

$$\Longleftrightarrow \frac{\mathcal{S}(\Upsilon\varrho, \Upsilon\varrho, \Upsilon\varpi) - \mathcal{S}(\Upsilon\varrho, \Upsilon\varrho, \varrho)}{\mathcal{S}(\Upsilon\varpi, \Upsilon\varpi, \varpi) - \mathcal{S}(\Upsilon\varrho, \Upsilon\varrho, \Upsilon\varpi)} < \delta < \infty$$

$$\Longrightarrow \mathcal{S}(\Upsilon\varrho, \Upsilon\varrho, \varrho) \leq \mathcal{S}(\Upsilon\varrho, \Upsilon\varrho, \Upsilon\varpi) \leq \mathcal{S}(\Upsilon\varpi, \Upsilon\varpi, \varpi), \tag{3.1}$$

where

$$\mathcal{S}(\Upsilon\varrho, \Upsilon\varrho, \Upsilon\varpi) \leq \frac{1}{2}[\mathcal{S}(\Upsilon\varrho, \Upsilon\varrho, \varrho) + \mathcal{S}(\Upsilon\varpi, \Upsilon\varpi, \varpi)]$$
$$\Longrightarrow \text{inequality (3.1)}$$

or

$$\mathcal{S}(\Upsilon\varrho,\Upsilon\varrho,\Upsilon\varpi) \geq \frac{1}{2}[\mathcal{S}(\Upsilon\varrho,\Upsilon\varrho,\varrho)+\mathcal{S}(\Upsilon\varpi,\Upsilon\varpi,\varpi)]$$
$$\Longrightarrow \text{inequality (3.1)}.$$

Case 2: Let $\delta = 1$. For each $\varrho, \varpi \in \mathcal{N}$, we have

$$\frac{\mathcal{S}(\Upsilon\varrho,\Upsilon\varrho,\Upsilon\varpi)}{\mathcal{S}(\Upsilon\varrho,\Upsilon\varrho,\varrho)+\mathcal{S}(\Upsilon\varpi,\Upsilon\varpi,\varpi)} \leq \gamma < \frac{1}{2}$$

$$\Longrightarrow \mathcal{S}(\Upsilon\varrho,\Upsilon\varrho,\Upsilon\varpi) < \frac{1}{2}[\mathcal{S}(\Upsilon\varrho,\Upsilon\varrho,\varrho)+\mathcal{S}(\Upsilon\varpi,\Upsilon\varpi,\varpi)]$$

$$\Longleftrightarrow \frac{\mathcal{S}(\Upsilon\varrho,\Upsilon\varrho,\Upsilon\varpi)-\mathcal{S}(\Upsilon\varrho,\Upsilon\varrho,\varrho)}{\mathcal{S}(\Upsilon\varpi,\Upsilon\varpi,\varpi)-\mathcal{S}(\Upsilon\varrho,\Upsilon\varrho,\Upsilon\varpi)} < 1,$$

or

$$\frac{\mathcal{S}(\Upsilon\varpi,\Upsilon\varpi,\varpi)-\mathcal{S}(\Upsilon\varrho,\Upsilon\varrho,\Upsilon\varpi)}{\mathcal{S}(\Upsilon\varrho,\Upsilon\varrho,\Upsilon\varpi)-\mathcal{S}(\Upsilon\varrho,\Upsilon\varrho,\varrho)} > 1,$$

provided

$$\mathcal{S}(\Upsilon\varrho,\Upsilon\varrho,\varrho) \leq \mathcal{S}(\Upsilon\varrho,\Upsilon\varrho,\Upsilon\varpi) \leq \mathcal{S}(\Upsilon\varpi,\Upsilon\varpi,\varpi).$$

Case 3: Let $\delta \in (0,1)$. For each $\varrho, \varpi \in \mathcal{N}$, we have

$$\frac{\mathcal{S}(\Upsilon\varrho,\Upsilon\varrho,\Upsilon\varpi)}{\mathcal{S}(\Upsilon\varrho,\Upsilon\varrho,\varrho)+\frac{\mathcal{S}(\Upsilon\varpi,\Upsilon\varpi,\varpi)}{\delta}} \leq \gamma < \frac{\delta}{1+\delta},$$

$$\Longleftrightarrow [\mathcal{S}(\Upsilon\varrho,\Upsilon\varrho,\Upsilon\varpi)-\mathcal{S}(\Upsilon\varrho,\Upsilon\varrho,\varrho)]\,\delta$$
$$< [\mathcal{S}(\Upsilon\varpi,\Upsilon\varpi,\varpi)-\mathcal{S}(\Upsilon\varrho,\Upsilon\varrho,\Upsilon\varpi)],$$

where

$$\mathcal{S}(\Upsilon\varrho,\Upsilon\varrho,\Upsilon\varpi) \leq \mathcal{S}(\Upsilon\varrho,\Upsilon\varrho,\varrho),$$

or

$$\mathcal{S}(\Upsilon\varrho,\Upsilon\varrho,\varrho) < \mathcal{S}(\Upsilon\varrho,\Upsilon\varrho,\Upsilon\varpi).$$

For each $\varrho, \varpi \in \mathcal{N}$, *we get*

$$\frac{\mathcal{S}(\Upsilon\varrho, \Upsilon\varrho, \Upsilon\varpi)}{\mathcal{S}(\Upsilon\varrho, \Upsilon\varrho, \varrho) + \frac{\mathcal{S}(\Upsilon\varpi, \Upsilon\varpi, \varpi)}{\delta}} \leq \gamma < \frac{\delta}{1+\delta}$$

$$\Longleftrightarrow \delta < \frac{\mathcal{S}(\Upsilon\varpi, \Upsilon\varpi, \varpi) - \mathcal{S}(\Upsilon\varrho, \Upsilon\varrho, \Upsilon\varpi)}{\mathcal{S}(\Upsilon\varrho, \Upsilon\varrho, \Upsilon\varpi) - \mathcal{S}(\Upsilon\varrho, \Upsilon\varrho, \varrho)}$$

$$\Longrightarrow \mathcal{S}(\Upsilon\varrho, \Upsilon\varrho, \varrho) \leq \mathcal{S}(\Upsilon\varrho, \Upsilon\varrho, \Upsilon\varpi) \leq \mathcal{S}(\Upsilon\varpi, \Upsilon\varpi, \varpi), \tag{3.2}$$

where

$$\mathcal{S}(\Upsilon\varrho, \Upsilon\varrho, \Upsilon\varpi) \leq \frac{1}{2}\left[\mathcal{S}(\Upsilon\varrho, \Upsilon\varrho, \varrho) + \mathcal{S}(\Upsilon\varpi, \Upsilon\varpi, \varpi)\right]$$
$$\Longrightarrow \text{inequality (3.2)}$$

or

$$\mathcal{S}(\Upsilon\varrho, \Upsilon\varrho, \Upsilon\varpi) \geq \frac{1}{2}\left[\mathcal{S}(\Upsilon\varrho, \Upsilon\varrho, \varrho) + \mathcal{S}(\Upsilon\varpi, \Upsilon\varpi, \varpi)\right]$$
$$\Longrightarrow \text{inequality (3.2)}.$$

(4) *If the S-metric is generated by d, then the notion of a* δ_S*-Kannan-type map coincides with the notion of* δ*-Kannan map introduced in [11].*

Applying the analogous reasoning presented in [11, 12], we provide the subsequent illustration:

Example 3.1 Let $\mathcal{N} = [0, 1]$ be an S-metric space with the S-metric defined as in 2.1. Let us consider the self-mapping $\Upsilon : \mathcal{N} \to \mathcal{N}$ defined as

$$\Upsilon\varrho = \begin{cases} \frac{\varrho}{3} & ; \quad \varrho \in [0,1) \\ \frac{1}{6} & ; \quad \varrho = 1 \end{cases},$$

for all $\varrho \in \mathcal{N}$ *[12]. Then* Υ *is a* δ_S*-Kannan-type map, but it is not Kannan-type map.*

Without sacrificing the generality, think about $\varrho < \varpi$. *Let's examine the following scenarios:*

Case 1: For $\gamma \in [0, \infty)$, $\delta \in (0, \infty)$ and each $\varrho, \varpi \in [0, 1)$, let

$$\max\left\{\frac{\varrho}{\delta} + \varpi, \varrho + \frac{\varpi}{\delta}\right\} = \begin{cases} \frac{\varrho}{\delta} + \varpi & ; \quad \delta \in (1, \infty) \\ \varrho + \frac{\varpi}{\delta} & ; \quad \delta \in (0, 1) \end{cases}.$$

Then we have

$$\frac{|\varrho-\varpi|}{2\max\left\{\frac{\varrho}{\delta}+\varpi,\varrho+\frac{\varpi}{\delta}\right\}}\leq\gamma.$$

Case 2: For $\gamma\in[0,\infty)$, $\delta\in(0,\infty)$ and each $\varrho\in[0,1)$, $\varpi=1$, let

$$\max\left\{\frac{4\varrho}{\delta}+5,4\varrho+\frac{5}{\delta}\right\}=\begin{cases}\frac{4\varrho}{\delta}+5 & ;\ \delta\in(1,\infty)\\ 4\varrho+\frac{5}{\delta} & ;\ \delta\in(0,1)\end{cases}.$$

Then we have

$$\frac{|2\varrho-1|}{\max\left\{\frac{4\varrho}{\delta}+5,4\varrho+\frac{5}{\delta}\right\}}\leq\gamma.$$

Hence, for $\varrho,\varpi\in\mathcal{N}$, we get

$$\frac{|\Upsilon\varrho-\Upsilon\varpi|}{\max\left\{\frac{|\varrho-\Upsilon\varrho|}{\delta}+|\varpi-\Upsilon\varpi|,|\varrho-\Upsilon\varrho|+\frac{|\varpi-\Upsilon\varpi|}{\delta}\right\}}\leq\gamma<\frac{\delta}{1+\delta},$$

that is, Υ is a δ_S-Kannan-type map.

If we take $\varrho=0$ and $\varpi=\frac{1}{3}$, it can be easily that Υ is not a Kannan-type map.

The contraction provided in [13] is utilized to demonstrate the subsequent theorem:

Theorem 3.1 *Assume that Υ satisfies the condition*

$$\mathcal{S}(\Upsilon\varrho,\Upsilon\varrho,\Upsilon\varpi)\leq\ell\mathcal{S}(\varrho,\varrho,\varpi),$$

for some $\ell\in[0,1)$ and for all $\varrho,\varpi\in\mathcal{N}$.

(1) *If $\ell\in\left[0,\frac{\delta}{1+2\delta}\right)$, where $\delta\in(0,1]$, then Υ is a δ_S-Kannan-type map with*

$$\gamma=\frac{2\ell}{1-2\ell}.$$

(2) *If $\ell\in\left[0,\frac{1}{2+\delta}\right)$, where $\delta\in[1,\infty)$, then Υ is a δ_S-Kannan-type map with*

$$\gamma'=\frac{2\ell\delta}{1-2\ell}.$$

Proof. Using the hypothesis, Lemma 2.1, and the condition (**S2**), we obtain

$$\begin{aligned}\mathcal{S}(\Upsilon\varrho,\Upsilon\varrho,\Upsilon\varpi) &\leq \ell\mathcal{S}(\varrho,\varrho,\varpi)\\ &\leq \ell\left[2\mathcal{S}(\varrho,\varrho,\Upsilon\varrho)+\mathcal{S}(\varpi,\varpi,\Upsilon\varrho)\right]\\ &\leq \ell\left[2\mathcal{S}(\varrho,\varrho,\Upsilon\varrho)+2\mathcal{S}(\varpi,\varpi,\Upsilon\varpi)+\mathcal{S}(\Upsilon\varrho,\Upsilon\varrho,\Upsilon\varpi)\right]\\ &\leq \ell\left[2\mathcal{S}(\varrho,\varrho,\Upsilon\varrho)+2\mathcal{S}(\varpi,\varpi,\Upsilon\varpi)+2\mathcal{S}(\Upsilon\varrho,\Upsilon\varrho,\Upsilon\varpi)\right]\\ &= 2\ell\left[\mathcal{S}(\Upsilon\varrho,\Upsilon\varrho,\varrho)+\mathcal{S}(\Upsilon\varrho,\Upsilon\varrho,\Upsilon\varpi)+\mathcal{S}(\Upsilon\varpi,\Upsilon\varpi,\varpi)\right],\end{aligned}$$

for each $\varrho,\varpi\in\mathcal{N}$.

(1) We have

$$\mathcal{S}(\Upsilon\varrho,\Upsilon\varrho,\Upsilon\varpi)-2\ell\mathcal{S}(\Upsilon\varrho,\Upsilon\varrho,\Upsilon\varpi)\leq 2\ell\left[\mathcal{S}(\Upsilon\varrho,\Upsilon\varrho,\varrho)+\mathcal{S}(\Upsilon\varpi,\Upsilon\varpi,\varpi)\right]$$

$$\Longrightarrow (1-2\ell)\mathcal{S}(\Upsilon\varrho,\Upsilon\varrho,\Upsilon\varpi)\leq 2\ell\left[\mathcal{S}(\Upsilon\varrho,\Upsilon\varrho,\varrho)+\mathcal{S}(\Upsilon\varpi,\Upsilon\varpi,\varpi)\right]$$

$$\Longrightarrow \mathcal{S}(\Upsilon\varrho,\Upsilon\varrho,\Upsilon\varpi)\leq \frac{2\ell}{1-2\ell}\left[\mathcal{S}(\Upsilon\varrho,\Upsilon\varrho,\varrho)+\mathcal{S}(\Upsilon\varpi,\Upsilon\varpi,\varpi)\right]$$

$$\Longrightarrow \mathcal{S}(\Upsilon\varrho,\Upsilon\varrho,\Upsilon\varpi)\leq \gamma\left[\frac{\mathcal{S}(\Upsilon\varrho,\Upsilon\varrho,\varrho)}{\delta}+\mathcal{S}(\Upsilon\varpi,\Upsilon\varpi,\varpi)\right]$$

$$\Longrightarrow \mathcal{S}(\Upsilon\varrho,\Upsilon\varrho,\Upsilon\varpi)\leq \gamma\max\left\{\begin{array}{l}\left[\frac{\mathcal{S}(\Upsilon\varrho,\Upsilon\varrho,\varrho)}{\delta}+\mathcal{S}(\Upsilon\varpi,\Upsilon\varpi,\varpi)\right],\\ \left[\mathcal{S}(\Upsilon\varrho,\Upsilon\varrho,\varrho)+\frac{\mathcal{S}(\Upsilon\varpi,\Upsilon\varpi,\varpi)}{\delta}\right]\end{array}\right\},$$

for each $\varrho,\varpi\in\mathcal{N}$ and $\delta\in(0,1]$.

(2) We have

$$\begin{aligned}\mathcal{S}(\Upsilon\varrho,\Upsilon\varrho,\Upsilon\varpi) &\leq \gamma\left[\mathcal{S}(\Upsilon\varrho,\Upsilon\varrho,\varrho)+\mathcal{S}(\Upsilon\varpi,\Upsilon\varpi,\varpi)\right]\\ &\leq \gamma\left[\mathcal{S}(\Upsilon\varrho,\Upsilon\varrho,\varrho)+\delta\mathcal{S}(\Upsilon\varpi,\Upsilon\varpi,\varpi)\right]\\ &\leq \gamma'\left[\frac{\mathcal{S}(\Upsilon\varrho,\Upsilon\varrho,\varrho)}{\delta}+\mathcal{S}(\Upsilon\varpi,\Upsilon\varpi,\varpi)\right]\end{aligned}$$

$$\Longrightarrow \mathcal{S}(\Upsilon\varrho,\Upsilon\varrho,\Upsilon\varpi)\leq \gamma'\max\left\{\begin{array}{l}\left[\frac{\mathcal{S}(\Upsilon\varrho,\Upsilon\varrho,\varrho)}{\delta}+\mathcal{S}(\Upsilon\varpi,\Upsilon\varpi,\varpi)\right],\\ \left[\mathcal{S}(\Upsilon\varrho,\Upsilon\varrho,\varrho)+\frac{\mathcal{S}(\Upsilon\varpi,\Upsilon\varpi,\varpi)}{\delta}\right]\end{array}\right\},$$

for each $\varrho,\varpi\in\mathcal{N}$ and $\delta\in[1,\infty)$.

Thus, for situations (1) and (2), Υ is a $\delta_{\mathcal{S}}$-Kannan-type map.□

Definition 3.2 Suppose that there exists a $\delta \in \mathbb{R}-\{0\}$ such that Υ satisfies

$$\mathcal{S}\left(\Upsilon\varrho, \Upsilon\varrho, \Upsilon\varpi\right) \leq \gamma \max\left\{ \begin{array}{c} \left[\frac{\mathcal{S}(\Upsilon\varrho,\Upsilon\varrho,\varrho)}{|\delta|} + \mathcal{S}\left(\Upsilon\varpi, \Upsilon\varpi, \varpi\right)\right], \\ \left[\mathcal{S}\left(\Upsilon\varrho, \Upsilon\varrho, \varrho\right) + \frac{\mathcal{S}(\Upsilon\varpi,\Upsilon\varpi,\varpi)}{|\delta|}\right] \end{array} \right\},$$

for each $\varrho, \varpi \in \mathcal{N}$, where

$$\gamma \in [0, \infty) : \gamma(1 + |\delta|) < |\delta| .$$

Then Υ is called a δ_S-modulus Kannan-type map.

Remark 3.2 (1) Since $\delta \in \mathbb{R}-\{0\}$ and $\gamma(1 + \delta) < \delta$, we have

$$0 \leq \gamma < \frac{|\delta|}{1 + |\delta|} \Longrightarrow \gamma \in [0, 1) .$$

(2) *If we take $|\delta| = 1$, then a Kannan-type map, first proposed in [14], and a δ_S-modulus Kannan-type map are equivalent.*

(3) *If $\delta \in (-\infty, 0)$, then Υ is an inverse of δ_S-Kannan-type map as Υ satisfies*

$$\mathcal{S}\left(\Upsilon\varrho, \Upsilon\varrho, \Upsilon\varpi\right) \leq \gamma \max\left\{ \begin{array}{c} \left[\mathcal{S}\left(\Upsilon\varrho, \Upsilon\varrho, \varrho\right) - \frac{\mathcal{S}(\Upsilon\varpi,\Upsilon\varpi,\varpi)}{\delta}\right], \\ \left[\mathcal{S}\left(\Upsilon\varpi, \Upsilon\varpi, \varpi\right) - \frac{\mathcal{S}(\Upsilon\varrho,\Upsilon\varrho,\varrho)}{\delta}\right] \end{array} \right\},$$

for each $\varrho, \varpi \in \mathcal{N}$, where

$$\gamma \in [0, \infty) : \gamma\left(1 - \frac{1}{\delta}\right) < 1.$$

(4) *If the S-metric is generated by d, then the notion of a δ_S-modulus Kannan-type map coincides with the notion of modulus Kannan map introduced in [11].*

Theorem 3.2 *Assume that $(\mathcal{N}, \mathcal{S})$ is complete and Υ is a δ_S-modulus Kannan-type map. Then*

(1) *Υ is a S-Picard operator.*

(2) *Υ is continuous at ϱ_Υ.*

Proof. (1) Let $\varrho, \varpi \in \mathcal{N}$. For some $\delta \in \mathbb{R} - \{0\}$: $|\delta| < 1$ or $|\delta| > 1$, we have

$$\mathcal{S}\left(\Upsilon\varrho, \Upsilon\varrho, \Upsilon^2\varrho\right) \leq \min\left\{\frac{\gamma}{1 - \frac{\gamma}{|\delta|}}, \frac{\frac{\gamma}{|\delta|}}{1-\gamma}\right\}\mathcal{S}\left(\Upsilon\varrho, \Upsilon\varrho, \varrho\right).$$

Let us consider a sequence of $\{\Upsilon^n\varrho\}$. For any $j \in \mathbb{N}$, we have

$$\mathcal{S}\left(\Upsilon^j\varrho, \Upsilon^j\varrho, \Upsilon^{j+1}\varrho\right) \leq \alpha^j\mathcal{S}\left(\Upsilon\varrho, \Upsilon\varrho, \varrho\right),$$

where

$$\alpha = \min\left\{\frac{\frac{\gamma}{|\delta|}}{1-\gamma}, \frac{\gamma}{1 - \frac{\gamma}{|\delta|}}\right\} < 1.$$

It implies that Υ is asymptotically Υ-regular. It follows that $\{\Upsilon^n\varrho\}$ is Cauchy. Since $\mathcal{N}$ is complete, there exists

$$\vartheta = \lim_{n\to\infty} \Upsilon^n\varrho.$$

We claim that

$$\varrho_\Upsilon = \vartheta.$$

Let $\mathcal{S}\left(\Upsilon\vartheta, \Upsilon\vartheta, \vartheta\right) > 0$. Then, using the notion of a δ_S-modulus Kannan-type map, as $n \to \infty$, we get

$$\begin{aligned}
\mathcal{S}\left(\Upsilon^{n+1}\varrho, \Upsilon^{n+1}\varrho, \Upsilon\vartheta\right) &\leq \gamma \max\left\{\begin{array}{c}\left[\frac{\mathcal{S}\left(\Upsilon^{n+1}\varrho, \Upsilon^{n+1}\varrho, \Upsilon^n\varrho\right)}{|\delta|} + \mathcal{S}\left(\Upsilon\vartheta, \Upsilon\vartheta, \vartheta\right)\right], \\ \left[\mathcal{S}\left(\Upsilon^{n+1}\varrho, \Upsilon^{n+1}\varrho, \Upsilon^n\varrho\right) + \frac{\mathcal{S}\left(\Upsilon\vartheta, \Upsilon\vartheta, \vartheta\right)}{|\delta|}\right]\end{array}\right\} \\
&= \gamma \max\left\{\mathcal{S}\left(\Upsilon\vartheta, \Upsilon\vartheta, \vartheta\right), \frac{\mathcal{S}\left(\Upsilon\vartheta, \Upsilon\vartheta, \vartheta\right)}{|\delta|}\right\}.
\end{aligned}$$

Case 1: Let $|\delta| < 1$. Then we have

$$\begin{aligned}
\mathcal{S}\left(\Upsilon\vartheta, \Upsilon\vartheta, \vartheta\right) = \mathcal{S}\left(\vartheta, \vartheta, \Upsilon\vartheta\right) &\leq \gamma\frac{\mathcal{S}\left(\Upsilon\vartheta, \Upsilon\vartheta, \vartheta\right)}{|\delta|} \\
&< \frac{|\delta|}{1+|\delta|}\frac{\mathcal{S}\left(\Upsilon\vartheta, \Upsilon\vartheta, \vartheta\right)}{|\delta|} \\
&= \frac{\mathcal{S}\left(\Upsilon\vartheta, \Upsilon\vartheta, \vartheta\right)}{1+|\delta|} < \mathcal{S}\left(\Upsilon\vartheta, \Upsilon\vartheta, \vartheta\right),
\end{aligned}$$

a contradiction.

Case 2: Let $|\delta| > 1$. Then we have

$$\mathcal{S}(\Upsilon\vartheta, \Upsilon\vartheta, \vartheta) \leq \gamma\mathcal{S}(\Upsilon\vartheta, \Upsilon\vartheta, \vartheta), \gamma \in [0, 1),$$

a contradiction.

It should be $\mathcal{S}(\Upsilon\vartheta, \Upsilon\vartheta, \vartheta) = 0$, that is, $\Upsilon\vartheta = \vartheta$. Also, the δ_S-modulus Kannan-type map has a unique fixed point. Indeed, suppose that ϑ^* is another fixed point of Υ such that $\vartheta \neq \vartheta^*$. Using the notion of δ_S-modulus Kannan-type map, we get

$$\begin{aligned}
\mathcal{S}(\Upsilon\vartheta, \Upsilon\vartheta, \Upsilon\vartheta^*) &= \mathcal{S}(\vartheta, \vartheta, \vartheta^*) \\
&\leq \gamma \max\left\{ \begin{array}{c} \left[\frac{\mathcal{S}(\Upsilon\vartheta, \Upsilon\vartheta, \vartheta)}{|\delta|} + \mathcal{S}(\Upsilon\vartheta^*, \Upsilon\vartheta^*, \vartheta^*)\right], \\ \left[\mathcal{S}(\Upsilon\vartheta, \Upsilon\vartheta, \vartheta) + \frac{\mathcal{S}(\Upsilon\vartheta^*, \Upsilon\vartheta^*, \vartheta^*)}{|\delta|}\right] \end{array} \right\} \\
&= 0,
\end{aligned}$$

a contradiction. It ought to be $\vartheta = \vartheta^*$.

(2) We acquire

$$\begin{aligned}
\mathcal{S}(\Upsilon\varrho_n, \Upsilon\varrho_n, \Upsilon\varrho_\Upsilon) &= \mathcal{S}(\Upsilon\varrho_n, \Upsilon\varrho_n, \varrho_\Upsilon) \\
&\leq \gamma \max\left\{ \begin{array}{c} \left[\frac{\mathcal{S}(\Upsilon\varrho_n, \Upsilon\varrho_n, \varrho_n)}{|\delta|} + \mathcal{S}(\Upsilon\varrho_\Upsilon, \Upsilon\varrho_\Upsilon, \varrho_\Upsilon)\right], \\ \left[\mathcal{S}(\Upsilon\varrho_n, \Upsilon\varrho_n, \varrho_n) + \frac{\mathcal{S}(\Upsilon\varrho_\Upsilon, \Upsilon\varrho_\Upsilon, \varrho_\Upsilon)}{|\delta|}\right] \end{array} \right\} \\
&= \gamma \max\left\{ \frac{\mathcal{S}(\Upsilon\varrho_n, \Upsilon\varrho_n, \varrho_n)}{|\delta|}, \mathcal{S}(\Upsilon\varrho_n, \Upsilon\varrho_n, \varrho_n) \right\},
\end{aligned}$$

as $n \to \infty$, we have

$$\lim_{n\to\infty} \Upsilon\varrho_n = \Upsilon\varrho_\Upsilon = \varrho_\Upsilon.$$

□

4.3.2 A New Form of Kannan-Type Map on b-Metric Spaces

This subsection derives a new version of the Kannan-type map on b-metric spaces via Theorem 2.1. Let $(\mathcal{N}, d^S)$ be a b-metric space generated by any S-metric $\mathcal{S}$.

Definition 3.3 Suppose that there exists a $\delta \in (0, \infty)$ such that Υ satisfies

$$d^S(\Upsilon\varrho, \Upsilon\varpi) \leq \gamma \max \left\{ \begin{array}{c} \left[\frac{d^S(\Upsilon\varrho,\varrho)}{\delta} + d^S(\Upsilon\varpi, \varpi)\right], \\ \left[d^S(\Upsilon\varrho, \varrho) + \frac{d^S(\Upsilon\varpi,\varpi)}{\delta}\right] \end{array} \right\},$$

for each $\varrho, \varpi \in \mathcal{N}$, where

$$\gamma \in [0, \infty) : \gamma(1 + \delta) < \delta.$$

Then Υ is called a δ_{d^S}-Kannan-type map.

Definition 3.4 Suppose that there exists a $\delta \in \mathbb{R} - \{0\}$ such that Υ satisfies

$$d^S(\Upsilon\varrho, \Upsilon\varpi) \leq \gamma \max \left\{ \begin{array}{c} \left[\frac{d^S(\Upsilon\varrho,\varrho)}{|\delta|} + d^S(\Upsilon\varpi, \varpi)\right], \\ \left[d^S(\Upsilon\varrho, \varrho) + \frac{d^S(\Upsilon\varpi,\varpi)}{|\delta|}\right] \end{array} \right\},$$

for each $\varrho, \varpi \in \mathcal{N}$, where

$$\gamma \in [0, \infty) : \gamma(1 + |\delta|) < |\delta|.$$

Then Υ is called a δ_{d^S}-modulus Kannan-type map.

4.4 INTEGRAL APPROACH TO KANNAN-TYPE MAP

Fixed-point theory has been studied recently for a variety of contractive circumstances. For instance, several examined fixed-point solutions were modified to include contractive conditions of the integral kind. Consequently, broader fixed-point theorems were derived.

Through the whole paper we assume that φ: $[0, \infty) \to [0, \infty)$ is a Lebesgue-integrable mapping which is summable (i.e., with finite integral) on each compact subset of $[0, \infty)$, nonnegative, and such that for each $\epsilon > 0$,

$$\int_0^{\epsilon} \varphi(t)\, dt > 0. \tag{4.1}$$

The following theorem shows that Branciari investigated a fixed-point theorem for a general contractive condition of integral type on a complete metric space [3].

Theorem 4.1 [3] Let $(\mathcal{N}, d)$ be a complete metric space, $\ell \in (0,1)$, the function $\varphi\colon [0,\infty) \to [0,\infty)$ be defined as in (4.1), and $\Upsilon\colon \mathcal{N} \to \mathcal{N}$ be a self-mapping such that

$$\int_0^{d(\Upsilon\varrho,\Upsilon\varpi)} \varphi(t)\,dt \leq \ell \int_0^{d(\varrho,\varpi)} \varphi(t)\,dt,$$

for all $\varrho, \varpi \in \mathcal{N}$, then Υ has a unique fixed point $\vartheta \in \mathcal{N}$ such that

$$\lim_{n\to\infty} \Upsilon^n \varrho = \vartheta,$$

for each $\varrho \in \mathcal{N}$.

We provide four new integral Kannan-type map contraction definitions using this method, which are as follows:

Definition 4.1 Suppose that there exists a $\delta \in (0,\infty)$ such that Υ satisfies

$$\int_0^{\mathcal{S}(\Upsilon\varrho,\Upsilon\varrho,\Upsilon\varpi)} \varphi(t)\,dt \leq \gamma \int_0^{M_1(\varrho,\varpi)} \varphi(t)\,dt,$$

for each $\varrho, \varpi \in \mathcal{N}$, where

$$\gamma \in [0,\infty) : \gamma(1+\delta) < \delta$$

and

$$M_1(\varrho,\varpi) = \max\left\{ \begin{array}{l} \left[\frac{\mathcal{S}(\Upsilon\varrho,\Upsilon\varrho,\varrho)}{\delta} + \mathcal{S}(\Upsilon\varpi,\Upsilon\varpi,\varpi)\right], \\ \left[\mathcal{S}(\Upsilon\varrho,\Upsilon\varrho,\varrho) + \frac{\mathcal{S}(\Upsilon\varpi,\Upsilon\varpi,\varpi)}{\delta}\right] \end{array} \right\}.$$

Then Υ is called a $\delta_{\mathcal{S}}$-integral Kannan-type map.

Remark 4.1 If we take $\varphi(t) = 1$ for each $t \in [0,\infty)$, the notion of a $\delta_{\mathcal{S}}$-integral Kannan-type map coincides with the notion of a $\delta_{\mathcal{S}}$-Kannan-type map.

Definition 4.2 Suppose that there exists a $\delta \in \mathbb{R} - \{0\}$ such that Υ satisfies

$$\int_0^{\mathcal{S}(\Upsilon\varrho,\Upsilon\varrho,\Upsilon\varpi)} \varphi(t)\,dt \leq \gamma \int_0^{M_2(\varrho,\varpi)} \varphi(t)\,dt,$$

for each $\varrho, \varpi \in \mathcal{N}$, where

$$\gamma \in [0, \infty) : \gamma (1 + |\delta|) < |\delta|$$

and

$$M_2 (\varrho, \varpi) = \max \left\{ \begin{array}{c} \left[\frac{\mathcal{S}(\Upsilon\varrho, \Upsilon\varrho, \varrho)}{|\delta|} + \mathcal{S}(\Upsilon\varpi, \Upsilon\varpi, \varpi) \right], \\ \left[\mathcal{S}(\Upsilon\varrho, \Upsilon\varrho, \varrho) + \frac{\mathcal{S}(\Upsilon\varpi, \Upsilon\varpi, \varpi)}{|\delta|} \right] \end{array} \right\}.$$

Then Υ is called a $\delta_{\mathcal{S}}$-integral modulus Kannan-type map.

Remark 4.2 If we take $\varphi(t) = 1$ for each $t \in [0, \infty)$, the notion of a $\delta_{\mathcal{S}}$-integral modulus Kannan-type map coincides with the notion of a $\delta_{\mathcal{S}}$-modulus Kannan-type map.

Definition 4.3 Suppose that there exists a $\delta \in (0, \infty)$ such that Υ satisfies

$$\int_0^{d^S(\Upsilon\varrho, \Upsilon\varpi)} \varphi(t)\, dt \leq \gamma \int_0^{M_3(\varrho, \varpi)} \varphi(t)\, dt,$$

for each $\varrho, \varpi \in \mathcal{N}$, where

$$\gamma \in [0, \infty) : \gamma (1 + \delta) < \delta$$

and

$$M_3 (\varrho, \varpi) = \max \left\{ \begin{array}{c} \left[\frac{d^S(\Upsilon\varrho, \varrho)}{\delta} + d^S(\Upsilon\varpi, \varpi) \right], \\ \left[d^S(\Upsilon\varrho, \varrho) + \frac{d^S(\Upsilon\varpi, \varpi)}{\delta} \right] \end{array} \right\}.$$

Then Υ is called a δ_{d^S}-integral Kannan-type map.

Remark 4.3 If we take $\varphi(t) = 1$ for each $t \in [0, \infty)$, the notion of a δ_{d^S}-integral Kannan-type map coincides with the notion of a δ_{d^S}-Kannan-type map.

Definition 4.4 Suppose that there exists a $\delta \in \mathbb{R} - \{0\}$ such that Υ satisfies

$$\int_0^{d^S(\Upsilon\varrho, \Upsilon\varpi)} \varphi(t)\, dt \leq \gamma \int_0^{M_4(\varrho, \varpi)} \varphi(t)\, dt,$$

for each $\varrho, \varpi \in \mathcal{N}$, where

$$\gamma \in [0, \infty) : \gamma (1 + |\delta|) < |\delta|$$

and

$$M_4 (\varrho, \varpi) = \max \left\{ \begin{array}{c} \left[\frac{d^S(\Upsilon\varrho, \varrho)}{|\delta|} + d^S (\Upsilon\varpi, \varpi) \right], \\ \left[d^S (\Upsilon\varrho, \varrho) + \frac{d^S(\Upsilon\varpi, \varpi)}{|\delta|} \right] \end{array} \right\}.$$

Then Υ is called a δ_{d^S}-integral modulus Kannan-type map.

Remark 4.4 If we take $\varphi (t) = 1$ for each $t \in [0, \infty)$, the notion of a δ_{d^S}-integral modulus Kannan-type map coincides with the notion of a δ_{d^S}-modulus Kannan-type map.

REFERENCES

1. Bakhtin IA. The contraction principle in quasimetric spaces. *Functional Analysis Ulianowsk Gos Ped Ins* 1989; 30: 26–37.
2. Banach S. Sur les operations dans les ensembles abstraits et leur application aux equations integrals. *Fundamenta Mathematicae* 1922; 2: 133–181.
3. Branciari A. A fixed point theorem for mappings satisfying a general contractive condition of integral type. *International Journal of Mathematics and Mathematical Sciences* 2002; 29 (9): 531–536.
4. Chouhan P, Malviya N. Some fixed point theorems for asymptotically regular sequences and maps. *International Journal of Mathematical Sciences* 2013; 33 (2): 1164–1167.
5. Gupta A. Cyclic contraction on S-metric space. *International Journal of Analysis and Applications* 2013; 3 (2): 119–130.
6. Hieu NT, Ly NT, Dung NV. A generalization of Ciric quasi-contractions for maps on S-metric spaces. *Thai Journal of Mathematics* 2015; 13 (2): 369–380.
7. Kannan R. Some results on fixed points. *Bulletin of the Calcutta Mathematical Society* 1968; 60: 71–76.
8. Kannan R. Some results on fixed points II. *The American Mathematical Monthly* 1969; 76: 405–408.

9. Özgür NY, Taş N. Some new contractive mappings on S-metric spaces and their relationships with the mapping ($S25$). *Mathematical Sciences* 2017; 11 (1): 7–16.
10. Özgür N, Taş N. On S-metric spaces with some topological aspects. *Electronic Journal of Mathematical Analysis and Applications* 2023; 11 (2): 1–8.
11. Prithvi BV, Katiyar SK. A revisit of the Kannan map. *The Journal of Analysis* 2024. https://doi.org/10.1007/s41478-023-00715-y
12. Reich S. Some remarks concerning contraction mappings. *Canadian Mathematical Bulletin* 1971; 14: 121–124.
13. Sedghi S, Shobe N, Aliouche A. A generalization of fixed point theorems in S-metric spaces. *Matematicki Vesnik* 2012; 64 (3): 258–266.
14. Sedghi S, Dung NV. Fixed point theorems on S-metric spaces. *Matematicki Vesnik* 2014; 66 (1): 113–124.

CHAPTER 5

Some Applications of a Kittaneh Inequality to Operator-Valued Integrals on Hilbert Spaces

Silvestru Sever Dragomir

5.1 INTRODUCTION

The *numerical radius* $w(T)$ of an operator T on H is given by

$$\omega(T) = sup\{|\langle Tx, x\rangle|, \|x\| = 1\}. \tag{1.1}$$

Obviously, by (1.1), for any $x \in H$ one has

$$|\langle Tx, x\rangle| \leq w(T)\|x\|^2. \tag{1.2}$$

It is well known that $w(\cdot)$ is a norm on the Banach algebra $B(H)$ of all bounded linear operators $T : H \to H$, i.e.,

(1) $\omega(T) \geq 0$ for any $T \in B(H)$ and $\omega(T) = 0$ if and only if $T = 0$;

(2) $\omega(\lambda T) = |\lambda|\,\omega(T)$ for any $\lambda \in \mathbb{C}$ and $T \in B(H)$;

(3) $\omega(T + V) \leq \omega(T) + \omega(V)$ for any $T, V \in B(H)$.

This norm is equivalent to the operator norm. In fact, the following more precise result holds:

$$\omega(T) \leq \|T\| \leq 2\omega(T) \tag{1.3}$$

DOI: 10.1201/9781003530602-5

for any $T \in B(H)$.

F. Kittaneh, in 2003 [14], showed that for any operator $T \in B(H)$ we have the following refinement of the first inequality in (1.3):

$$\omega(T) \leq \frac{1}{2}\left(\|T\| + \|T^2\|^{1/2}\right). \tag{1.4}$$

Utilizing the Cartesian decomposition for operators, F. Kittaneh in [15] improved the inequality (1.3) as follows:

$$\frac{1}{4}\|T^*T + TT^*\| \leq \omega^2(T) \leq \frac{1}{2}\|T^*T + TT^*\| \tag{1.5}$$

for any operator $T \in B(H)$.

For powers of the absolute value of operators, one can state the following results obtained by El-Haddad Kittaneh in 2007, [11]:

If for an operator $T \in B(H)$ we denote $|T| := (T^*T)^{1/2}$, then

$$\omega^r(T) \leq \frac{1}{2}\left\||T|^{2\alpha r} + |T^*|^{2(1-\alpha)r}\right\| \tag{1.6}$$

and

$$\omega^{2r}(T) \leq \left\|\alpha|T|^{2r} + (1-\alpha)|T^*|^{2r}\right\|, \tag{1.7}$$

where $\alpha \in (0,1)$ and $r \geq 1$.

If we take $\alpha = \frac{1}{2}$ and $r = 1$ we get from (1.6) that

$$\omega(T) \leq \frac{1}{2}\||T| + |T^*|\| \tag{1.8}$$

and from (1.7) that

$$\omega^2(T) \leq \frac{1}{2}\left\||T|^2 + |T^*|^2\right\|. \tag{1.9}$$

For more related results, see the recent books on inequalities for numerical radii [8, 7].

Let $T = U|T|$ be the *polar decomposition* of the bounded linear operator T. The *Aluthge transform* $\tilde{T}$ of T is defined by $\tilde{T} := |T|^{1/2}U|T|^{1/2}$; see [4].

The following properties of $\tilde{T}$ are as follows:

(i) $\|\tilde{T}\| \leq \|T\|$,

(ii) $w(\tilde{T}) \leq \omega(T)$,

(iii) $r(\tilde{T}) = \omega(T)$,

(iv) $\omega(\tilde{T}) \leq \|T^2\|^{1/2} \ (\leq \|T\|)$; [17].

Utilizing this transform T. Yamazaki [17] obtained in 2007 the following refinement of Kittaneh's inequality (1.4):

$$\omega(T) \leq \frac{1}{2}\left(\|T\| + \omega(\tilde{T})\right) \leq \frac{1}{2}\left(\|T\| + \|T^2\|^{1/2}\right) \tag{1.10}$$

for any operator $T \in B(H)$.

We remark that if $\tilde{T} = 0$, then obviously $w(T) = \frac{1}{2}\|T\|$.

Abu-Omar and Kittaneh [5] improved on inequality (1.10) using generalized Aluthge transform to prove that

$$\omega(T) \leq \frac{1}{2}\left(\|T\| + \min_{t\in[0,1]} \omega(\Delta_t(T))\right).$$

For $t = 1$ this also gives the following result for the *Dougal transform*

$$\omega(T) \leq \frac{1}{2}\left(\|T\| + \omega(\hat{T})\right). \tag{1.11}$$

In [6], Bunia et al. also proved that

$$\omega(T) \leq \min_{t\in[0,1]} \left\{\frac{1}{2}\omega(\Delta_t(T)) + \frac{1}{4}\left(\|T\|^{2t} + \|T\|^{2(1-t)}\right)\right\},$$

which for $t = 1/2$ gives (1.10) as well.

In 1988, F. Kittaneh obtained the following generalization of Schwarz inequality [13]:

Theorem 1 *Assume that f and g are non-negative functions on* $[0, \infty)$ *which are continuous and satisfy the relation* $f(t)g(t) = t$ *for all* $t \in [0, \infty)$. *Let T,V be operators in* $V(H)$ *such that* $|T|\,V = V^*\,|T|$, *then*

$$|\langle TVx, y\rangle| \leq r(V)\,\|f(|T|)\,x\|\,\|g(|T^*|)\,y\| \tag{1.12}$$

for all x,y $\in H$, *where* $r(V)$ *denotes the spectral radius of V.*

If we take $f(t) = t^{\alpha}$ and $g(t) = t^{1-\alpha}$ for $\alpha \in [0, 1]$ and $t > 0$,

$$|\langle TVx, y\rangle| \leq r(V)\,\||T|^{\alpha}x\|\||T^*|^{1-\alpha}y\| \tag{1.13}$$

for all $x,y \in H$.

Let $\mathcal{F}(B; E, \mathcal{A}, \mu)$ be the linear space of functions $x(t)$, $t \in E$, with values in a real or complex Banach space B, given on a measurable space $(E, \mathcal{A}, \mu)$ endowed with a countably-additive scalar measure μ on a σ-algebra $\mathcal{A}$ of subsets of E.

A function $x_0 \in \mathcal{F}$ is called *simple* if can be defined as, see [16],

$$x_0(t) := \begin{cases} x_i \in B, & t \in A_i \in \mathcal{A}, \quad \mu(A_i) < \infty, i \in \{1, ..., n\} \\ & A_k \cap A_j = \varnothing, \quad k \neq j, k, j \in \{1, ..., n\}, \\ & \\ 0, & t \in E \setminus \cup_{i=1}^{n} A_i, n \in \mathbb{N}. \end{cases}$$

A function $x \in \mathcal{F}$ is called *strongly measurable* if there exists a sequence $\{x_n\}$ of simple functions with $\|x_n - x\| \to 0$ almost-everywhere with respect to the measure μ on E. As a consequence of this, the scalar function $\|x\|$ is $\mathcal{A}$-measurable.

For the simple function $x_0 \in \mathcal{F}$ as above we define the integral by

$$\int_E x_0(t)\, d\mu(t) := \sum_{i=1}^{n} x_i \mu(A_i).$$

A function $x \in \mathcal{F}$ is said to be *Bochner integrable* if it is strongly measurable and if for some approximating sequence $\{x_n\}$ of simple functions we have

$$\lim_{n\to\infty} \int_E \|x(t) - x_n(t)\|\, d\mu(t) = 0.$$

The *Bochner integral* of such a function over a set $A \in \mathcal{A}$ is defined as

$$\int_A x(t)\, d\mu(t) = \lim_{n\to\infty} \int_E \chi_A(t)\, x_n(t)\, d\mu(t),$$

where χ_A is the *characteristic function* of A, and the limit is understood in the sense of strong convergence in the Banach space E. This limit exists, and is independent of the choice of the approximation sequence of simple functions.

It is well-known that, for a strongly-measurable function to be Bochner integrable, it is necessary and sufficient for the norm of this function to be integrable, i.e.

$$\int_A \|x(t)\|\, d\mu(t) < \infty.$$

The set of Bochner-integrable functions forms a vector subspace $\mathcal{L}(B;E,\mathcal{A},\mu)$ of $\mathcal{F}(B;E,\mathcal{A},\mu)$, and the Bochner integral is a linear operator on this subspace.

Some fundamental properties of Bochner integrals are as follows [16]; see also [1, 9, 11, 12, 18]:

(1) For any $x \in \mathcal{L}(B;E,\mathcal{A},\mu)$ we have the norm inequality

$$\left\| \int_A x(t)\, d\mu(t) \right\| \leq \int_A \|x(t)\| d\mu(t).$$

(2) Bochner integral is a countably-additive μ-absolutely continuous set-function on the σ-algebra $\mathcal{A}$, i.e.

$$\int_{\cup_{i=1}^{\infty} A_i} x(t)\, d\mu(t) = \sum_{i=1}^{\infty} \int_{A_i} x(t)\, d\mu(t)$$

if $A_i \in \mathcal{A}, \mu(A_i) < \infty, i \in \mathbb{N}, A_k \cap A_j = \emptyset, k \neq j, k, j \in \mathbb{N}$, and

$$\left\| \int_A x(t)\, d\mu(t) \right\| \to 0 \quad \text{if} \quad \mu(A) \to 0,$$

uniformly over $A \in \mathcal{A}$.

(3) If $x_n \in F, x_n \to x$ almost-everywhere with respect to the measure μ on $A \in \mathcal{A}$, if $\|x_n\| \leq f$ almost-everywhere with respect to μ on A, and if $\int_A f(t)\, d\mu(t) < \infty$, then $x \in \mathcal{L}(B;E,\mathcal{A},\mu)$ and

$$\int_A x_n(t)\, d\mu(t) \to \int_A x(t)\, d\mu(t).$$

(4) The space is complete with respect to the norm

$$\|x\| := \int_A \|x(t)\| d\mu(t).$$

(5) If T is a closed linear operator from a Banach space X into a Banach space Y and if $x \in \mathcal{L}(X;E,\mathcal{A},\mu)$ and $Tx \in \mathcal{L}(Y;E,\mathcal{A},\mu)$, then

$$\int_A Tx(t)\, d\mu(t) = T\left(\int_A x(t)\, d\mu(t) \right).$$

If T is bounded, the condition $Tx \in \mathcal{L}(Y;E,\mathcal{A},\mu)$ is automatically satisfied.

Motivated by the above results, in this chapter we establish several upper bounds for the quantities

$$\left|\left\langle\left(\int_E p(t)\, B_t T_t V_t A_t d\mu(t)\right) x, y\right\rangle\right|, x, y \in H$$

and

$$\left|\left\langle\left(\int_E p(t)\, B_t T_t V_t A_t d\mu(t)\right) x, x\right\rangle\right|, x \in H$$

for the fields of operators $V_{(\cdot)}, T_{(\cdot)}, A_{(\cdot)}, B_{(\cdot)}, : E \to \mathcal{B}(H)$ with $|T_t|\, V_t = V_t^* \,|T_t|$ for $t \in E$ and $p : E \to [0, \infty)$ a μ-measurable function on E with $\int_E p(t)\, d\mu(t) = 1$. Applications for operator norm and numerical radius are also provided.

5.2 MAIN RESULTS

We recall the following vector inequality for positive operators $A \geq 0$, obtained by C. A. McCarthy in [3]

$$\langle Ax, x\rangle^p \leq \langle A^p x, x\rangle, p \geq 1$$

for $x \in H$, $\|x\| = 1$ and *Buzano's inequality* [2],

$$|\langle x, e\rangle \langle e, y\rangle| \leq \frac{1}{2}\left[\|x\|\|y\| + |\langle x, y\rangle|\right] \tag{2.1}$$

that holds for any $x,y,e \in H$ with $\|e\| = 1$.

If we replace x by $\frac{y}{\|y\|}, y \neq 0$, we get

$$\left\langle A\frac{y}{\|y\|}, \frac{y}{\|y\|}\right\rangle^p \leq \left\langle A^p \frac{y}{\|y\|}, \frac{y}{\|y\|}\right\rangle, p \geq 1,$$

namely

$$\langle Ay, y\rangle^p \leq \|y\|^{2(p-1)} \langle A^p y, y\rangle, p \geq 1, \tag{2.2}$$

for all $y \in H$.

Our first result is as follows:

Theorem 2 *Assume that f_t and g_t,$t \in E$, are non-negative functions on $[0, \infty)$ which are continuous and satisfy the relation $f_t(u)_t g(u) = u$ for all $u \in$*

$[0,\infty)$, $t \in E$. *Let* $V_{t(\cdot)}, T_{(\cdot)}, A_{(\cdot)}, B_{(\cdot)}, : E \to \mathcal{B}(H)$ *with* $|T_t| V_t = V_t^* |T_t|$ *for* $t \in E$ *and* $p : E \to [0,\infty)$ *be* μ*-measurable functions on* E *with* $\int_E p(t)\, d\mu(t) = 1$. *If* $p(\cdot) r(V_{(\cdot)}) |f_{(\cdot)}(|T_{(\cdot)}|) A_{(\cdot)}|^2$, *and* $p(\cdot) r(V_{(\cdot)}) |g_{(\cdot)}(|T_{(\cdot)}^*|) B_{(\cdot)}^*|^2$ *are Bochner integrable, then*

$$\left|\left\langle\left(\int_E p(t) B_t T_t V_t A_t d\mu(t)\right) x, y\right\rangle\right|^2 \le \left\langle\left(\int_E p(t) r(V_t) |f_t(|T_t|) A_t|^2 d\mu(t)\right) x, x\right\rangle \times \left\langle\left(\int_E p(t) r(V_t) |g_t(|T_t^*|) B_t^*|^2 d\mu(t)\right) y, y\right\rangle \tag{2.3}$$

for all $x,y \in H$.

We also have the norm inequality

$$\left\|\int_E p(t) B_t T_t V_t A_t d\mu(t)\right\|^2 \le \left\|\int_E p(t) r(V_t) |f_t(|T_t|) A_t|^2 d\mu(t)\right\| \times \left\|\int_E p(t) r(V_t) |g_t(|T_t^*|) B_t^*|^2 d\mu(t)\right\|. \tag{2.4}$$

Proof. Let $t \in E$. Observe that by (1.12) we have

$$\begin{aligned}|\langle T_t V_t x, y\rangle|^2 &\le r^2(V_t) \|f_t(|T_t|) x\|^2 \left\|g_t(|T_t^*|) y\right\|^2 \\ &= r^2(V_t) \langle f_t(|T_t|) x, f_t(|T_t|) x\rangle \langle g_t(|T_t^*|) y, g_t(|T_t^*|) y\rangle \\ &= r^2(V_t) \langle f_t^2(|T_t|) x, x\rangle \langle g_t^2(|T_t^*|) y, y\rangle\end{aligned}$$

for all $x,y \in H$.

If we take $A_t x$ instead of x and $B_t^* y$ instead of y, then we get

$$\begin{aligned}|\langle T_t V_t A_t x, B_t^* y\rangle|^2 &\le r^2(V_t) \langle f_t^2(|T_t|) A_t x, A_t x\rangle \langle g_t^2(|T_t^*|) B_t^* y, B_t^* y\rangle \\ &= r^2(V_t) \langle A_t^* f_t^2(|T_t|) A_t x, x\rangle \langle B_t g_t^2(|T_t^*|) B_t^* y, y\rangle \\ &= r^2(V_t) \langle (f_t(|T_t|) A_t x)^* f_t(|T_t|) A_t x, x\rangle \\ &\quad \times \langle (g_t(|T_t^*|) B_t^*)^* g_t(|T_t^*|) B_t^* y, y\rangle \\ &= r^2(V_t) \langle |f_t(|T_t|) A_t|^2 x, x\rangle \langle |g_t(|T_t^*|) B_t^*|^2 y, y\rangle,\end{aligned}$$

namely

$$|\langle B_t T_t V_t A_t x, y\rangle|^2 \le r^2(V_t)\left\langle |f_t(|T_t|)A_t|^2 x, x\right\rangle \left\langle |g_t(|T_t^*|)B_t^*|^2 y, y\right\rangle \quad (2.5)$$

for all $x,y \in H$ and $t \in E$.

By taking the square root in (2.5), then we get

$$\left|\left\langle B_t T_t V_t A_t x, y\right\rangle\right| \le r(V_t)\left\langle |f_t(|T_t|)A_t|^2 x, x\right\rangle^{1/2}\left\langle |g_t(|T_t^*|)B_t^*|^2 y, y\right\rangle^{1/2} \quad (2.6)$$

for all $x,y \in H$ and $t \in E$.

If we multiply by $p(t) \ge 0$ and integrate, then we get

$$\int_E p(t)\,|\langle B_t T_t V_t A_t x, y\rangle|\,d\mu(t) \le \int_E p(t)\,r(V_t)\left\langle |f_t(|T_t|)A_t|^2 x, x\right\rangle^{1/2}\left\langle |g_t(|T_t^*|)B_t^*|^2 y, y\right\rangle^{1/2} d\mu(t). \quad (2.7)$$

By the triangle inequality we have

$$\left|\left\langle\left(\int_E p(t)\,B_t T_t V_t A_t d\mu(t)\right)x, y\right\rangle\right| \le \int_E p(t)\,|\langle B_t T_t V_t A_t x, y\rangle|\,d\mu(t) \quad (2.8)$$

and by the Cauchy-Bunyakowsky-Schwarz integral weighted inequality, we also have

$$\begin{aligned}
&\int_E p(t)\,r(V_t)\left\langle |f_t(|T_t|)A_t|^2 x, x\right\rangle^{1/2}\left\langle |g_t(|T_t^*|)B_t^*|^2 y, y\right\rangle^{1/2} d\mu(t)\\
&\le \left(\int_E p(t)\,r(V_t)\left(\left\langle |f_t(|T_t|)A_t|^2 x, x\right\rangle^{1/2}\right)^2 d\mu(t)\right)^{1/2}\\
&\quad\times\left(\int_E p(t)\,r(V_t)\left(\left\langle |g_t(|T_t^*|)B_t^*|^2 y, y\right\rangle^{1/2}\right)^2 d\mu(t)\right)^{1/2}\\
&= \left\langle\left(\int_E p(t)\,r(V_t)\,|f_t(|T_t|)A_t|^2 d\mu(t)\right)x, x\right\rangle^{1/2}\\
&\quad\times\left\langle\left(\int_E p(t)\,r(V_t)\,|g_t(|T_t^*|)B_t^*|^2 d\mu(t)\right)y, y\right\rangle^{1/2}
\end{aligned} \quad (2.9)$$

for all $x,y \in H$. By (2.7)–(2.9) we get (2.3).

By taking the supremum over $\|x\| = \|y\| = 1$ in (2.3), we get

$$
\begin{aligned}
&\left\| \int_E p(t) B_t T_t V_t A_t d\mu(t) \right\| \\
&= \sup_{\|x\|=\|y\|=1} \left| \left\langle \left(\int_E p(t) B_t T_t V_t A_t d\mu(t) \right) x, y \right\rangle \right|^2 \\
&\leq \sup_{\|x\|=\|y\|=1} \left[\left\langle \left(\int_E p(t) r(V_t) |f_t(|T_t|) A_t|^2 d\mu(t) \right) x, x \right\rangle \right. \\
&\quad \left. \times \left\langle \left(\int_E p(t) r(V_t) | g_t(| T_t^* |) B_t^* |^2 d\mu(t) \right) y, y \right\rangle \right] \\
&= \sup_{\|x\|=1} \left\langle \left(\int_E p(t) r(V_t) \, |f_t(|T_t|) A_t|^2 d\mu(t) \right) x, x \right\rangle \\
&\quad \times \sup_{\|y\|=1} \left\langle \left(\int_E p(t) r(V_t) \, |g_t(|T_t^*|) B_t^*|^2 d\mu(t) \right) y, y \right\rangle \\
&= \left\| \int_E p(t) r(V_t) \, |f_t(|T_t|) A_t|^2 d\mu(t) \right\| \\
&\quad \left\| \int_E p(t) r(V_t) \, |g_t(|T_t^*|) B_t^*|^2 d\mu(t) \right\|,
\end{aligned}
$$

which proves (2.4).

Corollary 1 *With the assumptions of Theorem 2 we have*

$$
\begin{aligned}
&\left| \left\langle \left(\int_E p(t) B_t T_t V_t A_t d\mu(t) \right) x, x \right\rangle \right|^2 \\
&\leq \left\langle \left(\int_E p(t) r(V_t) \, |f_t(|T_t|) A_t|^2 d\mu(t) \right) x, x \right\rangle \\
&\quad \times \left\langle \left(\int_E p(t) r(V_t) \, |g_t(|T_t^*|) B_t^*|^2 d\mu(t) \right) x, x \right\rangle \\
&\leq \left\langle \left(\int_E p(t) r(V_t) \left(\frac{|f_t(|T_t|) A_t|^2 + |g_t(|T_t^*|) B_t^*|^2}{2} \right) d\mu(t) \right) x, x \right\rangle
\end{aligned}
\tag{2.10}
$$

for all $x \in H$.

We also have the numerical radius inequality

$$\omega^2\left(\int_E p(t)\, B_t T_t V_t A_t d\mu(t)\right)$$
$$\leq \left\| \int_E p(t)\, r(V_t) \left(\frac{|f_t(|T_t|) A_t|^2 + |g_t(|T_t^*|) B_t^*|^2}{2} \right) d\mu(t) \right\|. \tag{2.11}$$

Proof. The inequality (2.10) follows by (2.3) for $y = x$ and using the *A-G means inequality*

$$\sqrt{ab} \leq \frac{a+b}{2},\ a, b \geq 0.$$

By taking the supremum over $\|x\| = 1$ in (2.10), we get

$$\omega^2\left(\int_E p(t)\, B_t T_t V_t A_t d\mu(t)\right)^2$$
$$= \sup_{\|x\|=1} \left| \left\langle \left(\int_E p(t)\, B_t T_t V_t A_t d\mu(t) \right) x, x \right\rangle \right|^2$$
$$\leq \sup_{\|x\|=1} \left\langle \left(\int_E p(t)\, r(V_t) \left(\frac{|f_t(|T_t|) A_t|^2 + |g_t(|T_t^*|) B_t^*|^2}{2} \right) d\mu(t) \right) x, x \right\rangle$$
$$= \left\| \int_E p(t)\, r(V_t) \left(\frac{|f_t(|T_t|) A_t|^2 + |g_t(|T_t^*|) B_t^*|^2}{2} \right) d\mu(t) \right\|,$$

which proves (2.11).

Remark 1 Assume that $\alpha : E \to [0, 1]$ is measurable. By taking $f_t(u) = u^{\alpha(t)}$ and $g_t(u) = u^{1-\alpha(t)}$, $u \geq 0, t \in E$ in Theorem 2, we obtain

$$\left| \left\langle \left(\int_E p(t)\, B_t T_t V_t A_t d\mu(t) \right) x, y \right\rangle \right|^2$$
$$\leq \left\langle \left(\int_E p(t)\, r(V_t) \left| |T_t|^{\alpha(t)} A_t \right|^2 d\mu(t) \right) x, x \right\rangle$$
$$\times \left\langle \left(\int_E p(t)\, r(V_t) \left| |T_t^*|^{1-\alpha(t)} B_t^* \right|^2 d\mu(t) \right) y, y \right\rangle \tag{2.12}$$

for all $x, y \in H$, *provided that* $p(\cdot)r(V_{(\cdot)})\left|T_{(\cdot)}^{\alpha(\cdot)}A_{(\cdot)}\right|^2$ *and* $p(\cdot)r(V_{(\cdot)})$ $\left|\left|T_{(\cdot)}^*\right|^{1-\alpha(\cdot)}B_{(\cdot)}^*\right|^2$ *are Bochner integrable.*

We also have the norm inequality

$$\left\|\int_E p(t)B_tT_tV_tA_t d\mu(t)\right\|^2 \le \left\|\int_E p(t)r(V_t)\left||T_t|^{\alpha(t)}A_t\right|^2 d\mu(t)\right\| \times \left\|\int_E p(t)r(V_t)\left||T_t^*|^{1-\alpha(t)}B_t^*\right|^2 d\mu(t)\right\|. \tag{2.13}$$

From Corollary 1 we derive

$$\begin{aligned}
&\left|\left\langle\left(\int_E p(t)B_tT_tV_tA_t d\mu(t)\right)x, x\right\rangle\right|^2 \\
&\le \left\langle\left(\int_E p(t)r(V_t)\left||T_t|^{\alpha(t)}A_t\right|^2 d\mu(t)\right)x, x\right\rangle \\
&\quad \times \left\langle\left(\int_E p(t)r(V_t)\left||T_t^*|^{1-\alpha(t)}B_t^*\right|^2 d\mu(t)\right)x, x\right\rangle \\
&\le \left\langle\int_E p(t)r(V_t)\frac{\left||T_t|^{\alpha(t)}A_t\right|^2 + \left||T_t^*|^{1-\alpha(t)}B_t^*\right|^2}{2} d\mu(t)x, x\right\rangle
\end{aligned} \tag{2.14}$$

for all $x \in H$.

We also have the numerical radius inequality

$$\begin{aligned}
&\omega^2\left(\int_E p(t)B_tT_tV_tA_t d\mu(t)\right) \\
&\le \left\|\int_E p(t)r(V_t)\frac{\left||T_t|^{\alpha(t)}A_t\right|^2 + \left||T_t^*|^{1-\alpha(t)}B_t^*\right|^2}{2} d\mu(t)\right\|.
\end{aligned} \tag{2.15}$$

The second result is as follows:

Theorem 3 *Assume that* f_t *and* $g_t, t \in E$ *are non-negative functions on* $[0, \infty)$ *which are continuous and satisfy the relation* $f_t(u)_t g(u) = u$ *for all* $u \in [0, \infty)$, $t \in E$. *Let* $V_{t(\cdot)}, T_{(\cdot)}, A_{(\cdot)}, B_{(\cdot)}, : E \to \mathcal{B}(H)$ *with*

$|T_t| V_t = V_t^* |T_t|$ for $t \in E$ and $p : E \to [0, \infty)$ be μ-measurable functions on E with $\int_E p(t)\, d\mu(t) = 1$. If $p(\cdot)\, r^2(V_{(\cdot)}) \left|f_{(\cdot)}(|T_{(\cdot)}|)\, A_{(\cdot)}\right|^{2s}$, and $p(\cdot)\, r^2(V_{(\cdot)}) \left|g_{(\cdot)}\left(\left|T_{(\cdot)}^*\right|\right) B_{(\cdot)}^*\right|^{2q}$ are Bochner integrable for $s,\ q > 1$ with $\frac{1}{s} + \frac{1}{q} = 1$, then

$$\left|\left\langle\left(\int_E p(t)\, B_t T_t V_t A_t d\mu(t)\right) x, y\right\rangle\right|^2 \leq \|x\|^{2/q} \|y\|^{2/s} \left\langle\left(\int_E p(t)\, r^2(V_t) \left|f_t(|T_t|)\, A_t\right|^{2s} d\mu(t)\right) x, x\right\rangle^{1/s} \times \left\langle\left(\int_E p(t)\, r^2(V_t) \left|g_t(|T_t^*|)\, B_t^*\right|^{2q} d\mu(t)\right) y, y\right\rangle^{1/q} \tag{2.16}$$

for all $x, y \in H$.

We also have the norm inequality

$$\left\|\int_E p(t)\, B_t T_t V_t A_t d\mu(t)\right\|^2 \leq \left\|\int_E p(t)\, r^2(V_t) \left|f_t(|T_t|)\, A_t\right|^{2s} d\mu(t)\right\|^{1/s} \times \left\|\int_E p(t)\, r^2(V_t) \left|g_t(|T_t^*|)\, B_t^*\right|^{2q} d\mu(t)\right\|^{1/q}. \tag{2.17}$$

Proof. If we multiply (2.5) by $p(t) \geq 0$ and integrate, then we get

$$\int_E p(t) \left|\langle B_t T_t V_t A_t x, y\rangle\right|^2 d\mu(t) \leq \int_E p(t)\, r^2(V_t) \left\langle \left|f_t(|T_t|)\, A_t\right|^2 x, x\right\rangle \left\langle \left|g_t(|T_t^*|)\, B_t^*\right|^2 y, y\right\rangle d\mu(t). \tag{2.18}$$

By Cauchy-Bunyakowsky-Schwarz weighted integral inequality we have

$$\left|\left\langle \int_E p(t)\, B_t T_t V_t A_t x, y\right\rangle d\mu(t)\right|^2 \leq \int_E p(t) \left|\langle B_t T_t V_t A_t x, y\rangle\right|^2 d\mu(t) \tag{2.19}$$

while by Hölder's weighted integral inequality for $s, q > 1$ with $\frac{1}{s} + \frac{1}{q} = 1$ we have

$$\begin{aligned}
&\int_E p(t)\, r^2(V_t) \left\langle |f_t(|T_t|) A_t|^2 x, x \right\rangle \left\langle |g_t(|T_t^*|) B_t^*|^2 y, y \right\rangle d\mu(t) \\
&\leq \left(\int_E p(t)\, r^2(V_t) \left\langle |f_t(|T_t|) A_t|^2 x, x \right\rangle^s d\mu(t) \right)^{1/s} \\
&\times \left(\int_E p(t)\, r^2(V_t) \left\langle |g_t(|T_t^*|) B_t^*|^2 y, y \right\rangle^q d\mu(t) \right)^{1/q} \qquad (2.20)
\end{aligned}$$

for all $x,y \in H$.

By (2.18)–(2.20) we get

$$\begin{aligned}
&\left| \left\langle \left(\int_E p(t) B_t T_t V_t A_t d\mu(t) \right) x, y \right\rangle \right|^2 \\
&\leq \left(\int_E p(t)\, r^2(V_t) \left\langle |f_t(|T_t|) A_t|^2 x, x \right\rangle^s d\mu(t) \right)^{1/s} \\
&\times \left(\int_E p(t)\, r^2(V_t) \left\langle |g_t(|T_t^*|) B_t^*|^2 y, y \right\rangle^q d\mu(t) \right)^{1/q} \qquad (2.21)
\end{aligned}$$

for all $x,y \in H$.

By McCarthy's inequality (2.2) we get

$$\left\langle |f_t(|T_t|) A_t|^2 x, x \right\rangle^s \leq \|x\|^{2(s-1)} \left\langle |f_t(|T_t|) A_t|^{2s} x, x \right\rangle, \; s > 1,$$

and

$$\left\langle |g_t(|T_t^*|) B_t^*|^2 y, y \right\rangle^q \leq \|y\|^{2(q-1)} \left\langle |g_t(|T_t^*|) B_t^*|^{2q} y, y \right\rangle, \; q > 1.$$

By employing (2.21) we then get

$$\begin{aligned}
&\left| \left\langle \left(\int_E p(t) B_t T_t V_t A_t d\mu(t) \right) x, y \right\rangle \right|^2 \\
&\leq \|x\|^{2(1-1/s)} \left\langle \left(\int_E p(t)\, r^2(V_t)\, |f_t(|T_t|) A_t|^{2s} d\mu(t) \right) x, x \right\rangle^{1/s} \\
&\times \|y\|^{2(1-1/q)} \left\langle \left(\int_E p(t)\, r^2(V_t)\, |g_t(|T_t^*|) B_t^*|^{2q} d\mu(t) \right) y, y \right\rangle^{1/q}
\end{aligned}$$

$$=\|x\|^{2/q}\left\langle\left(\int_E p(t)\,r^2(V_t)\,|f_t(|T_t|)\,A_t|^{2s}d\mu(t)\right)x,x\right\rangle^{1/s}$$

$$\times\|y\|^{2/s}\left\langle\left(\int_E p(t)\,r^2(V_t)\,|g_t(|T_t^*|)\,B_t^*|^{2q}d\mu(t)\right)y,y\right\rangle^{1/q}$$

for all $x,y \in H$, which proves (2.16).

By taking the supremum over $\|x\| = \|y\| = 1$ in (2.16), we get

$$\left\|\int_E p(t)\,B_tT_tV_tA_td\mu(t)\right\|^2$$

$$=\sup_{\|x\|=\|y\|=1}\left|\left\langle\int_E p(t)\,B_tT_tV_tA_tx,y\right\rangle d\mu(t)\right|^2$$

$$\leq\sup_{\|x\|=\|y\|=1}\left\{\|x\|^{2/q}\,\|y\|^{2/s}\left\langle\left(\int_E p(t)\,r^2(V_t)\middle|f_t(|T_t|)\,A_t|^{2s}d\mu(t)\right)x,x\right\rangle^{1/s}\right.$$

$$\left.\times\left\langle\left(\int_E p(t)\,r^2(V_t)\middle|g_t(|T_t^*|)\,B_t^*|^{2q}d\mu(t)\right)y,y\right\rangle^{1/q}\right\}$$

$$=\sup_{\|x\|=1}\left\langle\left(\int_E p(t)\,r^2(V_t)\,|f_t(|T_t|)\,A_t|^{2s}d\mu(t)\right)x,x\right\rangle^{1/s}$$

$$\times\sup_{\|y\|=1}\left\langle\left(\int_E p(t)\,r^2(V_t)\middle|g_t(|T_t^*|)\,B_t^*|^{2q}d\mu(t)\right)y,y\right\rangle^{1/q}$$

$$=\left\|\int_E p(t)\,r^2(V_t)\,|f_t(|T_t|)\,A_t|^{2s}d\mu(t)\right\|^{1/s}$$

$$\times\left\|\int_E p(t)\,r^2(V_t)\,|g_t(|T_t^*|)\,B_t^*|^{2q}d\mu(t)\right\|^{1/q},$$

which proves (2.17).

Corollary 2 *With the assumptions of Theorem 3 we have*

$$\left|\left\langle\left(\int_E p(t)\,B_tT_tV_tA_td\mu(t)\right)x,x\right\rangle\right|^2$$

$$\leq\|x\|^2\left\langle\left(\int_E p(t)\,r^2(V_t)\,|f_t(|T_t|)\,A_t|^{2s}d\mu(t)\right)x,x\right\rangle^{1/s}$$

$$\times\left\langle\left(\int_E p(t)\, r^2(V_t)\left|g_t\left(\left|T_t^*\right|\right)B_t^*\right|^{2q} d\mu(t)\right)x, x\right\rangle^{1/q}$$

$$\leq \|x\|^2$$

$$\times\left\langle\left(\int_E p(t)\, r^2(V_t)\left[\frac{1}{p}\left|f_t\left(\left|T_t\right|\right)A_t\right|^{2s} + \frac{1}{q}\left|g_t\left(\left|T_t^*\right|\right)B_t^*\right|^{2q}\right] d\mu(t)\right)x, x\right\rangle \tag{2.22}$$

for all $x \in H$.

We also have the numerical radius inequality

$$\omega^2\left(\int_E p(t)\, B_t T_t V_t A_t d\mu(t)\right)^2$$

$$\leq \left\|\int_E p(t)\, r^2(V_t)\left[\frac{1}{p}\left|f_t\left(\left|T_t\right|\right)A_t\right|^{2s} + \frac{1}{q}\left|g_t\left(\left|T_t^*\right|\right)B_t^*\right|^{2q}\right] d\mu(t)\right\|. \tag{2.23}$$

Proof. The inequality (2.22) follows by (2.16) on taking $y = x$ and also using the Young's inequality

$$ab \leq \frac{1}{s}a^s + \frac{1}{q}b^q,\ a, b \geq 0$$

for $s, q > 1$ with $\frac{1}{s} + \frac{1}{q} = 1$.

By taking the supremum over $\|x\| = 1$ in (2.22) we get

$$\omega^2\left(\int_E p(t)\, B_t T_t V_t A_t d\mu(t)\right)^2$$

$$= \sup_{\|x\|=1}\left|\left\langle\left(\int_E p(t)\, B_t T_t V_t A_t d\mu(t)\right)x, x\right\rangle\right|^2$$

$$\leq \sup_{\|x\|=1}\left\langle\left(\int_E p(t)\, r^2(V_t)\left[\frac{1}{p}\left|f_t\left(\left|T_t\right|\right)A_t\right|^{2s} + \frac{1}{q}\left|g_t\left(\left|T_t^*\right|\right)B_t^*\right|^{2q}\right] d\mu(t)\right)x, x\right\rangle$$

$$= \left\|\int_E p(t)\, r^2(V_t)\left[\frac{1}{p}\left|f_t\left(\left|T_t\right|\right)A_t\right|^{2s} + \frac{1}{q}\left|g_t\left(\left|T_t^*\right|\right)B_t^*\right|^{2q}\right] d\mu(t)\right\|,$$

which proves (2.23).

Remark 2 If we take $s = q = 2$ in Theorem 2 and assume that

$$p(\cdot)\, r^2\left(V_{(\cdot)}\right)\left|f_{(\cdot)}\left(\left|T_{(\cdot)}\right|\right)A_{(\cdot)}\right|^4$$

and

$$p(\cdot)\,r^2\left(V_{(\cdot)}\right)\left|g_{(\cdot)}\left(\left|T^*_{(\cdot)}\right|\right)B^*_{(\cdot)}\right|^4$$

are Bochner integrable, then we get

$$\begin{aligned}&\left|\left\langle\left(\int_E p(t)\,B_tT_tV_tA_t d\mu(t)\right)x,y\right\rangle\right|^4\\&\le \|x\|^2\|y\|^2\left\langle\left(\int_E p(t)\,r^2(V_t)\,|f_t(|T_t|)\,A_t|^4 d\mu(t)\right)x,x\right\rangle\\&\quad\times\left\langle\left(\int_E p(t)\,r^2(V_t)\,|g_t(|T_t^*|)\,B_t^*|^4 d\mu(t)\right)y,y\right\rangle\end{aligned}\tag{2.24}$$

for all $x, y \in H$.

We also have the norm inequality

$$\begin{aligned}\left\|\int_E p(t)\,B_tT_tV_tA_t d\mu(t)\right\|^4 &\le \left\|\int_E p(t)\,r^2(V_t)\,|f_t(|T_t|)\,A_t|^4 d\mu(t)\right\|\\&\quad\times\left\|\int_E p(t)\,r^2(V_t)\,|g_t(|T_t^*|)\,B_t^*|^4 d\mu(t)\right\|.\end{aligned}\tag{2.25}$$

From Corollary 2 we also have

$$\begin{aligned}&\left|\left\langle\left(\int_E p(t)\,B_tT_tV_tA_t d\mu(t)\right)x,x\right\rangle\right|^2\\&\le \|x\|^2\left\langle\left(\int_E p(t)\,r^2(V_t)\,|f_t(|T_t|)\,A_t|^4 d\mu(t)\right)x,x\right\rangle^{1/2}\\&\quad\times\left\langle\left(\int_E p(t)\,r^2(V_t)\,|g_t(|T_t^*|)\,B_t^*|^4 d\mu(t)\right)x,x\right\rangle^{1/2}\\&\le \frac{1}{2}\|x\|^2\\&\quad\times\left\langle\left(\int_E p(t)\,r^2(V_t)\left[|f_t(|T_t|)\,A_t|^4+|g_t(|T_t^*|)\,B_t^*|^4\right]d\mu(t)\right)x,x\right\rangle\end{aligned}\tag{2.26}$$

for all $x \in H$.

We also have the numerical radius inequality

$$\omega^2\left(\int_E p(t)\, B_t T_t V_t A_t d\mu(t)\right)^2 \le \frac{1}{2}\left\|\int_E p(t)\, r^2(V_t)\left[|f_t(|T_t|)A_t|^4 + |g_t(|T_t^*|)B_t^*|^4\right] d\mu(t)\right\|. \tag{2.27}$$

Remark 3 With the assumptions from Remark 1 we have by Theorem 3 that

$$\begin{aligned}\left|\left\langle\left(\int_E p(t)\, B_t T_t V_t A_t d\mu(t)\right)x, y\right\rangle\right|^2 &\le \|x\|^{2/q}\|y\|^{2/s}\left\langle\left(\int_E p(t)\, r^2(V_t)\left||T_t|^{\alpha(t)}A_t\right|^{2s} d\mu(t)\right)x, x\right\rangle^{1/s}\\ &\times\left\langle\left(\int_E p(t)\, r^2(V_t)\left||T_t^*|^{1-\alpha(t)}B_t^*\right|^{2q} d\mu(t)\right)y, y\right\rangle^{1/q}\end{aligned} \tag{2.28}$$

for all $x, y \in H$, provided that

$$p(\cdot)\, r^2\left(V_{(\cdot)}\right)\left||T_{(\cdot)}|^{\alpha(t)}A_{(\cdot)}\right|^{2s}$$

and

$$p(\cdot)\, r^2\left(V_{(\cdot)}\right)\left||T_{(\cdot)}^*|^{1-\alpha(t)}B_{(\cdot)}^*\right|^{2q}$$

are Bochner integrable for $s, q > 1$ with $\frac{1}{s} + \frac{1}{q} = 1$. We also have the norm inequality

$$\begin{aligned}\left\|\int_E p(t)\, B_t T_t V_t A_t d\mu(t)\right\|^2 &\le \left\|\int_E p(t)\, r^2(V_t)\left||T_t|^{\alpha(t)}A_t\right|^{2s} d\mu(t)\right\|^{1/s}\\ &\times\left\|\int_E p(t)\, r^2(V_t)\left||T_t^*|^{1-\alpha(t)}B_t^*\right|^{2q} d\mu(t)\right\|^{1/q}.\end{aligned} \tag{2.29}$$

From Corollary 2 we also obtain

$$\left|\left\langle\left(\int_E p(t)\, B_t T_t V_t A_t d\mu(t)\right)x, x\right\rangle\right|^2$$
$$\leq \|x\|^2\left\langle\left(\int_E p(t)\, r^2(V_t)\left||T_t|^{\alpha(t)}A_t\right|^{2s} d\mu(t)\right)x, x\right\rangle^{1/s}$$
$$\times\left\langle\left(\int_E p(t)\, r^2(V_t)\left||T_t^*|^{1-\alpha(t)}B_t^*\right|^{2q} d\mu(t)\right)x, x\right\rangle^{1/q}$$
$$\leq \|x\|^2$$
$$\times\left\langle\left(\int_E p(t)\, r^2(V_t)\left[\frac{1}{p}\left||T_t|^{\alpha(t)}A_t\right|^{2s}+\frac{1}{q}\left||T_t^*|^{1-\alpha(t)}B_t^*\right|^{2q}\right]d\mu(t)\right)x, x\right\rangle \tag{2.30}$$

for all $x \in H$.

We also have the numerical radius inequality

$$\omega^2\left(\int_E p(t)\, B_t T_t V_t A_t d\mu(t)\right)^2$$
$$\leq \left\|\int_E p(t)\, r^2(V_t)\left[\frac{1}{p}\left||T_t|^{\alpha(t)}A_t\right|^{2s}+\frac{1}{q}\left||T_t^*|^{1-\alpha(t)}B_t^*\right|^{2q}\right]d\mu(t)\right\|. \tag{2.31}$$

If we take $s = q = 2$ in Remark 3, then we get the following inequalities as well

$$\left|\left\langle\left(\int_E p(t)\, B_t T_t V_t A_t d\mu(t)\right)x, y\right\rangle\right|^2$$
$$\leq \|x\|\|y\|\left\langle\left(\int_E p(t)\, r^2(V_t)\left||T_t|^{\alpha(t)}A_t\right|^{4} d\mu(t)\right)x, x\right\rangle^{1/2}$$
$$\times\left\langle\left(\int_E p(t)\, r^2(V_t)\left||T_t^*|^{1-\alpha(t)}B_t^*\right|^{4} d\mu(t)\right)y, y\right\rangle^{1/2}$$

for all $x,y \in H$, provided that

$$p(\cdot)\, r^2(V_{(\cdot)})\left||T_{(\cdot)}|^{\alpha(t)}A_{(\cdot)}\right|^4$$

and

$$p(\cdot)\, r^2(V_{(\cdot)})\left||T_{(\cdot)}^*|^{1-\alpha(t)}B_{(\cdot)}^*\right|^4$$

are Bochner integrable. We also have the norm inequality

$$\left\|\int_E p(t)\, B_t T_t V_t A_t d\mu(t)\right\|^2 \le \left\|\int_E p(t)\, r^2(V_t) \left||T_t|^{\alpha(t)} A_t\right|^4 d\mu(t)\right\|^{1/2} \times \left\|\int_E p(t)\, r^2(V_t) \left||T_t^*|^{1-\alpha(t)} B_t^*\right|^4 d\mu(t)\right\|^{1/2}.$$

Moreover,

$$\begin{aligned}&\left|\left\langle\left(\int_E p(t)\, B_t T_t V_t A_t d\mu(t)\right) x, x\right\rangle\right|^2 \\ &\le \|x\|^2 \left\langle\left(\int_E p(t)\, r^2(V_t) \left||T_t|^{\alpha(t)} A_t\right|^4 d\mu(t)\right) x, x\right\rangle^{1/2} \\ &\quad\times \left\langle\left(\int_E p(t)\, r^2(V_t) \left||T_t^*|^{1-\alpha(t)} B_t^*\right|^4 d\mu(t)\right) x, x\right\rangle^{1/2} \\ &\le \frac{1}{2}\|x\|^2 \\ &\quad\times \left\langle\left(\int_E p(t)\, r^2(V_t) \left[\left||T_t|^{\alpha(t)} A_t\right|^4 + \left||T_t^*|^{1-\alpha(t)} B_t^*\right|^4\right] d\mu(t)\right) x, x\right\rangle\end{aligned}$$

for all $x \in H$. We also have the numerical radius inequality

$$\begin{aligned}&\omega^2\left(\int_E p(t)\, B_t T_t V_t A_t d\mu(t)\right) \\ &\le \frac{1}{2}\left\|\int_E p(t)\, r^2(V_t) \left[\left||T_t|^{\alpha(t)} A_t\right|^4 + \left||T_t^*|^{1-\alpha(t)} B_t^*\right|^4\right] d\mu(t)\right\|.\end{aligned}$$

5.3 SOME RELATED RESULTS

We also have the following result:

Theorem 4 *Assume that f_t and $g_t, t \in E$ are non-negative functions on $[0,\infty)$ which are continuous and satisfy the relation $f_t(u)_t g(u) = u$ for all $u \in [0,\infty)$, $t \in E$. Let $V_{t(\cdot)}, T_{(\cdot)}, A_{(\cdot)}, B_{(\cdot)}, : E \to \mathcal{B}(H)$ with $|T_t|\, V_t = V_t^* |T_t|$ for $t \in E$ and $p : E \to [0,\infty)$ be μ-measurable functions on E with $\int_E p(t)\, d\mu(t) = 1$. If $p(\cdot)\, r^2(V_{(\cdot)}) |f_{(\cdot)}(|T_{(\cdot)}|)\, A_{(\cdot)}|^4$, and*

$p(\cdot)\, r^2 (V_{(\cdot)}) \left|g_{(\cdot)}\left(\left|T^*_{(\cdot)}\right|\right) B^*_{(\cdot)}\right|^4$ *are Bochner integrable, then*

$$\left|\left\langle\left(\int_E p(t)\, B_t T_t V_t A_t d\mu(t)\right) x, x\right\rangle\right|^2$$
$$\leq \int_E p(t)\, |\langle B_t T_t V_t A_t x, x\rangle|^2 d\mu(t)$$
$$\leq \frac{1}{2}\left\langle\left(\int_E p(t)\, r^2(V_t)\, |f_t(|T_t|)\, A_t|^4 d\mu(t)\right) x, x\right\rangle^{1/2}$$
$$\times \left\langle\left(\int_E p(t)\, r^2(V_t)\, |g_t(|T_t^*|)\, B_t^*|^4 d\mu(t)\right) x, x\right\rangle^{1/2}$$
$$+ \frac{1}{2}\int_E p(t)\, r^2(V_t) \left|\left\langle |g_t(|T_t^*|)\, B_t^*|^2 |f_t(|T_t|)\, A_t|^2 x, x\right\rangle\right| d\mu(t) \tag{3.1}$$

for all $x \in H$, $\|x\| = 1$.

We also have the numerical radius inequality

$$\omega^2\left(\int_E p(t)\, B_t T_t V_t A_t d\mu(t)\right)$$
$$\leq \frac{1}{2}\left\|\int_E p(t)\, r^2(V_t)\, |f_t(|T_t|)\, A_t|^4 d\mu(t)\right\|^{1/2}$$
$$\times \left\|\int_E p(t)\, r^2(V_t)\, |g_t(|T_t^*|)\, B_t^*|^4 d\mu(t)\right\|^{1/2}$$
$$+ \frac{1}{2}\int_E p(t)\, r^2(V_t)\, \omega\left(|g_t(|T_t^*|)\, B_t^*|^2 |f_t(|T_t|)\, A_t|^2\right) d\mu(t). \tag{3.2}$$

Proof. From (2.5) we have

$$|\langle B_t T_t V_t A_t x, x\rangle|^2 \leq r^2(V_t) \left\langle |f_t(|T_t|)\, A_t|^2 x, x\right\rangle \left\langle x, |g_t(|T_t^*|)\, B_t^*|^2 x\right\rangle \tag{3.3}$$

for all $x \in H$ and $t \in E$.

If we use Buzano's inequality (2.1) then we can state that

$$\left\langle |f_t(|T_t|)\, A_t|^2 x, x\right\rangle \left\langle x, |g_t(|T_t^*|)\, B_t^*|^2 x\right\rangle$$
$$\leq \frac{1}{2}\left\| |f_t(|T_t|)\, A_t|^2 x\right\| \left\| |g_t(|T_t^*|)\, B_t^*|^2 x\right\|$$

$$
\begin{aligned}
&+\frac{1}{2}\left|\left\langle\left|f_t\left(\left|T_t\right|\right)A_t\right|^2x, \left|g_t\left(\left|T_t^*\right|\right)B_t^*\right|^2x\right\rangle\right| \\
&=\frac{1}{2}\left\|\left|f_t\left(\left|T_t\right|\right)A_t\right|^2x\right\|\left\|\left|g_t\left(\left|T_t^*\right|\right)B_t^*\right|^2x\right\| \\
&+\frac{1}{2}\left|\left\langle\left|g_t\left(\left|T_t^*\right|\right)B_t^*\right|^2\left|f_t\left(\left|T_t\right|\right)A_t\right|^2x, x\right\rangle\right|
\end{aligned}
$$

for all $x \in H$, $\|x\| = 1$ and $t \in E$.

Therefore, by (3.3) we get

$$
\begin{aligned}
&\left|\left\langle B_tT_tV_tA_tx, x\right\rangle\right|^2 \\
&\leq \frac{1}{2}r^2\left(V_t\right)\left\|\left|f_t\left(\left|T_t\right|\right)A_t\right|^2x\right\|\left\|\left|g_t\left(\left|T_t^*\right|\right)B_t^*\right|^2x\right\| \\
&+\frac{1}{2}r^2\left(V_t\right)\left|\left\langle\left|g_t\left(\left|T_t^*\right|\right)B_t^*\right|^2\left|f_t\left(\left|T_t\right|\right)A_t\right|^2x, x\right\rangle\right| \qquad (3.4)
\end{aligned}
$$

for all $x \in H$, $\|x\| = 1$ and $t \in E$.

If we multiply (3.4) by $p(t) \geq 0$ and integrate, then we get

$$
\begin{aligned}
&\int_E p(t)\left|\left\langle B_tT_tV_tA_tx, x\right\rangle\right|^2 d\mu(t) \\
&\leq \int_E p(t)\left\langle\left|f_t\left(\left|T_t\right|\right)A_t\right|^2x, x\right\rangle\left\langle x, \left|g_t\left(\left|T_t^*\right|\right)B_t^*\right|^2x\right\rangle d\mu(t) \\
&\leq \frac{1}{2}\int_E p(t)r^2\left(V_t\right)\left\|\left|f_t\left(\left|T_t\right|\right)A_t\right|^2x\right\|\left\|\left|g_t\left(\left|T_t^*\right|\right)B_t^*\right|^2x\right\| d\mu(t) \\
&+\frac{1}{2}\int_E p(t)r^2\left(V_t\right)\left|\left\langle\left|g_t\left(\left|T_t^*\right|\right)B_t^*\right|^2\left|f_t\left(\left|T_t\right|\right)A_t\right|^2x, x\right\rangle\right| d\mu(t) \qquad (3.5)
\end{aligned}
$$

for all $x \in H$, $\|x\| = 1$.

By utilizing the Cauchy-Bunyakowsky-Schwarz weighted integral inequality, we have

$$
\begin{aligned}
&\int_E p(t)r^2\left(V_t\right)\left\|\left|f_t\left(\left|T_t\right|\right)A_t\right|^2x\right\|\left\|\left|g_t\left(\left|T_t^*\right|\right)B_t^*\right|^2x\right\| d\mu(t) \\
&\leq \left(\int_E p(t)r^2\left(V_t\right)\left\|\left|f_t\left(\left|T_t\right|\right)A_t\right|^2x\right\|^2 d\mu(t)\right)^{1/2} \\
&\times\left(\int_E p(t)r^2\left(V_t\right)\left\|\left|g_t\left(\left|T_t^*\right|\right)B_t^*\right|^2x\right\|^2 d\mu(t)\right)^{1/2}
\end{aligned}
$$

$$= \left(\int_E p(t)\, r^2(V_t) \left\langle \left| f_t(|T_t|) A_t \right|^4 x, x \right\rangle d\mu(t) \right)^{1/2}$$

$$\times \left(\int_E p(t)\, r^2(V_t) \left\langle \left| g_t(|T_t^*|) B_t^* \right|^4 x, x \right\rangle d\mu(t) \right)^{1/2}$$

$$= \left\langle \left(\int_E p(t)\, r^2(V_t) \left| f_t(|T_t|) A_t \right|^4 d\mu(t) \right) x, x \right\rangle^{1/2}$$

$$\times \left\langle \left(\int_E p(t)\, r^2(V_t) \left| g_t(|T_t^*|) B_t^* \right|^4 d\mu(t) \right) x, x \right\rangle^{1/2}.$$

and by (3.5) we derive the second inequality in (3.1).

The first inequality follows by Cauchy-Bunyakowsky-Schwarz weighted integral inequality.

By taking the supremum over $x \in H, \|x\| = 1$ in (3.1) we obtain the inequalities (3.2). We also have the following result:

Theorem 5 *Assume that f_t and $g_t, t \in E$ are non-negative functions on $[0,\infty)$ which are continuous and satisfy the relation $f_t(u)_t g(u) = u$ for all $u \in [0,\infty)$, $t \in E$. Let $V_{t(\cdot)}$, $T_{(\cdot)}$, $A_{(\cdot)}$, $B_{(\cdot)}, : E \to \mathcal{B}(H)$ with $|T_t|\, V_t = V_t^* |T_t|$ for $t \in E$ and $p : E \to [0,\infty)$ be μ-measurable functions on E with $\int_E p(t)\, d\mu(t) = 1$. If $p(\cdot)\, r^{2r}(V_{(\cdot)}) \left| f_{(\cdot)}(|T_{(\cdot)}|) A_{(\cdot)} \right|^{2sr}$, and $p(\cdot)\, r^{2r}(V_{(\cdot)}) \left| g_{(\cdot)}\left(\left|T_{(\cdot)}^*\right|\right) B_{(\cdot)}^* \right|^{2qr}$ for $s, q > 1$ with $\frac{1}{s} + \frac{1}{q} = 1$ and $r \geq 1$ are Bochner integrable, then for $sr, qr \geq 2$,*

$$\left| \left\langle \left(\int_E p(t)\, B_t T_t V_t A_t d\mu(t) \right) x, x \right\rangle \right|^{2r}$$

$$\leq \int_E p(t) \left| \langle B_t T_t V_t A_t x, x \rangle \right|^{2r} d\mu(t)$$

$$\leq \frac{1}{2} \left\langle \left(\int_E p(t)\, r^{2r}(V_t) \left| f_t(|T_t|) A_t \right|^{2sr} d\mu(t) \right) x, x \right\rangle^{1/s}$$

$$\times \left\langle \left(\int_E p(t)\, r^{2r}(V_t) \left| g_t(|T_t^*|) B_t^* \right|^{2qr} d\mu(t) \right) x, x \right\rangle^{1/q}$$

$$+ \frac{1}{2} \int_E p(t)\, r^{2r}(V_t) \left| \left\langle \left| g_t(|T_t^*|) B_t^* \right|^2 \left| f_t(|T_t|) A_t \right|^2 x, x \right\rangle \right|^r d\mu(t) \quad (3.6)$$

for $x \in H$, $\|x\| = 1$.

We also have the numerical radius inequalities

$$
\begin{aligned}
&\omega^{2r}\left(\int_E p(t)\, B_t T_t V_t A_t d\mu(t)\right) \\
&\leq \frac{1}{2}\left\|\int_E p(t)\, r^{2r}(V_t)\, |f_t(|T_t|)\, A_t|^{2sr} d\mu(t)\right\|^{1/s} \\
&\times \left\|\int_E p(t)\, r^{2r}(V_t)\, |g_t(|T_t^*|)\, B_t^*|^{2qr} d\mu(t)\right\|^{1/q} \\
&+ \frac{1}{2}\int_E p(t)\, r^{2r}(V_t)\, \omega^r\left(|g_t(|T_t^*|)\, B_t^*|^2 |f_t(|T_t|)\, A_t|^2\right) d\mu(t).
\end{aligned} \tag{3.7}
$$

Proof. If we take the power $r \geq 1$ in (3.4), then

$$
\begin{aligned}
&|\langle B_t T_t V_t A_t x, x\rangle|^{2r} \\
&\leq \left(\frac{1}{2} r^2(V_t)\left\||f_t(|T_t|)\, A_t|^2 x\right\|\left\||g_t(|T_t^*|)\, B_t^*|^2 x\right\|\right. \\
&\quad \left. + \frac{1}{2} r^2(V_t)\, |\langle |g_t(|\, T_t^*\, |)\, B_t^*|^2 |f_t(|\, T_t\, |)\, A_t|^2 x, x\rangle|\right) r
\end{aligned} \tag{3.8}
$$

for $x \in H$, $\|x\| = 1$ and $t \in E$.

By the convexity of the power function, then we get

$$
\begin{aligned}
&\left(\frac{1}{2} r^2(V_t)\left\||f_t(|T_t|)\, A_t|^2 x\right\|\left\||g_t(|T_t^*|)\, B_t^*|^2 x\right\|\right. \\
&\quad + \frac{1}{2} r^2(V_t)\, |\langle |g_t(|\, T_t^*\, |)\, B_t^*|^2 |f_t(|\, T_t\, |)\, A_t|^2 x, x\rangle|) r \\
&\leq \frac{1}{2} r^{2r}(V_t)\left\||f_t(|T_t|)\, A_t|^2 x\right\|^r \left\||g_t(|T_t^*|)\, B_t^*|^2 x\right\|^r \\
&\quad + \frac{1}{2} r^{2r}(V_t)\left|\left\langle |g_t(|T_t^*|)\, B_t^*|^2 |f_t(|T_t|)\, A_t|^2 x, x\right\rangle\right|^r
\end{aligned}
$$

for $x \in H$, $\|x\| = 1$, which, by (3.8), gives that

$$
\begin{aligned}
&|\langle B_t T_t V_t A_t x, x\rangle|^{2r} \\
&\leq \frac{1}{2} r^{2r}(V_t)\left\||f_t(|T_t|)\, A_t|^2 x\right\|^r \left\||g_t(|T_t^*|)\, B_t^*|^2 x\right\|^r \\
&\quad + \frac{1}{2} r^{2r}(V_t)\left|\left\langle |g_t(|T_t^*|)\, B_t^*|^2 |f_t(|T_t|)\, A_t|^2 x, x\right\rangle\right|^r
\end{aligned} \tag{3.9}
$$

for $x \in H$, $\|x\| = 1$ and $t \in E$.

If we multiply (3.9) by $p(t) \geq 0$ and integrate, then we get

$$\int_E p(t) |\langle B_t T_t V_t A_t x, x\rangle|^{2r} d\mu(t)$$
$$\leq \frac{1}{2}\int_E p(t) r^{2r}(V_t) \left\| |f_t(|T_t|) A_t|^2 x \right\|^r \left\| |g_t(|T_t^*|) B_t^*|^2 x \right\|^r d\mu(t)$$
$$+ \frac{1}{2}\int_E p(t) r^{2r}(V_t) \left|\left\langle |g_t(|T_t^*|) B_t^*|^2 |f_t(|T_t|) A_t|^2 x, x\right\rangle\right|^r d\mu(t) \quad (3.10)$$

for $x \in H$, $\|x\| = 1$.

By Hölder's weighted integral inequality we have

$$\int_E p(t) r^{2r}(V_t) \left\| |f_t(|T_t|) A_t|^2 x \right\|^r \left\| |g_t(|T_t^*|) B_t^*|^2 x \right\|^r d\mu(t)$$
$$\leq \left(\int_E p(t) r^{2r}(V_t) \left\| |f_t(|T_t|) A_t|^2 x \right\|^{sr} d\mu(t)\right)^{1/s}$$
$$\times \left(\int_E p(t) r^{2r}(V_t) \left\| |g_t(|T_t^*|) B_t^*|^2 x \right\|^{qr} d\mu(t)\right)^{1/q}$$
$$= \left(\int_E p(t) r^{2r}(V_t) \left\langle |f_t(|T_t|) A_t|^4 x, x\right\rangle^{\frac{sr}{2}} d\mu(t)\right)^{1/s}$$
$$\times \left(\int_E p(t) r^{2r}(V_t) \left\langle |g_t(|T_t^*|) B_t^*|^4 x, x\right\rangle^{\frac{qr}{2}} d\mu(t)\right)^{1/q} \quad (3.11)$$

for $x \in H$, $\|x\| = 1$.

By utilizing McCarthy inequality, we have for $\frac{sr}{2}, \frac{qr}{2} \geq 1$ that

$$\left\langle |f_t(|T_t|) A_t|^4 x, x\right\rangle^{\frac{sr}{2}} \leq \left\langle |f_t(|T_t|) A_t|^{2sr} x, x\right\rangle$$

and

$$\left\langle |g_t(|T_t^*|) B_t^*|^4 x, x\right\rangle^{\frac{qr}{2}} \leq \left\langle |g_t(|T_t^*|) B_t^*|^{2qr} x, x\right\rangle$$

for $x \in H$, $\|x\| = 1$ and $t \in E$.

Therefore, by (3.11) we derive

$$\begin{aligned}
&\int_E p(t)\, r^{2r}(V_t) \left\| |f_t(|T_t|) A_t|^2 x \right\|^r \left\| |g_t(|T_t^*|) B_t^*|^2 x \right\|^r d\mu(t) \\
&\le \left(\int_E p(t)\, r^{2r}(V_t) \left\langle |f_t(|T_t|) A_t|^{2sr} x, x \right\rangle d\mu(t) \right)^{1/s} \\
&\quad \times \left(\int_E p(t)\, r^{2r}(V_t) \left\langle |g_t(|T_t^*|) B_t^*|^{2qr} x, x \right\rangle d\mu(t) \right)^{1/q} \\
&= \left\langle \left(\int_E p(t)\, r^{2r}(V_t) |f_t(|T_t|) A_t|^{2sr} d\mu(t) \right) x, x \right\rangle^{1/s} \\
&\quad \times \left\langle \left(\int_E p(t)\, r^{2r}(V_t) |g_t(|T_t^*|) B_t^*|^{2qr} d\mu(t) \right) x, x \right\rangle^{1/q} \qquad (3.12)
\end{aligned}$$

for $x \in H$, $\|x\| = 1$.

By utilizing (3.10) and (3.12) we derive the second part of (3.6).

The first inequality holds by Jensen's inequality for the power function $2r$

$$\begin{aligned}
\left| \left\langle \left(\int_E p(t)\, B_t T_t V_t A_t d\mu(t) \right) x, x \right\rangle \right|^{2r} &= \left| \int_E p(t) \langle (B_t T_t V_t A_t) x, x \rangle \, d\mu(t) \right|^{2r} \\
&\le \left(\int_E p(t) \left| \langle B_t T_t V_t A_t x, x \rangle \right| d\mu(t) \right)^{2r} \\
&\le \int_E p(t) \left| \langle B_t T_t V_t A_t x, x \rangle \right|^{2r} d\mu(t).
\end{aligned}$$

The inequality (3.7) follows by (3.6) by taking the supremum over for $x \in H$, $\|x\| = 1$.

Remark 4 If we take $s = q = 2$ and $r \geq 1$ in Theorem 5, then we get

$$\begin{aligned}
&\left| \left\langle \left(\int_E p(t)\, B_t T_t V_t A_t d\mu(t) \right) x, x \right\rangle \right|^{2r} \\
&\le \int_E p(t) \left| \langle B_t T_t V_t A_t x, x \rangle \right|^{2r} d\mu(t) \\
&\le \frac{1}{2} \left\langle \left(\int_E p(t)\, r^{2r}(V_t) |f_t(|T_t|) A_t|^{4r} d\mu(t) \right) x, x \right\rangle^{1/2}
\end{aligned}$$

$$\times\left\langle\left(\int_E p(t)\, r^{2r}(V_t)\left|g_t\left(\left|T_t^*\right|\right)B_t^*\right|^{4r} d\mu(t)\right)x, x\right\rangle^{1/2}$$
$$+\frac{1}{2}\int_E p(t)\, r^{2r}(V_t)\left|\left\langle\left|g_t\left(\left|T_t^*\right|\right)B_t^*\right|^2\left|f_t\left(\left|T_t\right|\right)A_t\right|^2 x, x\right\rangle\right|^r d\mu(t) \quad (3.13)$$

for $x \in H$, $\|x\| = 1$.

We also have the numerical radius inequalities

$$\omega^{2r}\left(\int_E p(t)\, B_t T_t V_t A_t d\mu(t)\right)$$
$$\leq \frac{1}{2}\left\|\int_E p(t)\, r^{2r}(V_t)\left|f_t\left(\left|T_t\right|\right)A_t\right|^{4r} d\mu(t)\right\|^{1/2}$$
$$\times\left\|\int_E p(t)\, r^{2r}(V_t)\left|g_t\left(\left|T_t^*\right|\right)B_t^*\right|^{4r} d\mu(t)\right\|^{1/2}$$
$$+\frac{1}{2}\int_E p(t)\, r^{2r}(V_t)\,\omega^r\left(\left|g_t\left(\left|T_t^*\right|\right)B_t^*\right|^2\left|f_t\left(\left|T_t\right|\right)A_t\right|^2\right) d\mu(t). \quad (3.14)$$

For $r = 1$ we obtain

$$\left|\left\langle\left(\int_E p(t)\, B_t T_t V_t A_t d\mu(t)\right)x, x\right\rangle\right|^2$$
$$\leq \int_E p(t)\left|\left\langle B_t T_t V_t A_t x, x\right\rangle\right|^2 d\mu(t)$$
$$\leq \frac{1}{2}\left\langle\left(\int_E p(t)\, r^2(V_t)\left|f_t\left(\left|T_t\right|\right)A_t\right|^4 d\mu(t)\right)x, x\right\rangle^{1/2}$$
$$\times\left\langle\left(\int_E p(t)\, r^2(V_t)\left|g_t\left(\left|T_t^*\right|\right)B_t^*\right|^4 d\mu(t)\right)x, x\right\rangle^{1/2}$$
$$+\frac{1}{2}\int_E p(t)\, r^2(V_t)\left|\left\langle\left|g_t\left(\left|T_t^*\right|\right)B_t^*\right|^2\left|f_t\left(\left|T_t\right|\right)A_t\right|^2 x, x\right\rangle\right| d\mu(t) \quad (3.15)$$

for $x \in H$, $\|x\| = 1$.

We also have the numerical radius inequalities

$$\omega^2\left(\int_E p(t)\, B_t T_t V_t A_t d\mu(t)\right)$$
$$\leq \frac{1}{2}\left\|\int_E p(t)\, r^2(V_t)\left|f_t\left(\left|T_t\right|\right)A_t\right|^4 d\mu(t)\right\|^{1/2}$$

$$\times \left\| \int_E p(t)\, r^2(V_t) \left|g_t\left(\left|T_t^*\right|\right) B_t^*\right|^4 d\mu(t) \right\|^{1/2}$$
$$+ \frac{1}{2} \int_E p(t)\, r^2(V_t)\, \omega\left(\left|g_t\left(\left|T_t^*\right|\right) B_t^*\right|^2 \left|f_t(|T_t|) A_t\right|^2 \right) d\mu(t). \quad (3.16)$$

Also, if we take $r = 2$, then for $s, q > 1$ with $\frac{1}{s} + \frac{1}{q} = 1$ we derive

$$\left| \left\langle \left(\int_E p(t)\, B_t T_t V_t A_t d\mu(t) \right) x, x \right\rangle \right|^4$$
$$\leq \int_E p(t) \left|\langle B_t T_t V_t A_t x, x\rangle\right|^4 d\mu(t)$$
$$\leq \frac{1}{2} \left\langle \left(\int_E p(t)\, r^4(V_t) \left|f_t(|T_t|) A_t\right|^{4s} d\mu(t) \right) x, x \right\rangle^{1/s}$$
$$\times \left\langle \left(\int_E p(t)\, r^4(V_t) \left|g_t\left(\left|T_t^*\right|\right) B_t^*\right|^{4q} d\mu(t) \right) x, x \right\rangle^{1/q}$$
$$+ \frac{1}{2} \int_E p(t)\, r^4(V_t) \left| \left\langle \left|g_t\left(\left|T_t^*\right|\right) B_t^*\right|^2 \left|f_t(|T_t|) A_t\right|^2 x, x \right\rangle \right|^4 d\mu(t) \quad (3.17)$$

for $x \in H$, $\|x\| = 1$.

We also have the numerical radius inequalities

$$\omega^4 \left(\int_E p(t)\, B_t T_t V_t A_t d\mu(t) \right)$$
$$\leq \frac{1}{2} \left\| \int_E p(t)\, r^4(V_t) \left|f_t(|T_t|) A_t\right|^{4s} d\mu(t) \right\|^{1/s}$$
$$\times \left\| \int_E p(t)\, r^4(V_t) \left|g_t\left(\left|T_t^*\right|\right) B_t^*\right|^{4q} d\mu(t) \right\|^{1/q}$$
$$+ \frac{1}{2} \int_E p(t)\, r^4(V_t)\, \omega^2\left(\left|g_t\left(\left|T_t^*\right|\right) B_t^*\right|^2 \left|f_t(|T_t|) A_t\right|^2 \right) d\mu(t). \quad (3.18)$$

5.4 APPLICATIONS VIA POLAR DECOMPOSITION

Let $X = U|X|$ be the *polar decomposition* of the bounded linear operator X, with U a partial isometry. If we take $T = U|X|^{1-\lambda}$ and $V = |X|^{\lambda}$, with $\lambda \in [0,1]$, then we have

$$TV = U|X| = X, |T| = |X|^{1-\lambda} \text{and} \, |T^*| = |X^*|^{1-\lambda}$$

and since

$$|T|\, V = |X| = V^* \, |T|$$

and

$$r\left(|X|^{\lambda}\right) = \left\| |X|^{\lambda} \right\| = \|X\|^{\lambda}.$$

We also assume that $|X|^0 = I$.

Let $X_t = U_t\,|X_t|$, $t \in E$, be the *polar decomposition* of the bounded linear family of operators X_t, with U_t a partial isometry. If we take $T_t = U_t|X_t|^{1-\lambda}$ and $V_t = |X_t|^{\lambda}$, $t \in E$, then we have

$$T_tV_t = X_t, |T_t| = |X_t|^{1-\lambda} \text{ and } |T_t^*| = |X_t^*|^{1-\lambda},$$

and

$$|T_t|\, V_t = V_t^* \, |T_t| = |X_t|$$

for all $t \in E$. Also

$$r(V_t) = \|X_t\|^{\lambda},$$

for all $t \in E$.

If we use Theorem 2 for $T_t = U_t|X_t|^{1-\lambda}$ and $V_t = |X_t|^{\lambda}$, $\quad t \in E$,

$$\begin{aligned}
&\left|\left\langle\left(\int_E p(t)\, B_t X_t A_t d\mu(t)\right) x, y\right\rangle\right|^2 \\
&\leq \left\langle\left(\int_E p(t)\, \|X_t\|^{\lambda} \left|f_t\left(|X_t|^{1-\lambda}\right) A_t\right|^2 d\mu(t)\right) x, x\right\rangle \\
&\times \left\langle\left(\int_E p(t)\, \|X_t\|^{\lambda} \left|g_t\left(|X_t^*|^{1-\lambda}\right) B_t^*\right|^2 d\mu(t)\right) y, y\right\rangle \qquad (4.1)
\end{aligned}$$

for all $x, y \in H$, provided

$$p(\cdot)\,\|X_{(\cdot)}\|^{\lambda}\left|f_{(\cdot)}\left(|X_{(\cdot)}|^{1-\lambda}\right)A_{(\cdot)}\right|^{2}$$

and

$$p(\cdot)\,\|X_{(\cdot)}\|^{\lambda}\left|g_{(\cdot)}\left(|X_{(\cdot)}^{*}|^{1-\lambda}\right)B_{(\cdot)}^{*}\right|^{2}$$

are Bochner integrable on E.

We also have the norm inequality

$$\left\|\int_E p(t)\,B_tX_tA_t d\mu(t)\right\|^2 \le \left\|\int_E p(t)\,\|X_t\|^{\lambda}\left|f_t\left(|X_t|^{1-\lambda}\right)A_t\right|^2 d\mu(t)\right\| \times \left\|\int_E p(t)\,\|X_t\|^{\lambda}\left|g_t\left(|X_t^{*}|^{1-\lambda}\right)B_t^{*}\right|^2 d\mu(t)\right\|. \tag{4.2}$$

As above, assume that $\alpha : E \to [0,1]$ is measurable. By taking $f_t(u) = u^{\alpha(t)}$ and $g_t(u) = u^{1-\alpha(t)}$, $u \ge 0$, $t \in E$ in (4.1) and (4.2), we get

$$\begin{aligned}\left|\left\langle\left(\int_E p(t)\,B_tX_tA_t d\mu(t)\right)x, y\right\rangle\right|^2 &\le \left\langle\left(\int_E p(t)\,\|X_t\|^{\lambda}\left|\,|X_t|^{\alpha(t)(1-\lambda)}A_t\right|^2 d\mu(t)\right)x, x\right\rangle \\ &\quad\times\left\langle\left(\int_E p(t)\,\|X_t\|^{\lambda}\left|\,|X_t^{*}|^{(1-\alpha(t))(1-\lambda)}B_t^{*}\right|^2 d\mu(t)\right)y, y\right\rangle \end{aligned} \tag{4.3}$$

for all $x, y \in H$, and the norm inequality

$$\left\|\int_E p(t)\,B_tX_tA_t d\mu(t)\right\|^2 \le \left\|\int_E p(t)\,\|X_t\|^{\lambda}\left|\,|X_t|^{\alpha(t)(1-\lambda)}A_t\right|^2 d\mu(t)\right\| \times \left\|\int_E p(t)\,\|X_t\|^{\lambda}\left|\,|X_t^{*}|^{(1-\alpha(t))(1-\lambda)}B_t^{*}\right|^2 d\mu(t)\right\|. \tag{4.4}$$

From Corollary 1 we also get

$$\begin{aligned}
&\left|\left\langle\left(\int_E p(t)\, B_t X_t A_t d\mu(t)\right) x, x\right\rangle\right|^2 \\
&\quad \le \left\langle\left(\int_E p(t)\, \|X_t\|^{\lambda} \left|f_t\left(|X_t|^{1-\lambda}\right) A_t\right|^2 d\mu(t)\right) x, x\right\rangle \\
&\quad\quad \times \left\langle\left(\int_E p(t)\, \|X_t\|^{\lambda} \left|g_t\left(|X_t^*|^{1-\lambda}\right) B_t^*\right|^2 d\mu(t)\right) x, x\right\rangle \\
&\quad \le \left\langle\left(\int_E p(t)\, \|X_t\|^{\lambda} \frac{\left|f_t\left(|X_t|^{1-\lambda}\right) A_t\right|^2 + \left|g_t\left(|X_t^*|^{1-\lambda}\right) B_t^*\right|^2}{2} d\mu(t)\right) x, x\right\rangle
\end{aligned} \tag{4.5}$$

for all $x \in H$. We also have the numerical radius inequality

$$\begin{aligned}
&\omega^2\left(\int_E p(t)\, B_t X_t A_t d\mu(t)\right) \\
&\quad \le \left\|\int_E p(t)\, \|X_t\|^{\lambda} \frac{\left|f_t\left(|X_t|^{1-\lambda}\right) A_t\right|^2 + \left|g_t\left(|X_t^*|^{1-\lambda}\right) B_t^*\right|^2}{2} d\mu(t)\right\|.
\end{aligned} \tag{4.6}$$

By taking $f_t(u) = u^{\alpha(t)}$ and $g_t(u) = u^{1-\alpha(t)}$, $u \ge 0$, $t \in E$ in (4.5) and (4.6), we get

$$\begin{aligned}
&\left|\left\langle\left(\int_E p(t)\, B_t X_t A_t d\mu(t)\right) x, x\right\rangle\right|^2 \\
&\quad \le \left\langle\left(\int_E p(t)\, \|X_t\|^{\lambda} \left||X_t|^{\alpha(t)(1-\lambda)} A_t\right|^2 d\mu(t)\right) x, x\right\rangle \\
&\quad\quad \times \left\langle\left(\int_E p(t)\, \|X_t\|^{\lambda} \left||X_t^*|^{(1-\alpha(t))(1-\lambda)} B_t^*\right|^2 d\mu(t)\right) x, x\right\rangle \\
&\quad \le \left\langle \int_E p(t)\, \|X_t\|^{\lambda} \frac{\left||X_t|^{\alpha(t)(1-\lambda)} A_t\right|^2 + \left||X_t^*|^{(1-\alpha(t))(1-\lambda)} B_t^*\right|^2}{2} d\mu(t)\, x, x\right\rangle
\end{aligned} \tag{4.7}$$

for all $x \in H$. We also have the numerical radius inequality

$$\omega^2\left(\int_E p(t)\, B_t X_t A_t d\mu(t)\right) \leq \left\| \int_E p(t)\, \|X_t\|^{\lambda} \frac{\left| |X_t|^{\alpha(t)(1-\lambda)} A_t \right|^2 + \left| |X_t^*|^{(1-\alpha(t))(1-\lambda)} B_t^* \right|^2}{2} d\mu(t) \right\|. \tag{4.8}$$

If we use Theorem 3 for $T_t = U_t |X_t|^{1-\lambda}$ and $V_t = |X_t|^{\lambda}$, $t \in E$, for $s, q > 1$ with $\frac{1}{s} + \frac{1}{q} = 1$, then we get

$$\begin{aligned} &\left|\left\langle\left(\int_E p(t)\, B_t X_t A_t d\mu(t)\right) x, y\right\rangle\right|^2 \\ &\leq \|x\|^{2/q} \|y\|^{2/s} \left\langle\left(\int_E p(t)\, \|X_t\|^{2\lambda} \left| f_t\left(|X_t|^{1-\lambda}\right) A_t \right|^{2s} d\mu(t)\right) x, x\right\rangle^{1/s} \\ &\times \left\langle\left(\int_E p(t)\, \|X_t\|^{2\lambda} \left| g_t\left(|X_t^*|^{1-\lambda}\right) B_t^* \right|^{2q} d\mu(t)\right) y, y\right\rangle^{1/q} \end{aligned} \tag{4.9}$$

for all $x, y \in H$, provided

$$p(\cdot)\, \|X_{(\cdot)}\|^{2\lambda} \left| f_{(\cdot)}\left(|X_{(\cdot)}|^{1-\lambda}\right) A_{(\cdot)} \right|^{2s}$$

and

$$p(\cdot)\, \|X_{(\cdot)}\|^{2\lambda} \left| g_{(\cdot)}\left(|X_{(\cdot)}^*|^{1-\lambda}\right) B_{(\cdot)}^* \right|^{2q}$$

are Bochner integrable on E.

We also have the norm inequality

$$\begin{aligned} &\left\| \int_E p(t)\, B_t X_t A_t d\mu(t) \right\|^2 \\ &\leq \left\| \int_E p(t)\, \|X_t\|^{2\lambda} \left| f_t\left(|X_t|^{1-\lambda}\right) A_t \right|^{2s} d\mu(t) \right\|^{1/s} \\ &\times \left\| \int_E p(t)\, \|X_t\|^{2\lambda} \left| g_t\left(|X_t^*|^{1-\lambda}\right) B_t^* \right|^{2q} d\mu(t) \right\|^{1/q}. \end{aligned} \tag{4.10}$$

By taking $f_t(u) = u^{\alpha(t)}$ and $g_t(u) = u^{1-\alpha(t)}$, $u \geq 0$, $t \in E$ in (4.1) and (4.2), we get

$$\begin{aligned}
&\left|\left\langle \left(\int_E p(t)\, B_t X_t A_t d\mu(t) \right) x, y \right\rangle\right|^2 \\
&\leq \|x\|^{2/q} \|y\|^{2/s} \left\langle \left(\int_E p(t) \|X_t\|^{2\lambda} \left| |X_t|^{\alpha(t)(1-\lambda)} A_t \right|^{2s} d\mu(t) \right) x, x \right\rangle^{1/s} \\
&\quad \times \left\langle \left(\int_E p(t) \|X_t\|^{2\lambda} \left| |X_t^*|^{(1-\alpha(t))(1-\lambda)} B_t^* \right|^{2q} d\mu(t) \right) y, y \right\rangle^{1/q} \qquad (4.11)
\end{aligned}$$

and

$$\begin{aligned}
&\left\| \int_E p(t)\, B_t X_t A_t d\mu(t) \right\|^2 \\
&\leq \left\| \int_E p(t) \|X_t\|^{2\lambda} \left| |X_t|^{\alpha(t)(1-\lambda)} A_t \right|^{2s} d\mu(t) \right\|^{1/s} \\
&\quad \times \left\| \int_E p(t) \|X_t\|^{2\lambda} \left| |X_t^*|^{(1-\alpha(t))(1-\lambda)} B_t^* \right|^{2q} d\mu(t) \right\|^{1/q}. \qquad (4.12)
\end{aligned}$$

From Corollary 2 we also have

$$\begin{aligned}
&\left|\left\langle \left(\int_E p(t)\, B_t X_t A_t d\mu(t) \right) x, x \right\rangle\right|^2 \\
&\leq \|x\|^2 \left\langle \left(\int_E p(t) \|X_t\|^{2\lambda} \left| f_t\left(|X_t|^{1-\lambda}\right) A_t \right|^{2s} d\mu(t) \right) x, x \right\rangle^{1/s} \\
&\quad \times \left\langle \left(\int_E p(t) \|X_t\|^{2\lambda} \left| g_t\left(|X_t^*|^{1-\lambda}\right) B_t^* \right|^{2q} d\mu(t) \right) x, x \right\rangle^{1/q} \\
&\leq \| x\|^2 \left\langle \int_E p(t) \| X_t\|^{2\lambda} \right. \\
&\quad \times \left. \left[\frac{1}{p} \left| f_t\left(|X_t|^{1-\lambda}\right) A_t \right|^{2s} + \frac{1}{q} \left| g_t\left(|X_t^*|^{1-\lambda}\right) B_t^* \right|^{2q} \right] d\mu(t)\, x, x \right\rangle \qquad (4.13)
\end{aligned}$$

for all $x \in H$. We also have the numerical radius inequality

$$\omega^2\left(\int_E p(t)\, B_t X_t A_t d\mu(t)\right)^2 \leq \left\|\int_E p(t) \| X_t\|^{2\lambda} \times \left[\frac{1}{p}\left|f_t\left(|X_t|^{1-\lambda}\right) A_t\right|^{2s} + \frac{1}{q}\left|g_t\left(|X_t^*|^{1-\lambda}\right) B_t^*\right|^{2q}\right] d\mu(t)\right\|. \quad (4.14)$$

By taking $f_t(u) = u^{\alpha(t)}$ and $g_t(u) = u^{1-\alpha(t)}$, $u \geq 0$, $t \in E$ in (4.13) and (4.14), we get

$$\begin{aligned}
&\left|\left\langle\left(\int_E p(t)\, B_t X_t A_t d\mu(t)\right) x, x\right\rangle\right|^2 \\
&\leq \|x\|^2 \left\langle\left(\int_E p(t) \|X_t\|^{2\lambda} \left||X_t|^{\alpha(t)(1-\lambda)} A_t\right|^{2s} d\mu(t)\right) x, x\right\rangle^{1/s} \\
&\quad \times \left\langle\left(\int_E p(t) \|X_t\|^{2\lambda} \left||X_t^*|^{(1-\alpha(t))(1-\lambda)} B_t^*\right|^{2q} d\mu(t)\right) x, x\right\rangle^{1/q} \\
&\leq \| x\|^2 \left\langle\int_E p(t) \| X_t\|^{2\lambda} \times \left[\frac{1}{p}\left||X_t|^{\alpha(t)(1-\lambda)} A_t\right|^{2s} + \frac{1}{q}\left||X_t^*|^{(1-\alpha(t))(1-\lambda)} B_t^*\right|^{2q}\right] d\mu(t) x, x\right\rangle
\end{aligned} \quad (4.15)$$

for all $x \in H$. We also have the numerical radius inequality

$$\omega^2\left(\int_E p(t)\, B_t X_t A_t d\mu(t)\right)^2 \leq \left\|\int_E p(t) \| X_t\|^{2\lambda} \times \left[\frac{1}{p}\left||X_t|^{\alpha(t)(1-\lambda)} At\right|^{2s} + \frac{1}{q}\left||X_t^*|^{(1-\alpha(t))(1-\lambda)} B_t^*\right|^{2q}\right] d\mu(t)\right\|. \quad (4.16)$$

If one uses the other general results from Theorem 4 and Theorem 5, that one can state some similar inequalities for three families of operators as above. We omit the details.

REFERENCES

1. S. Bochner, Integration von Funktionen, deren Werte die Elemente eines Vektorraumes sind, *Fund. Math.* **20** (1933), 262–276.
2. M. L. Buzano, Generalizzazione della diseguaglianza di Cauchy-Schwarz. (Italian), *Rend. Sem. Mat. Univ. e Politech. Torino*, **31** (1971/73), 405–409 (1974).
3. C. A. McCarthy, C_p, *Israel J. Math.* **5** (1967), 249–271.
4. A. Aluthge, Some generalized theorems on p-hyponormal operators, *Integr. Equat. Oper. Th.* **24** (1996), 497–501.
5. A. Abu-Omar and F. Kittaneh, A numerical radius inequality involving the generalized Aluthge transform, *Studia Math.* **216** (1) (2013), 69–75.
6. P. Bhunia, S. Bag, and K. Paul, Numerical radius inequalities and its applications in estimation of zeros of polynomials, *Linear Algebra and its App.* **573** (2019), 166–177.
7. P. Bhunia, S. S. Dragomir, M. S. Moslehian and K. Paul, *Lectures on Numerical Radius Inequalities*, Springer, Cham, 2022. https://doi.org/10.1007/978-3-031-13670-2.
8. S. S. Dragomir, *Inequalities for the Numerical Radius of Linear Operators in Hilbert Spaces*, SpringerBriefs in Mathematics, 2013. https://doi.org/10.1007/978-3-319-01448-7.
9. N. Dunford and J.T. Schwartz, *Linear Operators, Part I: General Theory*, Wiley-Interscience, New York, 1958.
10. M. El-Haddad and F. Kittaneh, Numerical radius inequalities for Hilbert space operators. II, *Studia Math.* **182** (2) (2007), 133–140.
11. T. H. Hildebrandt, Integration in abstract spaces, *Bull. Amer. Math. Soc.* **59** (1953), 111–139.
12. E. Hille and R. S. Phillips, Functional analysis and semi-groups, *Amer. Math. Soc.* **31** (1996), 808. Providence.
13. F. Kittaneh, Notes on some inequalities for Hilbert space operators, *Publ. Res. Inst. Math. Sci.* **24** (2) (1988), 283–293.
14. F. Kittaneh, A numerical radius inequality and an estimate for the numerical radius of the Frobenius companion matrix, *Studia Math.* **158** (1) (2003), 11–17.
15. F. Kittaneh, Numerical radius inequalities for Hilbert space operators, *Studia Math.* **168** (1) (2005), 73–80.

16. V.I. Sobolev, Bochner integral. Encyclopedia of Mathematics. http://www.encyclopediaofmath.org/index.php?title=Bochner_integraloldid=11334
17. T. Yamazaki, On upper and lower bounds of the numerical radius and an equality condition, *Studia Math.* **178** (1) (2007), 83–89.
18. K. Yosida, *Functional Analysis*, Springer, 1980, Chapt. 8, 209–231, §1.

CHAPTER 6

Frozen Derivative Iterative Methods of High Order for Equations

G. Deep and I. K. Argyros

6.1 INTRODUCTION

Let us consider a Fréchet-derivable operator $Q : E \subset T_1 \to T_2$, where T_1, T_2 are Banach spaces and $E\,(\neq \emptyset)$ is an open and convex set. In science and other practical fields, equations of the type

$$Q(x) = 0 \tag{1}$$

are regularly used to address a wide range of very complicated problems. The iterative method is defined for $k \geq 3$ standing for a natural number, and starting guess $x_0 \in E$ by

$$\begin{aligned}
Q'(x_0)\, u_j &= Q\left(x_{j-1}\right), j = 1, 2, 3, \\
x_j &= x_{j-1} - u_j, \\
Q'(x_0)\, u_4 &= Q(x_3)\, , j = 5, 6, 7, 8, \\
Q'(x_0)\, u_j &= Q'(x_3)\, u_{j-1} \\
x_4 &= x_3 - \frac{21}{4}u_4 + 11u_5 - \frac{23}{2}u_6 + 6u_7 - \frac{5}{4}u_8, \\
Q'(x_0)\, u_9 &= Q(x_4)\, , \\
Q'(x_0)\, u_j &= Q'(x_3)\, u_{j-1}, j = 10, 11, 12, 13,
\end{aligned}$$

DOI: 10.1201/9781003530602-6

$$x_5 = x_4 - \frac{11}{2}u_9 + 12u_{10} - 13u_{11} + 7u_{12} - \frac{3}{2}u_{13},$$

and

$$x_0 = x_k. \tag{2}$$

Note: Number of steps k, convergence order $5k - 11$, function evaluations k, Jacobian evaluations 2, LU-factorization 1, matrix-vector multiplications $4(k-2)$, vector-vector multiplications $5\,(k-3)$, and number of lower and upper triangular systems $5k-12$. The local convergence analysis was carried out in [1] when $T_1 = T_2 = \mathbb{R}^m$, where m is a natural number under the Taylor series expansion approach and assumptions reaching the $Q^{(9)}$ not on the method limiting its applicability to solving equations whose operator Q is at least that many times differentiable. But the method may converge otherwise. Let us present an academic and motivational function $f : E \to \mathbb{R}$ given by $f(t) = \begin{cases} \alpha t^2 \log t + \beta t^5 + \gamma t^4, & \text{if } t \neq 0 \\ 0, & t = 0, \end{cases}$ where $\alpha \neq 0$, and $\beta + \gamma = 0$ are given real parameters. Take $E = \left[-\frac{3}{2}, 2\right]$. Notice that by these definitions, $t^* = 1 \in E$ solves the equation $f(t) = 0$. But the function f'' is unbounded at the point $t = 0 \in E$. Hence, the convergence of the method (2) to t^* cannot be assured by the results in [1]. However, the method (2) converges to t^* if, e.g., we select $x_0 = 1.2 \in E$. Thus, the convergence results in [1] can be weakened. But there exist other issues: There are no computable bounds on $||x_k - q||$ or information about the uniqueness of the solution q. That is why in order to expand the applicability of the method, we provide a new local convergence using information only for operators Q, Q' appearing on it under generalized continuity assumptions [2, 3]. Other benefits of our approach include computable upper error bounds on the distances $||x_k - q||$ and uniqueness of the solution. Our approach can extend the applicability of other methods along the same lines [4–12]. The more challenging semi-local convergence is also given (not provided in [1]) by exchanging the role of the solution q by the initial point x_0 and using majorizing sequence. Finally, the analysis is carried out in the more general setting of a Banach space.

6.2 LOCAL CONVERGENCE

The convergence conditions are developed for $M = [0, +\infty)$ as follows. Assume:

(A_1) There exists a function which is continuous and nondecreasing (FCND) $\phi_0 : M \longrightarrow M$ such that the equation $\phi_0(t) - 1 = 0$ admits a smallest solution which is positive (SSP) and is denoted by ρ. Let $M_0 = [0, \rho)$.

(A_2) There exists a FCND $\phi : M_0 \longrightarrow M$. Define the real functions on the interval M_0 in turn by

$$h_1(t) = \frac{\int_0^1 \phi((1-\lambda)t)\, d\lambda}{1 - \phi_0(t)},$$

$$\phi_1(t) = \begin{cases} \phi((1 + h_1(t))t) \\ \text{or} \\ \phi_0(t) + \phi_0(h_1(t)t), \end{cases}$$

$$h_2(t) = \left[\frac{\int_0^1 \phi((1-\lambda)h_1(t)t)\, d\lambda}{1 - \phi_0(h_1(t)t)} + \frac{\phi_1(t)\left(1 + \int_0^1 \phi_0(\lambda h_1(t)t)\, d\lambda\right)}{(1-\phi_0(t))(1-\phi_0(h_1(t)t))}\right] h_1(t),$$

$$\phi_2(t) = \begin{cases} \phi((1 + h_2(t))t) \\ \text{or} \\ \phi_0(t) + \phi_0(h_2(t)t), \end{cases}$$

$$h_3(t) = \left[\frac{\int_0^1 \phi((1-\lambda)h_2(t)t)\, d\lambda}{1 - \phi_0(h_2(t)t)} + \frac{\phi_2(t)\left(1 + \int_0^1 \phi_0(\lambda h_2(t)t)\, d\lambda\right)}{(1-\phi_0(t))(1-\phi_0(h_2(t)t))}\right] h_2(t),$$

$$p_0(t) = \frac{1 + \phi_0(h_3(t)t)}{1 - \phi_0(t)}, \quad p_1(t) = \frac{1 + \int_0^1 \phi_0(\lambda h_3(t)t)\, d\lambda}{1 - \phi_0(t)},$$

$$p_m(t) = p_0(t)\, p_{m-1}(t), \; m = 2, 3, 4, 5,$$

$$h_4(t) = \Big(1 + \frac{21}{4} p_0(t) p_1(t) + 11 p_2(t) + \frac{23}{2} p_3(t) + 6 p_4(t) + \frac{5}{4} p_5(t)\Big) h_3(t),$$

$$q_1^{(s-1)}(t) = \frac{\left(1 + \int_0^1 \phi_0(\lambda h_{s-1}(t)\, t)\, d\lambda\right) g_{s-1}(t)}{1 - \phi_0(t)},$$

$$q_m^{(s-1)}(t) = p_0(t)\, q_{m-1}^{(s-1)}(t)$$

and

$$hs(t) = \Big(1 + \frac{11}{2} q_1^{(s-1)}(t) + 12\, q_2^{(s-1)}(t) + 13\, q_3^{(s-1)}(t) + 7\, q_4^{(s-1)}(t) + \frac{3}{2} q_5^{(s-1)}(t)\Big)\, hs - 1(t).$$

(A_3) The equations $h_\gamma(t) - 1 = 0, \gamma = 1, 2, ..., k$ have smallest solutions denoted by r_γ, respectively. Let

$$r = \min\{r_\gamma\}. \tag{3}$$

The real functions are associated to the operators on the method.

(A_4) There exists an invertible linear operator A, and a solution $q \in E$ of the equation $Q(x) = 0$ such that for each $v \in E$

$$\left\| A^{-1}(Q'(v) - Q'(q)) \right\| \leq \phi_0(\|v - q\|).$$

Let $E_0 = E \cap U(q, \rho)$.

(A_5)

$$\left\| A^{-1}(Q'(v_2) - Q'(v_1)) \right\| \leq \phi(\|v_2 - v_1\|)$$

for each $v_1, v_2 \in E_0$.
and

(A_6)

$$U[q, r] \subset E.$$

Theorem 1 *Assume that the conditions (A_1)–(A_6) are valid. Then, the sequence $\{x_k\}$ is convergent to q, provided that $x_0 \in U_0 := U(q, r) - \{q\}$ and $\|x_0 - q\| < r$.*

Proof. Notice that by (A_4) and the choice of r $\|A^{-1}(Q'(x_0) - Q'(q))\| \leq \phi_0(\|x_0 - q\|) < 1$, that $Q'(x_0)^{-1}$ exists and $\|Q'(x_0)^{-1}A\| \leq \frac{1}{1-\phi_0(\|x_0-q\|)}$ by Banach Lemma on invertible operators [6, 9]. The motivation for the introduction of the real functions is given by a series of estimates and induction:

$$x_1 - q = x_0 - q - u_1 = x_0 - q - Q'(x_0)^{-1}Q(x_0)$$
$$= \int_0^1 Q'(x_0)^{-1}(Q'(q + \lambda(x_0 - q)) - Q'(x_0))\, d\lambda\,(x_0 - q),$$
$$\|x_1 - q\| \leq \|Q'(x_0)^{-1}A\|$$
$$\left\|\int_0^1 A^{-1}(Q'(q + \lambda(x_0 - q)) - Q'(x_0))\, d\lambda\right\| \|x_0 - q\|$$
$$\leq \frac{\int_0^1 \phi((1-\lambda)\|x_0 - q\|)\, d\lambda \|x_0 - q\|}{1 - \phi_0(\|x_0 - q\|)} = h_1(\|x_0 - q\|)\|x_0 - q\|$$
$$\leq \|x_0 - q\| < r,$$

so the iterate $x_1 \in U_0$. Similarly,

$$x_2 - q = x_1 - q - u_2 = x_1 - q - Q'(x_0)^{-1}Q(x_1)$$
$$= x_1 - q - Q'(x_1)^{-1}Q(x_1) + Q'(x_1)^{-1}$$
$$(Q'(x_0) - Q'(x_1))\,Q'(x_0)^{-1}Q(x_1),$$
$$\|x_2 - q\| \leq \left[\frac{\int_0^1 \phi((1-\lambda)\|x_1 - q\|)\, d\lambda}{1 - \phi_0(\|x_1 - q\|)}\right.$$
$$\left. + \frac{\bar{\phi}_1\left(1 + \int_0^1 \phi_0(\lambda\|x_1 - q\|)\, d\lambda\right)}{(1 - \phi_0(\|x_0 - q\|))(1 - \phi_0(\|x_1 - q\|))}\right] \| x_1 - q \|$$
$$\leq h_2(\|x_1 - q\|)\|x_1 - q\| \leq \|x_1 - q\|,$$
$$x_3 - q = x_2 - q - u_3 = x_2 - q - Q'(x_0)^{-1}Q(x_2)$$
$$= x_2 - q - Q'(x_2)^{-1}Q(x_2) + Q'(x_2)^{-1}$$
$$(Q'(x_0) - Q'(x_2))\,Q'(x_0)^{-1}Q(x_2),$$

$$\|x_3 - q\| \le \left[\frac{\int_0^1 \phi((1-\lambda)\|x_2 - q\|)\, d\lambda}{1 - \phi_0(\|x_2 - q\|)} + \frac{\overline{\phi}_2 \left(1 + \int_0^1 \phi_0(\lambda \|x_2 - q\|)\, d\lambda\right)}{(1 - \phi_0(\|x_0 - q\|))(1 - \phi_0(\|x_2 - q\|))} \right] \| x_2 - q \|$$
$$\le h_3(\|x_2 - q\|)\|x_2 - q\| \le \|x_2 - q\|,$$

where we also used the estimates

$$\left\|A^{-1}(Q'(x_1) - Q'(x_0))\right\| \le \phi(\|x_1 - x_0\|) \le \phi(\|x_0 - q\| + \|x_1 - q\|) \le \overline{\phi}_1$$

or

$$\left\|A^{-1}(Q'(x_1) - Q'(x_0))\right\| \le \left\|A^{-1}(Q'(x_0) - A)\right\| + \left\|A^{-1}(Q'(x_1) - A)\right\| \le \phi_0(\|x_0 - q\|) + \phi_0(\|x_1 - q\|) \le \overline{\phi}_1, \tag{4}$$

where

$$\overline{\phi}_1 = \begin{cases} \phi(\|x_0 - q\| + \|x_1 - q\|) \\ \text{or} \\ \phi_0(\|x_0 - q\|) + \phi_0(\|x_1 - q\|) \end{cases} \tag{5}$$

and

$$\left\|A^{-1}(Q'(x_2) - Q'(x_0))\right\| \le \phi(\|x_2 - x_0\|) \le \phi(\|x_0 - q\| + \|x_2 - q\|) \le \overline{\phi}_2$$

or

$$\left\|A^{-1}(Q'(x_2) - Q'(x_0))\right\| \le \left\|A^{-1}(Q'(x_0) - A)\right\| + \left\|A^{-1}(Q'(x_2) - A)\right\| \le \phi_0(\|x_0 - q\|) + \phi_0(\|x_2 - q\|) \le \overline{\phi}_2,$$

where

$$\overline{\phi}_2 = \begin{cases} \phi(\|x_0 - q\| + \|x_2 - q\|) \\ \text{or} \\ \phi_0(\|x_0 - q\|) + \phi_0(\|x_2 - q\|). \end{cases}$$

Then, from the estimates

$$u_5 = Q'(x_0)^{-1} Q'(x_3)\, u_4 = Q'(x_0)^{-1} Q'(x_3)\, Q'(x_0)^{-1} Q(x_3),$$
$$u_6 = Q'(x_0)^{-1} Q'(x_3)\, u_5,$$
$$u_7 = Q'(x_0)^{-1} Q'(x_3)\, u_6,$$

and

$$u_8 = Q'(x_0)^{-1}\, Q'(x_3)\, u_7$$

we get in turn

$$\|u_4\| \leq \frac{\left(1 + \int_0^1 \phi_0(\lambda\|x_3 - q\|)\, d\lambda\right)\|x_3 - q\|}{1 - \phi_0(\|x_0 - q\|)} = \bar{p}_1\|x_3 - q\|,$$
$$\|u_5\| \leq \frac{(1 + \phi_0(\lambda\|x_3 - q\|))}{1 - \phi_0(\|x_0 - q\|)}\bar{p}_1\|x_3 - q\|,$$
$$= \bar{p}_0\bar{p}_1\|x_3 - q\| = \bar{p}_2\|x_3 - q\|,$$
$$\|u_6\| \leq \bar{p}_0\bar{p}_2\|x_3 - q\| = \bar{p}_3\|x_3 - q\|,$$
$$\|u_7\| \leq \bar{p}_0\bar{p}_3\|x_3 - q\| = \bar{p}_4\|x_3 - q\|,$$

and

$$\|u_8\| \leq \bar{p}_0\, \bar{p}_4\|x_3 - q\| = \bar{p}_5\, \|x_3 - q\|,$$

where we also used

$$Q(x_3) = Q(x_3) - Q(q) = \int_0^1 Q'(q + \lambda(x_3 - q))\, d\lambda\,(x_3 - q),$$
$$\left\|A^{-1}Q(x_3)\right\| = \left\|A^{-1}\left[\int_0^1 Q'(q + \lambda(x_3 - q)) - A + A\right] d\lambda\,(x_3 - q)\right\|$$
$$\leq \left(1 + \int_0^1 \phi_0(\lambda\|x_3 - q\|)\, d\lambda\right)\|x_3 - q\|$$

and

$$\left\|A^{-1}Q'(x_3)\right\| = \left\|A^{-1}(Q'(x_3) - A + A)\right\|$$
$$\leq 1 + \phi_0(\|x_3 - q\|).$$

Based on these calculations and the triangle inequality from

$$x_4 - q = x_3 - q - \frac{21}{4}u_4 + 11u_5 - \frac{23}{2}u_6 + 6u_7 - \frac{5}{4}u_8,$$

we get

$$\begin{aligned}\|x_4 - q\| &\le \left(1 + \frac{21}{4}\bar{p}_1 + 11\bar{p}_2 + \frac{23}{2}\bar{p}_3 + 6\bar{p}_4 + \frac{5}{4}\bar{p}_5\right)\|x_3 - q\| \\ &\le \bar{h}_4\left(\|x_3 - q\|\right)\|x_3 - q\| \le \|x_3 - q\|.\end{aligned}$$

Similarly, for $s = 5, 6, \ldots, k, j = 10, 11, 12, 13$, we obtain

$$\begin{aligned}\|u_9\| &\le \bar{q}_1^{(s-1)}\left(\|x_{s-1} - q\|\right)\|x_{s-1} - q\| \\ \|u_{10}\| &\le \bar{q}_m^{(s-1)}\|x_{s-1} - q\|, m = 2, 3, 4, 5,\end{aligned}$$

where

$$\bar{q}_1^{(s-1)}\left(\|x_{s-1} - q\|\right) = \frac{1 + \int_0^1 \phi_0\left(\lambda\|x_{s-1} - q\|\right)d\lambda}{1 - \phi_0\left(\|x_0 - q\|\right)}$$

and

$$\bar{q}_m^{(s-1)}\left(\|x_{s-1} - q\|\right) = \bar{p}_0\bar{q}_{m-1}^{(s-1)}\left(\|x_{s-1} - q\|\right).$$

Notice also that $\bar{q}_m^{(s-1)} \le q_m^{(s-1)}$, $m = 1, 2, 3, 4, 5$. By the preceding calculations and triangle inequality

$$\begin{aligned}\|x_s - q\| &\le \|x_{s-1} - q\| + \frac{11}{2}\|u_9\| + 12\|u_{10}\| + 13\|u_{11}\| + 7\|u_{12}\| + \frac{3}{2}\|u_{13}\| \\ &\le \left(1 + \frac{11}{2}\bar{q}_1^{(s-1)} + 12\bar{q}_2^{(s-1)} + 13\bar{q}_3^{(s-1)}\right. \\ &\quad \left. + 7\bar{q}_4^{(s-1)} + \frac{3}{2}\bar{q}_5^{(s-1)}\right)h_{s-1}\left(\| x_{s-1} - q \|\right)\| x_{s-1} - q \| \\ &\le \|x_{s-1} - q\|.\end{aligned}$$

Hence, $\{x_i\} \in U_0, i = 0, 1, 2, \ldots, k$. In particular, for $s = k$

$$\|x_k - q\| \le c\|x_0 - q\|,$$

where

$$c = h_k\left(\|x_0 - q\|\right) \in [0, 1).$$

Similarly, this process can not continue for $n = 0, 1, 2, \ldots$. Thus, we obtain $\|x_{k+n} - q\| \le c^{n+1}\|x_0 - q\|$ leading to $\{x_n\} \subset U_0$ and $\lim_{k\to+\infty} x_k = q$. ■

The uniqueness of the solution region is determined in the next result.

Proposition 1 *Suppose:*

- *There exists a solution* $q^* \in U(q, \kappa)$ *of the equation* $Q(x) = 0$ *for some* $\kappa > 0$.
- *The condition* (A_4) *is valid on the ball* $U(q, \kappa)$.
- *There exists some* $\kappa_1 \geq \kappa$ *such that*

$$\int_0^1 \phi_0(\theta\kappa_1)\, d\theta < 1.$$

Set $E_3 = E \cap U[q, \kappa_1]$. *Then, the equation (1) is uniquely solvable by* q *in the region* E_3.

Proof. Define the linear operator $\mathcal{H}$ by

$$\mathcal{H} = \int_0^1 Q'(q + \theta(q^* - q))\, d\theta.$$

It follows by (1)–(3) that

$$\begin{aligned}\left\|A^{-1}(\mathcal{H} - A)\right\| &\leq \int_0^1 \phi_0(\theta\|q^* - q\|)\, d\theta \\ &\leq \int_0^1 \phi_0(\theta\kappa_1)\, d\theta < 1,\end{aligned}$$

thus, $q^* - q = \mathcal{H}^{-1}(Q(q^*) - Q(q)) = \mathcal{H}^{-1}(0) = 0$, so $q^* = q$. ■

Remark 1

- The point ρ can replace r in the condition (A_6).
- Possible selections for the operator $A = I$ or $A = Q'(q)$. In the latter case, q is a simple solution of the equation $Q(x) = 0$. However, we do not make such an assumption in the Theorem 1. Therefore, the method (2) can be employed to find solutions of the equation $Q(x) = 0$ of multiplicity greater than one. Other choices for the operator A are possible as long as the conditions (A_4) and (A_5) hold.

6.3 CONCLUSION

A process is developed for expanding the applicability of the method (2) in the more general setting of a Banach space. The local convergence of the method given in [1] for $T_1 = T_2 = \mathbb{R}^m$ under Taylor series expansions requires high order derivatives not on the method (2). Other drawbacks are luck of computable error bounds on the distances $||q - x_k||$ and uniqueness of the solution results. These problems are positively addressed using only Q' that appears on the method which is controlled by generalized continuity conditions [4–6]. Computable upper error bounds are developed, and a set is found that contains only one solution. The same process can be applied to extend the applicability of other methods using Taylor series and inverses of linear operators [4–12].

REFERENCES

1. F. Ahmad, S. U. Rehman, M. Z. Ullah, H. M. Aljahdali, S. Ahmad, A. S. Alshomrani, J. A. Carrasco, S. Ahmad, and S. Sivasankaran, Qrozen Jocabian multistep iterative method for solving nonlinear IVPs and BVPs, *Complexity*, vol. 2017, pp. 1–30, 2017.
2. I.K.Argyros, G. Deep, and S, Regmi, Extended newton-like midpoint method for solving equations in banach space, *Foundations,* vol. 3, no. 1, pp. 82–98, 2023. https://doi.org/10.3390/foundations3010009
3. S. Regmi, I.K. Argyros, G. Deep, and L. Rathour, A newton-like midpoint method for solving equations in banach space, *Foundations,* vol. 3, pp. 154–166, 2023. https://doi.org/10.3390/foundations3020014
4. I.K. Argyros, *Convergence and Applications of Newton-Type Iterations*, Springer-Verlag, New York, 2008.
5. I.K. Argyros and S. George, On the complexity of extending the convergence region for Traub's method, *Journal of Complexity,* vol. 56, p. 101423, 2020.
6. I.K.Argyros and S. George, *Mathematical Modeling for the Solution of Equations and Systems of Equations with Applications*, Volume-IV, Nova Publishes, New York, 2024.
7. V. Arroyo, A. Cordero, and J. R. Torregrosa, Approximation of artificial satellites' preliminary orbits: The efficiency challenge, *Mathematical and Computer Modelling*, vol. 54, no. 7–8, pp. 1802–1807, 2011.

8. A. Cordero, M. Kansal, V. Kanwar, and J. R. Torregrosa, A stable class of improved second-derivative free Chebyshev-Halley type methods with optimal eighth order convergence, *Numerical Algorithms*, vol. 72, no. 4, pp. 937–958, 2016.
9. J. M. Ortega and W. C. Rheinboldt, *Iterative Solution of Nonlinear Equations in Several Variables*, Academic Press, New York, NY, 1970.
10. J. F. Traub, *Iterative Methods for the Solution of Equations, Prentice-Hall Series in Automatic Computation*, Prentice-Hall, Englewood Cliffs, NJ, 1964.
11. M. Z. Ullah, S. Serra-Capizzano, and F. Ahmad, An efficient multi-step iterative method for computing the numerical solution of systems of nonlinear equations associated with ODEs, *Applied Mathematics and Computation*, vol. 250, pp. 249–259, 2015.
12. G. Deep and I. K. Argyros, Improved higher order compositions for nonlinear equations, *Foundations,* vol. 3, no. 1, pp. 25–36, 2023. https://doi.org/10.3390/foundations3010003

CHAPTER 7

Application of Some Classes of Mittag-Leffler Functions in Solving Conformal Fractional Differential Equations

Arsalan Hojat Ansari, Snježana Maksimović, Hossam A. Nabwey, and Zoran D. Mitrović

7.1 INTRODUCTION

The Mittag-Leffler functions are introduced in [1–4], and their main properties are given in [5, 6]. The Mittag-Leffler functions have various applications in natural and engineering sciences, especially in physics, dynamical systems theory, disordered systems, and stochastic systems [7, 8]. Research in fractional calculus gave a new application of Mittag-Leffler functions, as Mittag-Leffler functions are used in the solution of fractional differential equations or fractional integral equations. Fractional calculus has various applications especially in physics, mechanics, engineering, and control theory of dynamic systems [9–12]. With the appearance and development of conformable differential calculus [13, 14], the interest in solving differential equations within the frame of conformable derivatives has grown [15, 16, 17].

DOI: 10.1201/9781003530602-7

In this chapter, we use the special case of Mittag-Leffler functions $T_{pj}, H_{pj} : \mathbb{R} \to \mathbb{R}, j = 0, 1, 2, ..., p-1, p \in \mathbb{N}$, defined as follows [18]:

$$T_{pj}(\tau) = \sum_{n=0}^{\infty} \frac{(-1)^n \tau^{pn+j}}{(pn+j)!}, \; H_{pj}(\tau) = \sum_{n=0}^{\infty} \frac{\tau^{pn+j}}{(pn+j)!}. \tag{1.1}$$

The aim of this chapter is to solve conformable fractional differential equations of the following type:

$$D^{np\lambda}y(\tau) + c_{n-1}D^{(n-1)p\lambda}y(\tau) + ... + c_1 D^{p\lambda}y(\tau) + c_0 y(\tau) = P_n(\tau^{\lambda}), \tag{1.2}$$

$\tau > 0$, $n = 1, 2, 3$. Also, we solve conformable fractional differential equations:

$$D^{(p+q)\lambda}y(\tau) - aD^{p\lambda}y(\tau) - bD^{q\lambda}y(\tau) + aby(t) - d = 0, \; \tau > 0, \tag{1.3}$$

where the a, b, d, c_i are constants, P_n is a polynomial of degree n, $p, q \in \mathbb{N}$, $p, q \geq 2$, and $0 < \lambda \leq 1$. Applying the fractional Laplace transform [19, 20] on (1.2) and (1.3), we obtain their solutions as a linear combination of functions (1.1).

7.2 PRELIMINARIES

In this section, we list some well-known results that we use. The following definition was introduced by Khalil et al. [14].

Definition 1 Let f: $[0, +\infty) \to \mathbb{R}$ be a function. We define the conformable fractional derivative with respect to τ of order λ as

$$D^{\lambda}f(\tau) = \lim_{\epsilon \to 0} \frac{f(\tau + \epsilon\tau^{1-\lambda}) - f(\tau)}{\epsilon}, \text{ for all } \quad \tau > 0, \lambda \in (0, 1]. \tag{2.1}$$

Remark 1 Note, if f is differentiable then

$$D^{\lambda}f(\tau) = \tau^{1-\lambda}f'(\tau), \text{ for all } \quad \tau > 0, \lambda \in (0, 1]. \tag{2.2}$$

We present the following two results obtained by Bashar et al. [16].

Theorem 1 *Let* $f_1, f_2, : [0, +\infty) \to \mathbb{R}$ *be two conformable differentiable functions in* $\tau > 0$ *and* $\lambda \in (0, 1]$, *then it holds*

(i) $D^{\lambda}(c) = 0$, where c is a constant,

(ii) $D^{\lambda}(\tau^{c}) = \lambda \tau^{c-\lambda}$, for all $c \in \mathbb{R}$,

(iii) $D^{\lambda}(a_1 f_1(\tau) + a_2 f_2(\tau)) = a_1 D^{\lambda} f_1(\tau) + a_2 D^{\lambda} f_2(\tau)$, for all $a_1, a_2 \in \mathbb{R}$,

(iv) $D^{\lambda}(f_1(\tau) f_2(\tau)) = D^{\lambda}(f_1(\tau)) f_2(\tau) + f_1(\tau) D^{\lambda}(f_2(\tau))$,

(v) $D^{\lambda}\left(\frac{f_1(\tau)}{f_2(\tau)}\right) = \frac{D^{\lambda}(f_1(\tau)) f_2(\tau) - f_1(\tau) D^{\lambda}(f_2(\tau))}{f_2^2(\tau)}, \quad f_2(\tau) \neq 0.$

Theorem 2 *If functions* $f_1, f_2 : [0, +\infty) \to \mathbb{R}$ *is twice differentiable and conformable differentiable, then the conformable derivative of the function* $f_1 \circ f_2$ *is*

$$D^{\lambda}(f_1 \circ f_2)(\tau) = \tau^{1-\lambda} f_1'(f_2(\tau)) f_2'(\tau). \tag{2.3}$$

In [13], Abdeljawad gives the following definition.

Definition 2 The fractional Laplace transform of the function $f : [0, +\infty) \to \mathbb{R}$ of an order $\lambda \in (0, 1]$ is defined by

$$\mathcal{L}_{\lambda}(f(\tau))(s) = F_{\lambda}(s) = \int_0^{+\infty} e^{-s\frac{\tau^{\lambda}}{\lambda}} f(\tau)\, \tau^{\lambda-1} d\tau, \; s \in \mathbb{C}. \tag{2.4}$$

Also in [13], Abdeljawad obtained the following two results.

Theorem 3 *Let* $f : [0, +\infty) \to \mathbb{R}$ *be a function such that* $\mathcal{L}_{\lambda}(f(\tau))(s) = F_{\lambda}(s)$ *exists. Then*

$$F_{\lambda}(s) = \mathcal{L}\left(f\left((\lambda\tau)^{\frac{1}{\lambda}}\right)\right)(s). \tag{2.5}$$

Theorem 4 *If* $\mathcal{L}_{\lambda}(f)(s)$ *exists for* $s > 0$, *then*

$$\mathcal{L}_{\lambda}(\tau^{c})(s) = \lambda^{\frac{c}{\lambda}} \frac{\Gamma\left(1 + \frac{c}{\lambda}\right)}{s^{1+\frac{c}{\lambda}}},$$

where c *is a constant and* Γ *is the Gamma function.*

We also give the results obtained in [15] (Theorem 5), [19] (Theorem 6), and [20] (Theorem 7).

Theorem 5 *Let $s, t > 0$, $0\lambda \in (0,1]$ and $p \in \mathbb{N}$, $p \geq 2$. The fractional Laplace of order λ of functions T_{pj} and H_{pj}, $j = 0, 1, ..., p-1$, is*

$$\mathcal{L}_\lambda\left(T_{pj}\left(c\frac{\tau^\lambda}{\lambda}\right)\right)(s) = \frac{s^{p-j-1}c^j}{s^p + c^p}, \quad \mathcal{L}_\lambda\left(H_{pj}\left(c\frac{\tau^\lambda}{\lambda}\right)\right)(s) = \frac{s^{p-j-1}c^j}{s^p - c^p}, \quad c \in \mathbb{R}. \tag{2.6}$$

Theorem 6 *Let f: $[0, +\infty) \to \mathbb{R}$ be a continuous, differentiable function and $\lambda \in (0,1]$. Then for any $n \in \mathbb{N}$ and $s > 0$, it holds*

$$\mathcal{L}_\lambda\left(D^{n\lambda}f(\tau)\right)(s) = s^n F_\lambda(s) - \sum_{j=0}^{n-1} s^j D^{(n-j-1)\lambda}f(0). \tag{2.7}$$

Theorem 7 *Let f_1, f_2: $[0, +\infty) \to \mathbb{R}$ and $\lambda \in (0,1]$. If $\mathcal{L}_\lambda(f_1)(s)$ and $\mathcal{L}_\lambda(f_2)(s)$ exist for $s > 0$, then*

$$\mathcal{L}_\lambda(f_1 * f_2)(s) = \mathcal{L}_\lambda(f_1)(s)\,\mathcal{L}_\lambda(f_2)(s). \tag{2.8}$$

7.3 MAIN RESULTS

In this section, we give our main results.

Theorem 8 *Let $p \in \{2, 3, ...\}$, $n \in \mathbb{N}$ such that $n \leq p-1$, $\tau > 0$ and $\lambda \in (0,1]$. Then the equation*

$$D^{p\lambda}y(\tau) - ay(\tau) = \sum_{k=0}^{n} b_k \tau^{\lambda k}, \tag{3.1}$$

with initial conditions $y(0) = a_0$, $D^\lambda y(0) = a_1 ...$, $D^{(p-1)\lambda}y(0) = a_{p-1}$ has a solution

$$y(\tau) = \sum_{j=0}^{p-1} \frac{a_j}{\sqrt[p]{a^j}} H_{pj}\left(\sqrt[p]{a}\frac{\tau^\lambda}{\lambda}\right) + \frac{1}{a}\left(\sum_{k=0}^{n} \frac{b_k \lambda^k k!}{\sqrt[p]{a^k}} H_{pk}\left(\sqrt[p]{a}\frac{\tau^\lambda}{\lambda}\right) - \sum_{k=0}^{n} b_k \tau^{\lambda k}\right), \tag{3.2}$$

for $a > 0$ *and*

$$y(\tau) = \sum_{j=0}^{p-1} \frac{a_j}{\sqrt[p]{(-a)^j}} T_{pj}\left(\sqrt[p]{-a}\,\frac{\tau^\lambda}{\lambda}\right)$$
$$+\frac{1}{a}\left(\sum_{k=0}^{n} \frac{b_k \lambda^k k!}{\sqrt[p]{(-a)^k}} T_{pk}\left(\sqrt[p]{-a}\,\frac{\tau^\lambda}{\lambda}\right) - \sum_{k=0}^{n} b_k t^{\lambda k}\right), \qquad (3.3)$$

for $a < 0$.

Proof. We will prove case (3.2), because the proof in case (3.3) is similar. If we apply the fractional Laplace transform on (3.1), by Theorem 6 and Theorem 4 we obtain

$$\mathcal{L}_\lambda\left(D^{p\lambda} y(\tau) - a y(\tau) - \sum_{k=0}^{n} b_k \tau^{\lambda k}\right)(s)$$
$$= s^p \mathcal{L}_\lambda (y(\tau))(s) - \sum_{j=0}^{p-1} s^{p-j-1} D^{j\lambda} y(0) - a \mathcal{L}_\lambda (y(\tau))(s)$$
$$-\sum_{k=0}^{n} b_k \frac{\lambda^k \Gamma(k+1)}{s^{k+1}} = 0,\ s > 0,$$

from which it follows

$$\mathcal{L}_\lambda (y(\tau))(s) = \sum_{j=0}^{p-1} \frac{a_j s^{p-j-1}}{s^p - a} + \sum_{k=0}^{n} \frac{b_k \lambda^k \Gamma(k+1)}{s^{k+1}(s^p - a)}.$$

Using an inverse fractional Laplace transform and Theorem 5 we obtain

$$y(\tau) = \sum_{j=0}^{p-1} a_j \mathcal{L}_\lambda^{-1}\left(\frac{s^{p-j-1}}{s^p - a}\right)(\tau) + \sum_{k=0}^{n} b_k \lambda^k \Gamma(k+1) \mathcal{L}_\lambda^{-1}\left(\frac{1}{s^{k+1}(s^p - a)}\right)(\tau)$$
$$= \sum_{j=0}^{p-1} \frac{a_j}{\sqrt[p]{a^j}} H_{pj}\left(\sqrt[p]{a}\,\frac{\tau^\lambda}{\lambda}\right) + \frac{1}{a}\sum_{k=0}^{n} b_k \lambda^k \Gamma(k+1) \mathcal{L}_\lambda^{-1}\left(\frac{s^{p-k-1}}{s^p - a}\right)(\tau)$$
$$-\frac{1}{a}\sum_{k=0}^{n} b_k \lambda^k \Gamma(k+1) \mathcal{L}_\lambda^{-1}\left(\frac{1}{s^{k+1}}\right)(\tau)$$
$$= \sum_{j=0}^{p-1} \frac{a_j}{\sqrt[p]{a^j}} H_{pj}\left(\sqrt[p]{a}\,\frac{\tau^\lambda}{\lambda}\right) + \frac{1}{a}\sum_{k=0}^{n} \frac{b_k \lambda^k k!}{\sqrt[p]{a^k}} H_{pk}\left(\sqrt[p]{a}\,\frac{\tau^\lambda}{\lambda}\right) - \frac{1}{a}\sum_{k=0}^{n} b_k \tau^{k\lambda}.$$

□

Theorem 9 *Let* $p \in \{2, 3, ...\}$, $a > 0$ *and* $\lambda \in (0, 1]$, *then the equation*

$$D^{2p\lambda}y(\tau) - aD^{p\lambda}y(\tau) + by(\tau) = \sum_{k=0}^{r} b_k\tau^{k\lambda}, r \leq p-1, \tau > 0, \tag{3.4}$$

with initial conditions $y(0) = a_0$, $D^{\lambda}y(0) = a_1, ..., D^{(2p-1)\lambda}y(0) = a_{2p-1}$, *where* $a = m + n$, $b = m{\cdot}n$, *has a solution*

$$\begin{aligned} y(\tau) = &\frac{1}{m-n}\sum_{j=0}^{p-1}\left(\frac{a_{p+j} - na_j}{\sqrt[p]{m^j}}H_{pj}\left(\sqrt[p]{m}\frac{\tau^{\lambda}}{\lambda}\right) + \frac{ma_j - a_{p+j}}{\sqrt[p]{n^j}}H_{pj}\left(\sqrt[p]{n}\frac{\tau^{\lambda}}{\lambda}\right)\right) \\ &+ \frac{1}{b}\left(\sum_{k=0}^{r} b_k\tau^{\lambda k} + \frac{1}{m-n}\sum_{k=0}^{r} b_k\lambda^k k! \right. \\ &\left.\times\left(\frac{nH_{pk}\left(\sqrt[p]{m}\frac{\tau^{\lambda}}{\lambda}\right)}{\sqrt[p]{m^k}} - \frac{mH_{pk}\left(\sqrt[p]{n}\frac{\tau^{\lambda}}{\lambda}\right)}{\sqrt[p]{n^k}}\right)\right), \end{aligned} \tag{3.5}$$

for $m > n > 0$ *and*

$$\begin{aligned} y(\tau) = &\sum_{j=0}^{p-1}\frac{a_j}{\sqrt[p]{m^j}}H_{pj}\left(\sqrt[p]{m}\frac{\tau^{\lambda}}{\lambda}\right) + \frac{a_{p+j} - ma_j}{\sqrt[p]{m^{p+j-1}}}H_{pp-1}\left(\sqrt[p]{m}\frac{\tau^{\lambda}}{\lambda}\right) * H_{pj}\left(\sqrt[p]{m}\frac{\tau^{\lambda}}{\lambda}\right) \\ &+ \frac{1}{m^2}\left(\sum_{k=0}^{r} b_k\tau^{k\lambda} + \sum_{k=0}^{r}\frac{b_k k!\lambda^k}{\sqrt[p]{m^k}}\left(\sqrt[p]{m}H_{pp-1}\left(\sqrt[p]{m}\frac{\tau^{\lambda}}{\lambda}\right) * H_{pk}\left(\sqrt[p]{m}\frac{\tau^{\lambda}}{\lambda}\right)\right.\right. \\ &\left.\left.-H_{pk}\left(\sqrt[p]{m}\frac{\tau^{\lambda}}{\lambda}\right)\right)\right), \end{aligned} \tag{3.6}$$

for $m = n > 0$.

Proof. In the case of $m > n > 0$, Let $s > 0$. If we apply the fractional Laplace transform on (3.4) we obtain

$$\begin{aligned} &\mathcal{L}_{\lambda}\left(D^{2p\lambda}y(\tau) - aD^{p\lambda}y(\tau) + by(\tau) - \sum_{k=0}^{r} b_k\tau^{k\lambda}\right)(s) \\ &= s^{2p}\mathcal{L}_{\lambda}(y(\tau))(s) - \sum_{j=0}^{2p-1} a_{2p-j-1}s^j \end{aligned}$$

$$
\begin{aligned}
&-as^p\mathcal{L}_\lambda\left(y(\tau)\right)(s)+a\sum_{j=0}^{p-1}a_{p-j-1}s^j+b\mathcal{L}_\lambda\left(y(\tau)\right)(s)\\
&-\sum_{k=0}^{r}b_k\frac{\lambda^k\Gamma(k+1)}{s^{k+1}}=0
\end{aligned}
$$

from which it follows

$$
\begin{aligned}
&\mathcal{L}_\lambda\left(y(\tau)\right)(s)\\
&=\frac{\sum_{j=0}^{p-1}\left(a_{2p-j-1}-aa_{p-j-1}\right)s^j+\sum_{j=p}^{2p-1}a_{2p-j-1}s^j+\sum_{k=0}^{r}b_k\frac{\lambda^k\Gamma(k+1)}{s^{k+1}}}{\left(s^p-m\right)\left(s^p-n\right)}.
\end{aligned}
$$

From

$$
\begin{aligned}
&\frac{\sum_{j=0}^{p-1}\left(a_{2p-j-1}-aa_{p-j-1}\right)s^j+\sum_{j=p}^{2p-1}a_{2p-j-1}s^j}{\left(s^p-m\right)\left(s^p-n\right)}\\
&\quad=\frac{\sum_{j=0}^{p-1}A_js^{p-j-1}}{s^p-m}+\frac{\sum_{j=0}^{p-1}B_js^{p-j-1}}{s^p-n}
\end{aligned}
$$

we obtain the system

$$
\begin{aligned}
A_j+B_j&=a_j\\
nA_j+mB_j&=aa_j-a_{p+j},
\end{aligned}
$$

$j=0,1,\ldots,p-1$, where the solution is

$$
A_j=\frac{a_j(m-a)+a_{p+j}}{m-n},\quad B_j=\frac{a_j(a-n)-a_{p+j}}{m-n},\quad j=0,1,\ldots,p-1.
$$

Using an inverse fractional Laplace transform, Theorem 5, and Theorem 4, we obtain

$$
\begin{aligned}
y(\tau)=&\frac{1}{m-n}\sum_{j=0}^{p-1}\left(\frac{a_j(m-a)+a_{p+j}}{\sqrt[p]{m^j}}H_{pj}\left(\sqrt[p]{m}\frac{\tau^\lambda}{\lambda}\right)\right.\\
&\left.+\frac{a_j(a-n)-a_{p+j}}{\sqrt[p]{n^j}}H_{pj}\left(\sqrt[p]{n}\frac{\tau^\lambda}{\lambda}\right)\right)\\
&+\frac{1}{m-n}\sum_{k=0}^{r}b_k\lambda^k\Gamma(k+1)\mathcal{L}_\lambda^{-1}\left(\frac{1}{s^{k+1}\left(s^p-m\right)}-\frac{1}{s^{k+1}\left(s^p-n\right)}\right)(\tau).
\end{aligned}
$$

From

$$\mathcal{L}_{\lambda}^{-1}\left(\frac{1}{s^{k+1}(s^p-m)}-\frac{1}{s^{k+1}(s^p-n)}\right)(\tau)$$
$$=\mathcal{L}_{\lambda}^{-1}\left(-\frac{1}{ms^{k+1}}+\frac{s^{p-k-1}}{m(s^p-m)}+\frac{1}{ns^{k+1}}-\frac{s^{p-k-1}}{n(s^p-n)}\right)(\tau)$$
$$=\frac{m-n}{mn}\cdot\frac{\tau^{\lambda}}{\lambda^k k!}+\frac{1}{m\sqrt[p]{m^k}}H_{pk}\left(\sqrt[p]{m}\frac{\tau^{\lambda}}{\lambda}\right)-\frac{1}{n\sqrt[p]{n^k}}H_{pk}\left(\sqrt[p]{n}\frac{\tau^{\lambda}}{\lambda}\right),$$

the assertion follows. Using case $m=n>0$, *if* we apply the fractional Laplace transform on (3.4), we obtain

$$\mathcal{L}_{\lambda}\left(D^{2p\lambda}y(\tau)-2mD^{p\lambda}y(\tau)+m^2y(\tau)-\sum_{k=0}^{r}b_k\tau^{k\lambda}\right)(s)$$
$$=s^{2p}\mathcal{L}_{\lambda}(y(\tau))(s)-\sum_{j=0}^{2p-1}a_{2p-j-1}s^j$$
$$-2ms^p\mathcal{L}_{\lambda}(y(\tau))(s)+2m\sum_{j=0}^{p-1}a_{p-j-1}s^j+m^2\mathcal{L}_{\lambda}(y(\tau))(s)$$
$$-\sum_{k=0}^{r}b_k\frac{\lambda^k\Gamma(k+1)}{s^{k+1}}=0,$$

from which it follows

$$\mathcal{L}_{\lambda}(y(\tau))(s)$$
$$=\frac{\sum_{j=0}^{p-1}\left(a_{2p-j-1}-2ma_{p-j-1}\right)s^j+\sum_{j=0}^{p-1}a_{p-j-1}s^{j+p}+\sum_{k=0}^{r}b_k\frac{\lambda^k\Gamma(k+1)}{s^{k+1}}}{\left(s^p-m\right)^2}.$$

From

$$\frac{\sum_{j=0}^{p-1}\left(a_{2p-j-1}-2ma_{p-j-1}\right)s^j++\sum_{j=0}^{p-1}a_{p-j-1}s^{j+p}}{\left(s^p-m\right)^2}$$
$$=\frac{\sum_{j=0}^{p-1}A_js^{p-j-1}}{s^p-m}+\frac{\sum_{j=0}^{p-1}B_js^{p-j-1}}{\left(s^p-m\right)^2},$$

we obtain the system

$$A_j=a_j$$
$$mA_j-B_j=2ma_j-a_{p+j},$$

$j = 0, 1, ..., p-1$, where the solution is

$$A_j = a_j, \quad B_j = a_{p+j} - ma_j, \quad j = 0, 1, ..., p-1.$$

Using an inverse fractional Laplace transform, Theorem 5, and Theorem 4, we obtain

$$y(\tau) = \sum_{j=0}^{p-1} \frac{a_j}{\sqrt[p]{m^j}} H_{pj}\left(\sqrt[p]{m}\frac{\tau^\lambda}{\lambda}\right) + \sum_{j=0}^{p-1} \frac{a_{p+j} - ma_j}{\sqrt[p]{m^{p+j-1}}} H_{pp-1}\left(\sqrt[p]{m}\frac{\tau^\lambda}{\lambda}\right)$$
$$* H_{pj}\left(\sqrt[p]{m}\frac{\tau^\lambda}{\lambda}\right) + \frac{1}{m}\sum_{k=0}^{r} b_k \lambda^k \Gamma(k+1)\, \mathcal{L}_\lambda^{-1}\left(\frac{m}{s^{k+1}(s^p - m)^2}\right)(\tau).$$

The assertion follows from Theorem 7 and

$$\mathcal{L}_\lambda^{-1}\left(\frac{m}{s^{k+1}(s^p - m)^2}\right)(\tau)$$
$$= \mathcal{L}_\lambda^{-1}\left(-\frac{1}{ms^{k+1}} + \frac{s^{p-k-1}}{m(s^p - m)} + \frac{s^{p-k-1}}{(s^p - m)^2}\right)(\tau)$$
$$= \frac{1}{m} \cdot \frac{\tau^\lambda}{\lambda^k k!} + \frac{1}{m\sqrt[p]{m^k}} H_{pk}\left(\sqrt[p]{m}\frac{\tau^\lambda}{\lambda}\right) + \frac{1}{\sqrt[p]{m^{p+j-1}}} H_{pp-1}$$
$$\times \left(\sqrt[p]{m}\frac{\tau^\lambda}{\lambda}\right) * H_{pk}\left(\sqrt[p]{m}\frac{\tau^\lambda}{\lambda}\right).$$

□

Remark 2 (i) If $n < 0$, then in (3.5), instead of $\sqrt[p]{n}$ we put $\sqrt[p]{(-n)}$, and instead of $H_{pj}\left(\sqrt[p]{n}\frac{\tau^\lambda}{\lambda}\right)$ we put $T_{pj}\left(\sqrt[p]{-n}\frac{\tau^\lambda}{\lambda}\right)$. (ii) If $m < 0$, then in (3.5), instead of $\sqrt[p]{m}$ we put $\sqrt[p]{(-m)}$, and instead of $H_{pj}\left(\sqrt[p]{m}\frac{\tau^\lambda}{\lambda}\right)$ we put $T_{pj}\left(\sqrt[p]{-m}\frac{\tau^\lambda}{\lambda}\right)$. (iii) If $m < 0$, then in (3.6), instead of $\sqrt[p]{m}$ we put $\sqrt[p]{(-m)}$, and instead of $H_{pj}\left(\sqrt[p]{m}\frac{\tau^\lambda}{\lambda}\right)$ we put $T_{pj}\left(\sqrt[p]{-m}\frac{\tau^\lambda}{\lambda}\right)$.

Theorem 10 *Let $p \in \mathbb{N}$, $p \geq 2$, $a, b, c > 0$ and $\lambda \in (0, 1]$. Then the equation*

$$D^{3p\lambda}y(\tau) - (a+b+c)\, D^{2p\lambda}y(\tau) + (ab+bc+ca)\, D^{p\lambda}y(\tau) - abcy(\tau)$$
$$= \sum_{k=0}^{n} b'_k t^{k\lambda}, \tau > 0, \tag{3.7}$$

with initial conditions $y(0) = a_0$, $D^{\lambda}y(0) = a_1, ..., D^{(3p-1)\lambda}y(0) = a_{3p-1}$ *has a solution*

$$y(\tau) = \sum_{j=0}^{p-1}\left(\frac{A_j}{\sqrt[p]{a^j}}H_{pj}\left(\sqrt[p]{a}\frac{\tau^{\lambda}}{\lambda}\right) + \frac{B_j}{\sqrt[p]{b^j}}H_{pj}\left(\sqrt[p]{b}\frac{\tau^{\lambda}}{\lambda}\right) + \frac{C_j}{\sqrt[p]{c^j}}H_{pj}\left(\sqrt[p]{c}\frac{\tau^{\lambda}}{\lambda}\right)\right)$$
$$+ \frac{1}{a(b-a)(c-a)}\left(\sum_{k=0}^{n}\frac{b_k\lambda^k k!}{\sqrt[p]{a^k}}H_{pk}\left(\sqrt[p]{a}\frac{\tau^{\lambda}}{\lambda}\right) - \sum_{k=0}^{n} b_k\tau^{k\lambda}\right)$$
$$+ \frac{1}{b(a-b)(c-b)}\left(\sum_{k=0}^{n}\frac{b_k\lambda^k k!}{\sqrt[p]{b^k}}H_{pk}\left(\sqrt[p]{b}\frac{\tau^{\lambda}}{\lambda}\right) - \sum_{k=0}^{n} b_k\tau^{k\lambda}\right)$$
$$+ \frac{1}{c(a-c)(b-c)}\left(\sum_{k=0}^{n}\frac{b_k\lambda^k k!}{\sqrt[p]{c^k}}H_{pk}\left(\sqrt[p]{c}\frac{\tau^{\lambda}}{\lambda}\right) - \sum_{k=0}^{n} b_k\tau^{k\lambda}\right), \quad (3.8)$$

for $a \neq b \neq c$, *where* $A_j, B_j, C_j, j = 0, 1, ..., p-1$, *are solutions of the system*

$$A_j + B_j + C_j = a_j$$
$$(b+c)A_j + (a+c)B_j + (a+b)C_j = (a+b+c)a_j - a_{p+j}$$
$$bcA_j + acB_j + abC_j = a_{2p+j} - (b+c)a_{p+j}$$
$$+ (ab+ac+bc)a_j, \quad (3.9)$$

$$y(\tau) = \sum_{j=0}^{p-1}\left(\frac{A_j}{\sqrt[p]{a^j}}H_{pj}\left(\sqrt[p]{a}\frac{\tau^{\lambda}}{\lambda}\right) + \frac{B_j}{\sqrt[p]{a^{p+j-1}}}H_{pp-1}\left(\sqrt[p]{a}\frac{\tau^{\lambda}}{\lambda}\right) * H_{pj}\left(\sqrt[p]{a}\frac{\tau^{\lambda}}{\lambda}\right)\right.$$
$$\left. + \frac{C_j}{\sqrt[p]{c^j}}H_{pj}\left(\sqrt[p]{c}\frac{\tau^{\lambda}}{\lambda}\right)\right) - \frac{1}{a(a-c)^2}\left(\sum_{k=0}^{n}\frac{b_k\lambda^k k!}{\sqrt[p]{a^k}}H_{pk}\left(\sqrt[p]{a}\frac{\tau^{\lambda}}{\lambda}\right) - \sum_{k=0}^{n} b_k\tau^{k\lambda}\right)$$
$$+ \frac{1}{a(a-c)}\left(\frac{1}{a}\sum_{k=0}^{n} b_k\tau^{k\lambda} - \frac{1}{a}\sum_{k=0}^{n}\frac{b_k\lambda^k k!}{\sqrt[p]{a^k}}H_{pk}\left(\sqrt[p]{a}\frac{\tau^{\lambda}}{\lambda}\right)\right.$$
$$\left. + \sum_{k=0}^{n}\frac{b_k\lambda^k k!}{\sqrt[p]{a^{p+k-1}}}H_{pp-1}\left(\sqrt[p]{a}\frac{\tau^{\lambda}}{\lambda}\right) * H_{pk}\left(\sqrt[p]{a}\frac{\tau^{\lambda}}{\lambda}\right)\right)$$
$$+ \frac{1}{c(a-c)^2}\left(\sum_{k=0}^{n}\frac{b_k\lambda^k k!}{\sqrt[p]{c^k}}H_{pk}\left(\sqrt[p]{c}\frac{\tau^{\lambda}}{\lambda}\right) - \sum_{k=0}^{n} b_k\tau^{k\lambda}\right), \quad (3.10)$$

for $a = b \neq c$, *where* A_j, B_j, C_j, $j = 0, 1, ..., p-1$, *are solutions of the system*

$$\begin{aligned} A_j + C_j &= a_j \\ (a+c)\,A_j - B_j + 2aC_j &= (2a+c)\,a_j - a_{p+j} \\ bcA_j + acB_j + abC_j &= a_{2p+j} - (a+c)\,a_{p+j} + a(a+2c)\,a_j. \end{aligned} \tag{3.11}$$

$$\begin{aligned} y(\tau) = {} & \sum_{j=0}^{p-1} \frac{a_j}{\sqrt[p]{a^j}} H_{pj}\left(\sqrt[p]{a}\frac{\tau^\lambda}{\lambda}\right) + \sum_{j=0}^{p-1} \frac{a_{p+j} - aa_j}{\sqrt[p]{a^{p+j-1}}} H_{pp-1}\left(\sqrt[p]{a}\frac{\tau^\lambda}{\lambda}\right) \\ & * H_{pj}\left(\sqrt[p]{a}\frac{\tau^\lambda}{\lambda}\right) + \sum_{j=0}^{p-1} \frac{a_{2p+j} - 2aa_{p+j} + a^2 a_j}{\sqrt[p]{a^{2p+j-2}}} H_{pp-1}\left(\sqrt[p]{a}\frac{\tau^\lambda}{\lambda}\right) \\ & * H_{pp-1}\left(\sqrt[p]{a}\frac{\tau^\lambda}{\lambda}\right) * H_{pj}\left(\sqrt[p]{a}\frac{\tau^\lambda}{\lambda}\right) \\ & + \frac{1}{a}\sum_{k=0}^{n} \frac{b_k \lambda^k k!}{\sqrt[p]{a^{2p+k-2}}} H_{pp-1}\left(\sqrt[p]{a}\frac{\tau^\lambda}{\lambda}\right) * H_{pp-1}\left(\sqrt[p]{a}\frac{\tau^\lambda}{\lambda}\right) * H_{pk}\left(\sqrt[p]{a}\frac{\tau^\lambda}{\lambda}\right) \\ & - \frac{1}{a^2}\sum_{k=0}^{n} \frac{b_k \lambda^k k!}{\sqrt[p]{a^{p+k-1}}} H_{pp-1}\left(\sqrt[p]{a}\frac{\tau^\lambda}{\lambda}\right) * H_{pk}\left(\sqrt[p]{a}\frac{\tau^\lambda}{\lambda}\right) \\ & + \frac{1}{a^3}\sum_{k=0}^{n} \frac{b_k \lambda^k k!}{\sqrt[p]{a^k}} H_{pk}\left(\sqrt[p]{a}\frac{\tau^\lambda}{\lambda}\right) - \frac{1}{a^3}\sum_{k=0}^{n} b_k \tau^{k\lambda}, \end{aligned} \tag{3.12}$$

for $a = b = c$.

Proof. We will prove the case of $a \neq b \neq c$, since the proof of $a = b \neq c$ and proof of $a = b = c$ is similar. Let $s > 0$. If we apply the fractional Laplace transform on (3.7), we obtain

$$\begin{aligned} & \mathcal{L}_\lambda(D^{3p\lambda} y(\tau) - (a+b+c)\, D^{2p\lambda} y(\tau) + (ab+bc+ca)\, D^{p\lambda} y(\tau) - abcy(\tau) \\ & \quad - \sum_{k=0}^{n} b_k \tau^{k\lambda})(s) \\ & = s^{3p} \mathcal{L}_\lambda(y(\tau))(s) - \sum_{j=0}^{3p-1} a_{3p-j-1} s^j - (a+b+c)(s^{2p}\mathcal{L}_\lambda(y(\tau))(s) \\ & \quad - \sum_{j=0}^{2p-1} a_{2p-j-1} s^j) + (ab+bc+ca)(s^p \mathcal{L}_\lambda(y(\tau))(s) - \sum_{j=0}^{p-1} a_{p-j-1} s^j) \\ & \quad - \sum_{k=0}^{n} \frac{b_k \lambda^k k!}{s^{k+1}} = 0 \end{aligned}$$

from which it follows

$$
\begin{aligned}
&\mathcal{L}_\lambda\left(y(\tau)\right)(s)\\
&= \frac{\sum_{j=0}^{p-1}\left(a_{3p-j-1}-(a+b+c)\,a_{2p-j-1}+a_{p-j-1}\,(ab+bc+ca)\right)s^j}{(s^p-a)\,(s^p-b)\,(s^p-c)}\\
&+\frac{\sum_{j=p}^{2p-1}\left(a_{3p-j-1}-(a+b+c)\,a_{2p-j-1}\right)s^j+\sum_{j=2p}^{3p-1}a_{3p-j-1}s^j+\sum_{k=0}^{n}\frac{b_k\lambda^k k!}{s^{k+1}}}{(s^p-a)\,(s^p-b)\,(s^p-c)}.
\end{aligned}
$$

From

$$
\begin{aligned}
&\frac{\sum_{j=0}^{p-1}\left(a_{3p-j-1}-(a+b+c)\,a_{2p-j-1}+a_{p-j-1}\,(ab+bc+ca)\right)s^j}{(s^p-a)\,(s^p-b)\,(s^p-c)}\\
&\quad+\frac{\sum_{j=p}^{2p-1}\left(a_{3p-j-1}-(a+b+c)\,a_{2p-j-1}\right)s^j+\sum_{j=2p}^{3p-1}a_{3p-j-1}s^j}{(s^p-a)\,(s^p-b)\,(s^p-c)}\\
&\quad=\frac{\sum_{j=0}^{p-1}A_j s^{p-j-1}}{s^p-a}+\frac{\sum_{j=0}^{p-1}B_j s^{p-j-1}}{s^p-b}+\frac{\sum_{j=0}^{p-1}C_j s^{p-j-1}}{s^p-c},
\end{aligned}
$$

we obtain system (3.9). Further, from

$$
\begin{aligned}
&\frac{1}{s^{k+1}\,(s^p-a)\,(s^p-b)\,(s^p-c)}\\
&\quad=\frac{A}{s^{k+1}\,(s^p-a)}+\frac{B}{s^{k+1}\,(s^p-b)}+\frac{C}{s^{k+1}\,(s^p-c)},
\end{aligned}
$$

it follows that

$$
A=\frac{1}{(b-a)\,(c-a)},\ B=\frac{1}{(a-b)\,(c-b)},\ C=\frac{1}{(a-c)\,(b-c)}.
$$

Applying the inverse fractional Laplace transform and using Theorem 5 we obtain the assertion. □

Remark 3 (i) If $a<0$, then in (3.8)–(3.12), instead of $\sqrt[p]{a}$ we put $\sqrt[p]{(-a)}$, and instead of $H_{pj}\left(\sqrt[p]{a}\,\frac{\tau^\lambda}{\lambda}\right)$ we put $T_{pj}\left(\sqrt[p]{-a}\,\frac{\tau^\lambda}{\lambda}\right)$. (ii) If $b<0$, then in (3.8), instead of $\sqrt[p]{b}$ we put $\sqrt[p]{(-b)}$, and instead of $H_{pj}\left(\sqrt[p]{b}\,\frac{\tau^\lambda}{\lambda}\right)$ we put $T_{pj}\left(\sqrt[p]{-b}\,\frac{\tau^\lambda}{\lambda}\right)$. (iii) If $c<0$, then in (3.8) and (3.10), instead of $\sqrt[p]{c}$ we put $\sqrt[p]{(-c)}$, and instead of $H_{pj}\left(\sqrt[p]{c}\,\frac{\tau^\lambda}{\lambda}\right)$ we put $T_{pj}\left(\sqrt[p]{-c}\,\frac{\tau^\lambda}{\lambda}\right)$.

Theorem 11 *Let $p \in \mathbb{N}$, $p+q \geq 4$. Then the equation*

$$D^{(p+q)\lambda}y(\tau) - aD^{p\lambda}y(\tau) - by^{q\lambda}y(\tau) + aby(\tau) - d = 0, \tau > 0, \quad (3.13)$$

with initial conditions $y(0) = a_0$, $D^{\lambda}y(0) = a_1, ..., D^{(p+q-1)\lambda}y(0) = a_{p+q-1}$, where $a, b > 0, a \neq b$, has a solution

$$\begin{aligned} y(\tau) = & \sum_{j=0}^{p-1} \frac{A_j}{\sqrt[p]{b^j}} H_{pj}\left(\sqrt[p]{b}\frac{\tau^{\lambda}}{\lambda}\right) + \sum_{j=0}^{q-1} \frac{B_j}{\sqrt[q]{a^j}} H_{qj}\left(\sqrt[q]{a}\frac{\tau^{\lambda}}{\lambda}\right) \\ & + d\sum_{j=1}^{p-1} \frac{C_{j-1}}{\sqrt[p]{b^j}} H_{pj}\left(\sqrt[p]{b}\frac{\tau^{\lambda}}{\lambda}\right) + d\sum_{j=1}^{q-1} \frac{D_{j-1}}{\sqrt[q]{a^j}} H_{qj}\left(\sqrt[q]{a}\frac{\tau^{\lambda}}{\lambda}\right) \\ & + \frac{dC_{p-1}}{b}\left(H_{p0}\left(\sqrt[p]{b}\frac{\tau^{\lambda}}{\lambda}\right) - 1\right) + \frac{dD_{q-1}}{a}\left(H_{q0}\left(\sqrt[q]{a}\frac{\tau^{\lambda}}{\lambda}\right) - 1\right), \end{aligned} \quad (3.14)$$

where A_j, $j = 0, ..., p-1$ and B_j, $j = 0, ..., q-1$ are solutions of the system

$$\begin{aligned} aA_{p-j-1} + bB_{q-j-1} &= aa_{p-j-1} + ba_{q-j-1} - a_{p+q-j-1}, \ j = 0, ...q-1, \\ A_{q-j-1} + B_{q-j-1} &= a_{q-j-1}, \ j = 0, ...q-1, \\ A_{p+q-j-1} - aA_{p-j-1} &= a_{p+q-j-1} - aa_{p-j-1}, \ j = q, ..., p-1, \end{aligned} \quad (3.15)$$

and C_j, $j = 0, ..., p-1$ and D_j, $j = 0, ..., q-1$ are solutions of the system

$$\begin{aligned} aC_{p-1} + bD_{q-1} &= -1 \\ aC_{p-j-1} + bD_{q-j-1} &= 0, \ j = 1, ..., q-1 \\ C_{q-j-1} + D_{q-j-1} &= 0, \ j = 0, ..., q-1 \\ C_{p+q-j-1} - aC_{p-j-1} &= 0, \ j = q, ..., p-1. \end{aligned} \quad (3.16)$$

Proof. If $p = q$, then we have Theorem 9. So, without a loosing a generality, we can assume $p > q$. If we apply the fractional Laplace transform on (3.13) we obtain

$$\begin{aligned} & \mathcal{L}_{\lambda}\left(D^{(p+q)\lambda}y(\tau) - aD^{p\lambda}y(\tau) - by^{q\lambda}y(\tau) + aby(\tau) - d\right)(s) \\ & = s^{p+q}\mathcal{L}_{\lambda}(y(\tau))(s) - \sum_{j=0}^{p+q-1} a_{p+q-j-1}s^j - as^p\mathcal{L}_{\lambda}(y(\tau))(s) + a\sum_{j=0}^{p-1} a_{p-j-1}s^j \\ & \quad - bs^q\mathcal{L}_{\lambda}(y(\tau))(s) + b\sum_{j=0}^{q-1} a_{q-j-1}s^j + ab\mathcal{L}_{\lambda}(y(\tau))(s) - \frac{d}{s} = 0 \end{aligned}$$

from which it follows

$$\mathcal{L}_\lambda(y(\tau))(s)$$
$$= \frac{\sum_{j=0}^{q-1}\left(a_{p+q-j-1} - aa_{p-j-1} - ba_{q-j-1}\right)s^j + \sum_{j=q}^{p-1}\left(a_{p+q-j-1} - aa_{p-j-1}\right)s^j}{(s^p - b)(s^q - a)}$$
$$+ \frac{\sum_{j=p}^{p+q-1} a_{p+q-j-1}s^j + \frac{d}{s}}{(s^p - b)(s^q - a)}.$$

From

$$\frac{\sum_{j=0}^{q-1}\left(a_{p+q-j-1} - aa_{p-j-1} - ba_{q-j-1}\right)s^j + \sum_{j=q}^{p-1}\left(a_{p+q-j-1} - aa_{p-j-1}\right)s^j}{(s^p - b)(s^q - a)}$$

$$+\frac{\sum_{j=p}^{p+q-1} a_{p+q-j-1}s^j}{(s^p - b)(s^q - a)} = \frac{\sum_{j=0}^{p-1} A_{p-j-1}s^j}{s^p - b} + \frac{\sum_{j=0}^{q-1} B_{p-j-1}s^j}{s^q - a},$$

we have

$$\sum_{j=0}^{q-1}\left(a_{p+q-j-1} - aa_{p-j-1} - ba_{q-j-1}\right)s^j + \sum_{j=q}^{p-1}\left(a_{p+q-j-1} - aa_{p-j-1}\right)s^j$$
$$+ \sum_{j=p}^{p+q-1} a_{p+q-j-1}s^j$$
$$= (s^q - a)\sum_{j=0}^{p-1} A_{p-j-1}s^j + (s^p - b)\sum_{j=0}^{q-1} B_{p-j-1}s^j$$
$$= -\sum_{j=0}^{q-1}\left(aA_{p-j-1} + bB_{q-j-1}\right)s^j + \sum_{j=q}^{p-1}\left(A_{p+q-j-1} - aA_{p-j-1}\right)s^j$$
$$+ \sum_{j=p}^{p+q-1}\left(A_{p+q-j-1} + B_{p+q-j-1}\right)s^j,$$

from which we have system (3.15). Similarly,

$$\frac{1}{(s^p - b)(s^q - a)} = \frac{\sum_{j=0}^{p-1} C_{p-j-1}s^j}{s^p - b} + \frac{\sum_{j=0}^{q-1} D_{q-j-1}s^j}{s^q - a}$$

implies

$$1 = -\sum_{j=0}^{q-1}\left(aC_{p-j-1} + bD_{q-j-1}\right)s^j + \sum_{j=q}^{p-1}\left(C_{p+q-j-1} - aC_{p-j-1}\right)s^j$$
$$+ \sum_{j=p}^{p+q-1}\left(C_{p+q-j-1} + D_{p+q-j-1}\right)s^j,$$

from which we obtain (3.16). Further,

$$\frac{1}{s\left(s^p-b\right)\left(s^q-a\right)} = \frac{\sum_{j=0}^{p-1} C_j s^{p-j-1}}{s\left(s^p-b\right)} + \frac{\sum_{j=0}^{q-1} D_j s^{q-j-1}}{s\left(s^q-a\right)}$$
$$= \frac{\sum_{j=0}^{p-2} C_j s^{p-j-2}}{s^p-b} + \frac{\sum_{j=0}^{q-2} D_j s^{q-j-2}}{s^q-a} + \frac{C_{p-1}}{s\left(s^p-b\right)} + \frac{D_{q-1}}{s\left(s^q-a\right)}$$
$$= \sum_{j=1}^{p-1}\frac{C_{j-1}s^{p-j-1}}{s^p-b} + \sum_{j=1}^{q-1}\frac{D_{j-1}s^{q-j-1}}{s^q-a} - \frac{C_{p-1}}{bs} + \frac{C_{p-1}s^{p-1}}{b\left(s^p-b\right)} - \frac{D_{q-1}}{as}$$
$$+ \frac{D_{q-1}s^{q-1}}{a\left(s^q-a\right)},$$

so using the inverse fractional Laplace transform and Theorem 5 we obtain the assertion. □

Remark 4 (i) If $a < 0$, then in (3.14) instead of $H_{qj}\left(\sqrt[q]{a}\frac{\tau^\lambda}{\lambda}\right)$ we put $T_{qj}\left(\sqrt[q]{-a}\frac{\tau^\lambda}{\lambda}\right)$. (ii) If $b < 0$, then in (3.14), instead of $H_{pj}\left(\sqrt[p]{b}\frac{\tau^\lambda}{\lambda}\right)$ we put $T_{pj}\left(\sqrt[p]{-b}\frac{\tau^\lambda}{\lambda}\right)$.

7.4 EXAMPLES

In this section, we present some examples that indicate how our results can be applied to concrete problems.

Example 1 Using Theorem 8, we obtain that the equation equation

$$D^{7\lambda}y(\tau) - y(\tau) = 3 + \tau^{2\lambda} + \tau^{6\lambda}, \tag{4.1}$$

with initial conditions

$$y(0) = 1, D^{\lambda}y(0) = 0, D^{2\lambda}y(0) = 0, D^{3\lambda}y(0) = 1, D^{4\lambda}y(0) = 2, D^{5\lambda}y(0) = 1,$$

$D^{6\lambda}y(0) = 3$, has solution

$$y(\tau) = 4H_{70}\left(\frac{\tau^{\lambda}}{\lambda}\right) + H_{73}\left(\frac{\tau^{\lambda}}{\lambda}\right) + 2H_{74}\left(\frac{\tau^{\lambda}}{\lambda}\right) + H_{75}\left(\frac{\tau^{\lambda}}{\lambda}\right) + 3H_{76}\left(\frac{\tau^{\lambda}}{\lambda}\right)$$
$$+ 2!\lambda^2 H_{72}\left(\frac{\tau^{\lambda}}{\lambda}\right) + 6!\lambda^6 H_{76}\left(\frac{\tau^{\lambda}}{\lambda}\right) - 3 - \tau^{2\lambda} - \tau^{7\lambda},$$

where $\tau > 0$ and $\lambda \in (0, 1]$.

Example 2 Using Theorem 9 we obtain that the solution of the fractional differential equation

$$D^{10\lambda}y(\tau) - 3D^{5\lambda}y(\tau) + 2y(\tau) - 4 = 0, \tag{4.2}$$

with initial conditions

$$y(0) = 1,\ D^{\lambda}y(0) = 1,\ D^{2\lambda}y(0) = 2,\ D^{3\lambda}y(0) = 1,\ D^{4\lambda}y(0) = 3,$$

$$D^{5\lambda}y(0) = 4,\ D^{6\lambda}y(0) = 2,\ D^{7\lambda}y(0) = 0,\ D^{8\lambda}y(0) = 1,\ D^{9\lambda}y(0) = 5$$

is

$$y(\tau) = 5H_{50}\left(\sqrt[5]{2}\frac{\tau^{\lambda}}{\lambda}\right) - \frac{2}{\sqrt[5]{4}}H_{52}\left(\sqrt[5]{2}\frac{\tau^{\lambda}}{\lambda}\right) + \frac{2}{\sqrt[5]{16}}H_{54}\left(\sqrt[5]{2}\frac{\tau^{\lambda}}{\lambda}\right)$$
$$- 6H_{50}\left(\frac{\tau^{\lambda}}{\lambda}\right) + 4H_{52}\left(\frac{\tau^{\lambda}}{\lambda}\right) + H_{53}\left(\frac{\tau^{\lambda}}{\lambda}\right) + H_{54}\left(\frac{\tau^{\lambda}}{\lambda}\right) + 2,$$

where $\tau > 0$ *and* $\lambda \in (0, 1]$.

Example 3 Using Theorem 9, we establish that the solution of the fractional differential equation

$$D^{6\lambda}y(\tau) - 2D^{3\lambda}y(\tau) + y(t) = 3 + \frac{\tau^{\lambda}}{\lambda}, \tag{4.3}$$

$$y(0) = 1,\ D^{\lambda}y(0) = 0,\ D^{2\lambda}y(0) = 0,\ D^{3\lambda}y(0) = 0,\ D^{4\lambda}y(0) = 0,\ D^{5\lambda}y(0) = 1, \tag{4.4}$$

is

$$y(\tau) = -2H_{30}\left(\frac{\tau^\lambda}{\lambda}\right) - H_{31}\left(\frac{\tau^\lambda}{\lambda}\right) + 2H_{32}\left(\frac{\tau^\lambda}{\lambda}\right) * H_{30}\left(\frac{\tau^\lambda}{\lambda}\right) + H_{32}\left(\frac{\tau^\lambda}{\lambda}\right)$$
$$* H_{32}\left(\frac{\tau^\lambda}{\lambda}\right) + H_{32}\left(\frac{\tau^\lambda}{\lambda}\right) * H_{31}\left(\frac{\tau^\lambda}{\lambda}\right) + 3 + \frac{\tau^\lambda}{\lambda},$$
$$\tau > 0, \lambda \in (0, 1].$$

Example 4 Using Theorem 10, we establish that the solution of the equation

$$D^{9\lambda}y(\tau) - 6D^{6\lambda}y(\tau) + 11D^{3\lambda}y(\tau) - 6y(\tau) = 1 \tag{4.5}$$

with initial conditions

$$y(0) = 1,\ D^\lambda y(0) = 0,\ D^{2\lambda}y(0) = 0,\ D^{3\lambda}y(0) = 0,\ D^{4\lambda}y(0) = 0,\ D^{5\lambda}y(0) = 0,$$
$$D^{6\lambda}y(0) = 0, D^{7\lambda}y(0) = 0, D^{8\lambda}y(0) = 1$$

is

$$y(\tau) = \frac{7}{2}H_{30}\left(\frac{\tau^\lambda}{\lambda}\right) - \frac{7}{2\sqrt[3]{2}}H_{30}\left(\sqrt[3]{2}\frac{\tau^\lambda}{\lambda}\right) + \frac{7}{6\sqrt[3]{3}}H_{30}\left(\sqrt[3]{3}\frac{\tau^\lambda}{\lambda}\right)$$
$$+ \frac{1}{2}H_{32}\left(\frac{\tau^\lambda}{\lambda}\right) - \frac{1}{2\sqrt[3]{2}}H_{32}\left(\sqrt[3]{2}\frac{\tau^\lambda}{\lambda}\right) + \frac{1}{2\sqrt[3]{3}}H_{32}\left(\sqrt[3]{3}\frac{\tau^\lambda}{\lambda}\right) - \frac{1}{6},$$

$\tau > 0, \lambda \in (0, 1]$.

Example 5 Using Theorem 11, we establish that the solution of the equation

$$D^{5\lambda}y(\tau) - 4D^{3\lambda}y(\tau) - 3D^{2\lambda}y(\tau) + 12y(\tau) - 5 = 0,$$
$$y(0) = 1,\ D^\lambda y(0) = 1,\ D^{2\lambda}y(0) = 2,\ D^{3\lambda}y(0) = 3,\ D^{4\lambda}y(0) = 4 \tag{4.6}$$

is

$$y(\tau) = \left(A_0 + \frac{5C_2}{3}\right)H_{30}\left(\sqrt[3]{3}\frac{\tau^\lambda}{\lambda}\right) + \left(A_1 + \frac{5C_0}{\sqrt[3]{3}}\right)H_{31}\left(\sqrt[3]{3}\frac{\tau^\lambda}{\lambda}\right)$$
$$+ \left(A_2 + \frac{5C_1}{\sqrt[3]{9}}\right)H_{32}\left(\sqrt[3]{3}\frac{\tau^\lambda}{\lambda}\right)$$
$$+ \left(B_0 + \frac{5D_1}{4}\right)H_{20}\left(2\frac{\tau^\lambda}{\lambda}\right) + \left(B_1 + \frac{5D_0}{2}\right)H_{21}\left(2\frac{\tau^\lambda}{\lambda}\right) - \frac{5C_2}{3} - \frac{5D_1}{4},$$

where $A_0 = \frac{51}{55}$, $A_1 = \frac{52}{55}$, $A_2 = \frac{94}{55}$, $B_0 = \frac{4}{55}$, $B_1 = \frac{3}{55}$, $C_0 = -\frac{4}{55}$, $C_1 = -\frac{3}{55}$, $C_2 = -\frac{16}{55}$, $D_0 = \frac{4}{55}$, $D_1 = \frac{3}{55}$, $\tau > 0$, $\lambda \in (0, 1]$.

7.5 APPLICATIONS

In this section, we provide some applications of the theorems from Section 7.3.

Theorem 12 *(Special case of Theorem 8, for $a = 0$) Let $p \in \mathbb{N}$, $p \geq 2$ and $\lambda \in (0,1]$. Then the solution of the equation*

$$D^{p\lambda}y(\tau) = \sum_{k=0}^{n} b_k \tau^{\lambda k}, \; n \in \mathbb{N}, \tau > 0, \tag{5.1}$$

with initial conditions $y(0) = a_0$, $D^{\lambda}y(0) = a_1, \ldots, D^{(p-1)\lambda}y(0) = a_{p-1}$ is

$$y(\tau) = \sum_{j=0}^{p-1} \frac{a_j}{\lambda^j j!}\tau^{j\lambda} + \frac{\tau^{p\lambda}}{\lambda^p} \sum_{k=0}^{n} \frac{b_k k!}{(p+k)!}\tau^{k\lambda}.$$

Proof. The proof follows from Theorems 4 and 5. □

Theorem 13 *(Special case of Theorem 9, for $b = 0$) Let $p \in \mathbb{N}$, $p \geq 2$ and $\lambda \in (0,1]$. Then the solution of the equation*

$$D^{2p\lambda}y(\tau) - (m+n)D^{\lambda}y(\tau) = \sum_{k=0}^{r} b_k t^{k\lambda}, \; r \leq p-1, \tau > 0, \tag{5.2}$$

with initial conditions $y(0) = a_0$, $D^{\lambda}y(0) = a_1, \ldots, D^{(2p-1)\lambda}y(0) = a_{2p-1}$, is

$$\begin{aligned} y(\tau) = &-\frac{1}{m}\sum_{j=0}^{p-1}\left(a_{2p-j-1} + m a_{p-j-1}\right)\frac{\tau^{(p-j-1)\lambda}}{\lambda^{p-j-1}(p-j-1)!} \\ &+ \frac{1}{m}\sum_{j=0}^{p-1}\frac{a_{p+j} + m a_j}{\sqrt[p]{m^j}} H_{pj}\left(\sqrt[p]{m}\frac{\tau^{\lambda}}{\lambda}\right) \\ &+ \sum_{j=0}^{p-1}\frac{a_j}{\sqrt[p]{m^j}} H_{pj}\left(\sqrt[p]{m}\frac{\tau^{\lambda}}{\lambda}\right) - \frac{1}{m\lambda^p}\sum_{k=0}^{r}\frac{b_k k!}{(p+k)!}\tau^{(p+k)\lambda} - \frac{1}{m^2}\sum_{k=0}^{r} b_k \tau^{k\lambda} \\ &+ \frac{1}{m^2}\sum_{k=0}^{r}\frac{b_k \lambda^k k!}{\sqrt[p]{m^k}} H_{pk}\left(\sqrt[p]{m}\frac{\tau^{\lambda}}{\lambda}\right), \end{aligned}$$

for $n = 0$, $m > 0$ and

$$y(\tau) = -\frac{1}{n}\sum_{j=0}^{p-1}\left(a_{2p-j-1} + na_{p-j-1}\right)\frac{\tau^{(p-j-1)\lambda}}{\lambda^{p-j-1}(p-j-1)!}$$
$$+\frac{1}{n}\sum_{j=0}^{p-1}\frac{a_{p+j}+na_j}{\sqrt[p]{n^j}}H_{pj}\left(\sqrt[p]{n}\frac{\tau^{\lambda}}{\lambda}\right)$$
$$+\sum_{j=0}^{p-1}\frac{a_j}{\sqrt[p]{n^j}}H_{pj}\left(\sqrt[p]{n}\frac{\tau^{\lambda}}{\lambda}\right) - \frac{1}{n\lambda^p}\sum_{k=0}^{r}\frac{b_k k!}{(p+k)!}t^{(p+k)\lambda} - \frac{1}{n^2}\sum_{k=0}^{r} b_k\tau^{k\lambda}$$
$$+\frac{1}{n^2}\sum_{k=0}^{r}\frac{b_k\lambda^k k!}{\sqrt[p]{n^k}}H_{pk}\left(\sqrt[p]{n}\frac{\tau^{\lambda}}{\lambda}\right),$$

for $m = 0$, $n > 0$.

Remark 5 In the case of $n = 0$, $m > 0$: If in (5.2) $m < 0$, $n = 0$, then instead of $\sqrt[p]{m}$ we put $\sqrt[p]{(-m)}$, and instead of $H_{pj}\left(\sqrt[p]{m}\frac{\tau^{\lambda}}{\lambda}\right)$ we put $T_{pj}\left(\sqrt[p]{-m}\frac{\tau^{\lambda}}{\lambda}\right)$. In the case of $m = 0$, $n > 0$: If in (5.2) $n < 0$, $m = 0$, then instead of $\sqrt[p]{n}$ we put $\sqrt[p]{(-n)}$, and instead of $H_{pj}\left(\sqrt[p]{n}\frac{\tau^{\lambda}}{\lambda}\right)$ we put $T_{pj}\left(\sqrt[p]{-n}\frac{\tau^{\lambda}}{\lambda}\right)$.

Theorem 14 *(Special case of Theorem 9, for $a = 0$) Let $p \in \mathbb{N}$, $p \geq 2$, and $\lambda \in (0,1]$. The the solution of the equation*

$$D^{2p\lambda}y(\tau) - m^2 y(\tau) = \sum_{k=0}^{r} b_k\tau^{k\lambda},\ r \leq p-1, \tau > 0, m \neq 0 \tag{5.3}$$

with initial conditions $y(0) = a_0$, $D^{\lambda}y(0) = a_1, ..., D^{(2p-1)\lambda}y(0) = a_{2p-1}$, is

$$y(\tau)$$
$$= \frac{1}{2m}\sum_{j=0}^{p-1}\left(\frac{a_{p+j}+ma_j}{\sqrt[p]{m^j}}H_{pj}\left(\sqrt[p]{m}\frac{\tau^{\lambda}}{\lambda}\right) + \frac{ma_j - a_{p+j}}{\sqrt[p]{m^j}}T_{pj}\left(\sqrt[p]{m}\frac{\tau^{\lambda}}{\lambda}\right)\right)$$
$$-\frac{1}{m^2}\left(\sum_{k=0}^{r} b_k\tau^{\lambda k} - \frac{1}{2}\sum_{k=0}^{r} b_k\lambda^k k!\left(\frac{H_{pk}\left(\sqrt[p]{m}\frac{\tau^{\lambda}}{\lambda}\right)}{\sqrt[p]{m^k}} + \frac{T_{pk}\left(\sqrt[p]{m}\frac{\tau^{\lambda}}{\lambda}\right)}{\sqrt[p]{m^k}}\right)\right).$$

Remark 6 If we put $a = b = 0$ in Theorem 9, then we obtain Theorem 12.

7.6 CONCLUSION

The aim of this chapter is to solve linear constant coefficient conformable fractional differential equations of forms:

$$D^{np\lambda}y(\tau) + c_{n-1}D^{(n-1)p\lambda}y(\tau) + ... + c_1D^{p\lambda}y(\tau) + c_0y(\tau) = P_n\left(\tau^{\lambda}\right), \tag{5.4}$$

$\tau > 0, n = 1, 2, 3$ and

$$D^{(p+q)\lambda}y(\tau) - aD^{p\lambda}y(\tau) - bD^{q\lambda}y(\tau) + aby(\tau) - d = 0, \tau > 0, \tag{5.5}$$

where the a, b, d, c_i are constants, P_n is a polynomial of degree $n, p, q \in \mathbb{N}$, $p, q \geq 2$, and $\lambda \in (0, 1]$. Applying the fractional Laplace transform on (5.4) and (5.5) we obtain their solutions as a linear combination of functions $T_{pj}, H_{pj} : \mathbb{R} \to \mathbb{R}, j = 0, 1, 2, ..., p-1, p \in \mathbb{N}$, defined as follows

$$T_{pj}(\tau) = \sum_{n=0}^{\infty} \frac{(-1)^n \tau^{pn+j}}{(pn+j)!}, \; H_{pj}(\tau) = \sum_{n=0}^{\infty} \frac{\tau^{pn+j}}{(pn+j)!}. \tag{5.6}$$

We believe that the methods presented in this paper provide an opportunity for further research in the field of fractional calculus.

7.7 FUNDING

This research received no external funding.

REFERENCES

1. Mittag-Leffler, G. M. Une generalisation de lá integrale de Laplace-Abel. *Comptes Rendus de lá Academie des Sciences Série II*, **1903**, *13*, 537–539.
2. Mittag-Leffler, G. M. Sur la nouvelle fonction $E_\lambda(z)$. *Comptes Rendus de lá Academie des Sciences*, **1903**, *137*, 554–558.

3. Mittag-Leffler, G. M. Sur la representation analytiqie dá une fonction monogene (cinquieme note). *Acta Mathematica*, **1905**, *29*(1), 101–181.
4. Wiman, A. Uber de fundamental satz in der theorie der funktionen $E_\lambda(z)$. *Acta Mathematica*, **1905**, *29*, 191–201.
5. Dzherbashyan, M. M. *Integral Transforms and Representations of Functions in the Complex Plane*, Nauka, Moscow, Russia, 1966.
6. Erdélyi, A.; Magnus, W.; Oberhettinger, F.; Tricomi, F. G. *Higher Transcendental Functions,* vol. 3, McGraw-Hill, New York, NY, 1955.
7. Gorenflo, R.; Kilbas, A. A.; Mainardi, F.; Rogosin, S. V. *Mittag-Leffler Functions, Related Topics and Applications*, Springer-Verlag, Berlin, Heidelberg, 2014.
8. Haubold, H. J.; Mathai, A. M.; Saxena, R. K. Mittag-leffler functions and their applications. *Journal of Applied Mathematics*, **2011**, *2011*, Article ID 298628.
9. Kilbas, A.; Srivastava, H.; Trujillo, J. *Theory and Applications of Fractional Differential Equations*, North-Holland Mathematics Studies, New York, 2006.
10. Miller, K. S. *An Introduction to Fractional Calculus and Fractional Differential Equations*, J. Wiley and Sons, New York, 1993.
11. Oldham, K.; Spanier, J. *The Fractional Calculus, Theory and Applications of Differentiation and Integration of Arbitrary Order*, Academic Press, Cambridge, 1974.
12. Podlubny, I. *Fractional Differential Equations. An Introduction to Fractional Derivatives, Fractional Differential Equations, Some Methods of their Solution and Some of their Applications*, Academic Press, New York, 1999.
13. Abdeljawad, T. On conformable fractional calculus. *Journal of Computational and Applied Mathematics*, **2015**, *279*, 57–66.
14. Khalil, R.; Al Horani, M.; Yousef, A.; Sababheh, M. A new definition of fractional derivative. *Journal of Computational and Applied Mathematics*, **2014**, *264*, 65–70.
15. Ansari, A. H.; Maksimović, S. Solving conformable fractional Sturm-Liouville equations using an one class of special polynomials and special functions. *Mathematical Analysis and its Contemporary Applications*, **2023**, *5*(1), 69–84.
16. Bashar, M. H.; Inc, M.; Islam, S. R.; Mahmoud, K. H.; Akbar, M. A. Soliton solutions and fractional effects to the time-fractional modified

equal width equation. *Alexandria Engineering Journal*, **2022**, *61*(12), 12539–12547.

17. Benkhettou, N.; Hassani, N. S.; Torres, D. F. M. A conformable fractional calculus on arbitrary time scales. *Journal of King Saud University - Science*, **2016**, *28*(1), 93–98.
18. Ansari, A. H. ; Liu, X.; Mishra, V. N. On Mittag-Leffler function and beyond. *Nonlinear Science Letters A*, **2017**, *8*(2), 187–199.
19. Bouchenak, A. Generalization of fractional Laplace transform for higher order and its application. *Journal of Innovative Applied Mathematics and Computational Sciences,* **2021**, *1*(1), 79–92.
20. Silva, F. S.; Moreira, D. M.; Moret, M. A. Conformable laplace transform of fractional differential equations. *Axioms*, **2018**, *7*, 55.

CHAPTER 8

The Non-Population Conserving SIR Model on Time Scales

Zahra Belarbi, Benaoumeur Bayour, and Delfim F. M. Torres

8.1 INTRODUCTION

The historical trajectory of employing mathematical models to understand and predict the dynamics of disease transmission can be traced back to the pivotal year of 1766 and the work of Daniel Bernoulli. Indeed, Bernoulli's work laid a foundational understanding of the application of mathematical principles in describing the intricate patterns of disease spread [11].

Fast forward to the year 1927, a crucial milestone emerged with the seminal publication of Kermack and McKendrick [13]. In their groundbreaking work, they introduced the SIR (Susceptible–Infected–Recovered/Removed) model for epidemics. This pioneering model not only provided a conceptual framework for comprehending the transmission dynamics of infectious diseases but also set the stage for the development of numerous subsequent models. In fact, SIR-type models, with their compartmental classification of individuals into different classes, became a cornerstone in the field of epidemiology, offering a versatile and widely adopted template for modeling various contagious diseases [2].

Since then, the field has witnessed a burgeoning array of mathematical models, each tailored to address specific nuances and challenges associated with different diseases. The continuous evolution of these models reflects

DOI: 10.1201/9781003530602-8

the ongoing commitment of researchers to refine and enhance our understanding of epidemic dynamics, ultimately contributing to more effective strategies for disease control and prevention [1].

Let $N(t) = S(t) + I(t) + R(t)$ denote the total population at time t. The classical and more standard SIR epidemic model assumes that there are no births or deaths during the period under study, based on the assumption that these are on a much slower time scale and can therefore be ignored. The combined dynamics is then given mathematically by

$$\begin{cases} \dot{S}(t) = -\lambda \frac{S(t)I(t)}{N(t)}, \\ \dot{I}(t) = \lambda \frac{S(t)I(t)}{N(t)} - \gamma I(t), \\ \dot{R}(t) = \gamma I(t), \end{cases} \tag{1}$$

where $\lambda > 0$ is the infection rate, and $\gamma > 0$ is the rate at which infected individuals recover. Clearly, model (1) assumes the total population under study to be constant: $(S(t) + I(t) + R(t))' = 0$. Some of its limitations stand out immediately: for diseases such as Ebola [3, 14] or COVID-19 [9, 15], where death rates are not negligible, we do not have a constant population, and model (1) ceases to be valid.

In [9], Borkar and Manjunath propose a variant of (1), called the SIR-NC model, that, unlike the standard SIR model (1), does not assume the conservation of the population. Surprisingly, by incorporating a nonzero death rate into the model, thus being more suitable for diseases like Ebola or COVID-19, the new SIR-NC model is analytically tractable [9, 10].

Calculus on time scales is a mathematical area that generalizes the traditional calculus by unifying continuous and discrete analysis on an arbitrary time scale. By combining both continuous and discrete elements, time scales allow for a more flexible and inclusive approach to modeling systems that exhibit both continuous and discrete behaviors (hybrid systems). The theory was introduced by Stefan Hilger from 1988 to 1990 as a special case of a general analysis on measure chains [12] (see also [4]).

The analysis of time scales allows one to generalize differential and difference equations, incorporating both continuous and discrete dynamics in a unified setting. Moreover, the new analysis holds in any nonempty closed set, such as the set of integers, rationals, or more complex structures like the Cantor set, offering a more comprehensive mathematical framework for analyzing and modeling systems with mixed continuous and discrete dynamics. This permits the extension of the applicability of traditional calculus to a broader range of scenarios. In particular, this is true in epidemic modeling, where analysis of time scales has allowed the modeling of noncontinuous disease dynamics, e.g., diseases where a virus remains unnoticed within the host for several years before continuing to spread [7, 8].

Here we investigate the SIR-NC model on an arbitrary time scale. In Section 8.2, we briefly recall the fundamentals of calculus on time scales. Our results are then given in Section 8.3: we formulate our SIR-NC model on time scales (cf. the dynamic system (2)), proving that there exists a unique solution (cf. Theorem 3.1). More than that, an explicit analytical formula for the solution is obtained (cf. (5)). We end with Section 8.4, discussing the SIR-NC model on time scales with imported infections. Under some conditions, a closed formula for the solution is also obtained (cf. Theorem 4.1). Along the text, two examples are given to illustrate the obtained results and to show how our results generalize those available in the literature (cf. Example 3.1 and Example 4.1).

8.2 PRELIMINARIES

A time scale $\mathbb{T}$ is an arbitrary nonempty closed subset of the real numbers $\mathbb{R}$. For $t \in \mathbb{T}$, we define the forward jump operator $\sigma : \mathbb{T} \to \mathbb{T}$ by

$$\sigma(t) = inf\{s \in \mathbb{T} : s > t\}$$

and the backward jump operator $\rho : \mathbb{T} \to \mathbb{T}$ is defined by

$$\rho(t) := sup\{s \in \mathbb{T} : s < t\}.$$

Then, one defines the graininess function $\mu : \mathbb{T} \to [0, +\infty[$ by $\mu(t) = \sigma(t) - t$.

If $\sigma(t) > t$, then we say that t is right-scattered; if $\rho(t) < t$, then t is left-scattered. Moreover, if $t < sup\mathbb{T}$ and $\sigma(t) = t$, then t is called right-dense; if $t > inf\,\mathbb{T}$ and $\rho(t) = t$, then t is called left-dense. If $\mathbb{T}$ has a left-scattered maximum m, then we define $\mathbb{T}^{\kappa} = \mathbb{T} \setminus \{m\}$; otherwise we set $\mathbb{T}^{\kappa} = \mathbb{T}$. If $f : \mathbb{T} \to \mathbb{R}$, then $f^{\sigma} : \mathbb{T} \to \mathbb{R}$ is given by $f^{\sigma}(t) = f(\sigma(t))$ for all $t \in \mathbb{T}$.

Definition 2.1 (The Hilger Derivative [6]) Let $f : \mathbb{T} \to \mathbb{R}$ and $t \in \mathbb{T}$. We define $f^{\Delta}(t)$ to be the number (provided it exists) with the property that given any $\epsilon > 0$ there is a neighborhood U of t (i.e., $U = (t - \delta, t + \delta) \cap \mathbb{T}$ for some $\delta > 0$) such that

$$|[f(\sigma(t)) - f(s)] - f^{\Delta}(t)[\sigma(t) - s]| \leq \epsilon |\sigma(t) - s|$$

for all $s \in U$. We call $f^{\Delta}(t)$ the Hilger (or delta) derivative of f at t.

We denote the set of *rd*-continuous (right-dense continuous) functions by $C_{rd} = C_{rd}(\mathbb{T}) = C_{rd}(\mathbb{T}, \mathbb{R})$. A function f is said to be regressive if

$$1 + \mu(t)f(t) \neq 0$$

holds for all $t \in \mathbb{T}$. The set of all regressive functions is denoted by

$$\mathcal{R} = \mathcal{R}(\mathbb{T}) = \mathcal{R}(\mathbb{T}, \mathbb{R}).$$

We also define the set $\mathcal{R}^+$ of all positively regressive elements by

$$\mathcal{R}^+ = \mathcal{R}^+(\mathbb{T}, \mathbb{R}) = \{f \in \mathcal{R} : 1 + \mu(t)f(t) > 0 \quad \text{for all } t \in \mathbb{T}\}.$$

Now let $p, q \in \mathcal{R}$. We define the circle plus addition $\oplus$ on $\mathcal{R}$ by

$$(p \oplus q)(t) := p(t) + q(t) + \mu(t)p(t)q(t) \quad \text{for all } t \in \mathbb{T},$$

while the circle minus subtraction $\ominus$ on $\mathcal{R}$ is given by

$$(p \ominus q)(t) := \frac{p(t) - q(t)}{1 + \mu(t)q(t)} \quad \text{for all } t \in \mathbb{T}.$$

Theorem 2.1 (See Theorem 2.33 of [5]) *If $p \in \mathcal{R}$ and $t_0 \in \mathbb{T}$, then the IVP*

$$y^{\Delta} = p(t)y, \quad y(t_0) = 1,$$

possesses a unique solution, which is denoted by $e_p(\cdot, t_0)$ and called the exponential function.

Some useful properties of the exponential function are the following.

Theorem 2.2 (See Theorem 2.36 of [5]) *If $p \in \mathcal{R}$, then*

- $e_0(t, s) = 1$ *and* $e_p(t, t) = 1$;
- $e_p(t, s) = \frac{1}{e_p(s,t)}$;
- $e_p(t, r)\, e_p(r, s) = e_p(t, s)$.

Theorem 2.3 (See Theorem 2.44 of [5]) *If* $p \in \mathcal{R}^+$ *and* $t_0 \in \mathbb{T}$, *then* $e_p(t, t_0) > 0$ *for all* $t \in \mathbb{T}$.

Corollary 2.1 (See [5]) *If* $p, q \in \mathcal{R}$, *then*

1. $e_{p \oplus q}(t, s) = e_p(t, s)\, e_q(t, s)$;
2. $e_{\ominus p}(t, s) = e_p(s, t) = \frac{1}{e_p(t,s)}$.

Theorem 2.4 (Variation of Constants, see Theorems 2.74 and 2.77 of [5]) *Suppose* $p \in \mathcal{R}$ *and* $f \in C_{rd}$. *If* $t_0 \in \mathbb{T}$ *and* $y_0 \in \mathbb{R}$ *are given, then the unique solution of the IVP*

$$y^{\Delta} = p(t)\, y + f(t), \quad y(t_0) = y_0,$$

is given by

$$y(t) = e_p(t, t_0)\, y_0 + \int_{t_0}^{t} e_p(t, \sigma(s)) f(s)\, \Delta s.$$

Similarly, the unique solution of the IVP

$$y^{\Delta} = -p(t)\, y + f(t), \quad y(t_0) = y_0,$$

is given by

$$y(t) = e_{\ominus p}(t, t_0)\, y_0 + \int_{t_0}^{t} e_{\ominus p}(t, s) f(s)\, \Delta s.$$

Theorem 2.5 (See Theorem 2.39 of [5]) *If* $p \in \mathbb{R}$ *and* $a, b, c \in \mathbb{T}$, *then*

$$\int_a^b p(t)\, e_p(t, c)\, \Delta t = e_p(b, c) - e_p(a, c)$$

and

$$\int_a^b p(t)\, e_p(c, \sigma(t))\, \Delta t = e_p(c, a) - e_p(c, b).$$

8.3 THE NON-POPULATION CONSERVING SIR MODEL ON TIME SCALES (SIR-NC)

On a given time scale $\mathbb{T}$, we propose the following SIR-NC model:

$$\begin{cases} S^{\Delta}(t) = -\lambda \frac{S(t)I^{\sigma}(t)}{S(t)+I(t)}, \\ I^{\Delta}(t) = \lambda \frac{S(t)I^{\sigma}(t)}{S(t)+I(t)} - \gamma I^{\sigma}(t), \\ R^{\Delta}(t) = \gamma I^{\sigma}(t), \end{cases} \tag{2}$$

where $S, I, R : \mathbb{T} \longrightarrow \mathbb{R}^{+}$ and $\lambda, \gamma > 0$, subject to given initial conditions

$$S(0) = S_0, \quad I(0) = I_0, \quad R(0) = R_0 \tag{3}$$

with $S_0 > 0$, $I_0 > 0$, and $R_0 \geq 0$. In the particular case of $\mathbb{T} = \mathbb{R}$, problems (2)–(3) are studied in [9].

We begin by remarking that it is enough to solve the first two equations of system (2). Indeed, by knowing $I(t)$, we immediately get $R(t)$ from the third equation of (2). For this reason, in the sequel, we restrict ourselves to the two-dimensional IVP

$$\begin{cases} S^{\Delta}(t) = -\lambda \frac{S(t)I^{\sigma}(t)}{S(t)+I(t)}, & S(0) = S_0, \\ I^{\Delta}(t) = \lambda \frac{S(t)I^{\sigma}(t)}{S(t)+I(t)} - \gamma I^{\sigma}(t), & I(0) = I_0. \end{cases} \tag{4}$$

Let $N_0 := I_0 + S_0$ (we could also add R_0, but since $R(t)$ does not affect the evolution of (4), we consider here, without loss of generality, that $R_0 = 0$).

It is easy to see that in our SIR-NC model, the condition for an epidemic outbreak is given by

$$I^{\Delta}(0) > 0 \Rightarrow \frac{\lambda}{\gamma} > 1 + \frac{I_0}{S_0}.$$

Theorem 3.1 *Let* $C = \frac{S(0)}{I(0)}$. *If* $\gamma - \lambda, p(t) \in \mathcal{R}$, *then the unique solution to (4) is given by*

$$\begin{cases} S(t) = e_{(\ominus p(t)) \oplus (\gamma - \lambda)}(t, 0)\, S(0), \\ I(t) = e_{\ominus p(t)}(t, 0)\, I(0), \end{cases} \tag{5}$$

where $p(t) = \gamma - \frac{\lambda C}{e_{\ominus(\gamma-\lambda)}(t,0)+C}$.

Proof. Let

$$\begin{cases} x(t) = \frac{S(t)}{S(t)+I(t)}, \\ y(t) = \frac{I(t)}{S(t)+I(t)}. \end{cases}$$

By the assumption that $I(0) > 0$, we have $x(t) + y(t) = 1$ and $x(0) < 1$. We can rewrite $x^{\Delta}(t)$ and $y^{\Delta}(t)$ as follows:

$$\begin{aligned} x^{\Delta}(t) &= \frac{S^{\Delta}(t)(S(t)+I(t)) - \left(S^{\Delta}(t)+I^{\Delta}(t)\right)S(t)}{(S(t)+I(t))(S^{\sigma}(t)+I^{\sigma}(t))} \\ &= (\gamma-\lambda)\,x(t)\,y^{\sigma}(t) \\ &= (\gamma-\lambda)(1-x^{\sigma}(t))\,x(t). \end{aligned} \tag{6}$$

Applying the substitution $z = \frac{1}{x}$, we obtain the linear first-order dynamic equation

$$z^{\Delta} = -(\gamma-\lambda)z^{\sigma} + (\gamma-\lambda).$$

Its solution is

$$z(t) = e_{\ominus(\gamma-\lambda)}(t,0)\,z(0) + \int_0^t e_{\ominus(\gamma-\lambda)}(t,s)(\gamma-\lambda)\,\Delta s. \tag{7}$$

Integrating yields

$$z(t) = e_{\ominus(\gamma-\lambda)}(t,0)(z(0)-1)+1$$

and, re-substituting $z = \frac{1}{x}$, we obtain that

$$x(t) = \frac{x(0)}{e_{\ominus(\gamma-\lambda)}(t,0)(1-x(0))+x(0)}.$$

We have

$$\begin{aligned} I^{\Delta}(t) &= I^{\sigma}(t)\left(\lambda\frac{S(t)}{S(t)+I(t)} - \gamma\right) \\ &= -(\gamma-\lambda x(t))\,I^{\sigma}(t). \end{aligned}$$

Letting $p(t) = \gamma - \lambda x(t) \in \mathcal{R}$ it follows that

$$I^{\Delta}(t) = -p(t)\,I^{\sigma}(t).$$

Clearly,

$$I(t) = e_{\ominus p(t)}(t,0)\,I(0), \tag{8}$$

where

$$p(t) = \gamma - \lambda x(t) = \gamma - \frac{\lambda x(0)}{e_{\ominus(\gamma-\lambda)}(t,0)(1-x(0)) + x(0)},$$

and

$$S(t) = \frac{-I(t)\,x(t)}{x(t)-1} = e_{(\ominus p(t))\oplus(\gamma-\lambda)}(t,0)\,S(0).$$

The proof for the time-dependent γ and λ goes exactly the same way. □

Remark 3.1 If $\gamma = \lambda$, then $\gamma - \lambda \in \mathcal{R}$ and, by Theorem 3.1, the solution of system (2) is

$$\begin{cases} S(t) = e_{\ominus p(t)}(t,0)\,S(0), \\ I(t) = e_{\ominus p(t)}(t,0)\,I(0), \end{cases} \tag{9}$$

where $p(t) = \frac{\lambda(t)}{1+C}$ with $C = \frac{S(0)}{I(0)}$.

As a corollary, we apply Theorem 3.1 to solve the discrete epidemic model

$$\begin{cases} S(t+1) = S(t) - \frac{\lambda(t)S(t)I(t+1)}{S(t)+I(t)}, \\ I(t+1) = I(t) + \frac{\lambda(t)S(t)I(t+1)}{S(t)+I(t)} - \gamma(t)\,I(t+1), \end{cases} \tag{10}$$

$t \in \mathbb{Z}$, with initial conditions $S(0) = S_0 > 0, I(0) = I_0 > 0, R(0) = R_0 \geq 0$. Note that for any $t \in \mathbb{Z}$ we have

$$\begin{aligned} p(t) &= \gamma(t) - \frac{\lambda(t)C}{e_{\ominus(\gamma(t)-\lambda(t))}(t,0) + C} \\ &= \gamma(t) - \frac{\lambda(t)C}{\frac{1}{e_{(\gamma(t)-\lambda(t))}(t,0)} + C} \\ &= \gamma(t) - \frac{\lambda(t)C}{\frac{1}{(1+(\gamma-\lambda(t)))^t} + C}, \end{aligned}$$

Corollary 3.1 *If* $1+\gamma(t)-\lambda(t), 1+\lambda(t) \neq 0$ *for all* $t \in \mathbb{Z}$, *then the unique solution to system (10) is given by*

$$\begin{cases} S(t) = \dfrac{S(0)\delta(t)(1+C\delta(t))^{t}}{(1+\gamma(t)+C\gamma(t)(1+\gamma(t)-\lambda(t)\gamma(t)))^{t}}, \\ I(t) = \dfrac{I(0)(1+C\delta(t))^{t}}{(1+\gamma(t)+C\gamma(t)(1+\gamma(t)-\lambda(t)\gamma(t)))^{t}}, \end{cases} \tag{11}$$

where $C = \frac{S(0)}{I(0)}$ and $\delta(t) = (1 + (\gamma - \lambda)(t))^{t}$.

A very simple particular case is obtained when the time scale is the set of real numbers. In that particular case, we obtain what is given in [9].

Example 3.1 Let $\mathbb{T} = \mathbb{R}$ and $\gamma, \lambda \in \mathbb{R}$ with $\gamma \neq \lambda$. Then, by Theorem 3.1, the solution to system (2) is given by

$$S(t) = S(0)\frac{e^{\int_0^t (\gamma-\lambda)ds}}{e^{\int_0^t p(s)ds}}$$

$$= S(0)e^{(\gamma-\lambda)t}e^{-\int_0^t \gamma ds}e^{\int_0^t \frac{\lambda C e^{(\gamma-\lambda)s}}{1+Ce^{(\gamma-\lambda)s}}ds}$$

and

$$I(t) = I(0)\,e^{-\int_0^t \frac{\gamma e^{-(\gamma-\lambda)s} + C(\gamma-\lambda)}{e^{-(\gamma-\lambda)s}+C}ds}.$$

8.4 THE SIR-NC MODEL WITH IMPORTED INFECTIONS

On a general time scale $\mathbb{T}$, we propose the following SIR-NC model with imported infections:

$$\begin{cases} S^{\Delta}(t) = -\lambda\dfrac{S(t)I^{\sigma}(t)}{S(t)+I(t)} - \nu S^{\sigma}(t), \\ I^{\Delta}(t) = \lambda\dfrac{S(t)I^{\sigma}(t)}{S(t)+I(t)} + \nu S^{\sigma}(t) - \gamma I^{\sigma}(t), \\ R^{\Delta}(t) = \gamma I^{\sigma}(t), \end{cases} \tag{12}$$

where $S, I, R : \mathbb{T} \longrightarrow \mathbb{R}^{+}$ and $\lambda, \gamma, \nu > 0$. Similarly to Section 8.3, we restrict our attention, without loss of generality, to the first two equations of (12), and we set $N(t) := S(t) + I(t)$.

We observe that here the condition for an epidemic to break out is given by

$$\lambda\frac{I(0)}{N(0)}+\nu>\gamma\frac{I(0)}{S(0)}.$$

Theorem 4.1 *If $\nu,\gamma-\lambda\in\mathcal{R}$, then the solution to the SIR-NC system (12) with imported infections is given as follows:*

$$\begin{cases} S(t)=\frac{x(t)}{1-x(t)}e_{\ominus g(t)}(t,0)\,I(0), \\ I(t)=e_{\ominus g(t)}(t,0)\,I(0), \end{cases} \tag{13}$$

$t\in\mathbb{T}$, where $S,I:\mathbb{T}\longrightarrow\mathbb{R}^{+}$, $\lambda,\gamma,\nu>0$, and

$$g(t)=-\lambda x(t)+\gamma-\frac{\nu x^{\sigma}(t)}{1-x^{\sigma}(t)},$$

$$x(t)=\frac{x(0)}{e_{\nu\oplus(\gamma-\lambda)}(t,0)\left(1+\frac{x(0)(\gamma-\lambda)}{\nu-\gamma+\lambda}\right)-\frac{x(0)(\gamma-\lambda)}{\nu-\gamma+\lambda}}.$$

Proof. Defining $x(t)=\frac{S(t)}{N(t)}$, we have

$$\begin{aligned} x^{\Delta}(t) &= \frac{S^{\Delta}(t)N(t)-N^{\Delta}(t)S(t)}{N(t)N^{\sigma}(t)} \\ &= -\lambda x(t)y^{\sigma}(t)-\nu x^{\sigma}(t)+\gamma x(t)y^{\sigma}(t) \\ &= (\gamma-\lambda)\left(\frac{\nu}{\lambda-\gamma}-x(t)\right)x^{\sigma}(t)-(\lambda-\gamma)x(t). \end{aligned} \tag{14}$$

Since $x(t)=x^{\sigma}(t)-\mu(t)x^{\Delta}(t)$, then

$$x^{\Delta}(t)=\frac{\gamma-\lambda}{1-\mu(t)(\lambda-\gamma)}\left(\frac{\nu}{\lambda-\gamma}-x(t)\right)x^{\sigma}(t)-\frac{\lambda-\gamma}{1-\mu(t)(\lambda-\gamma)}x^{\sigma}(t).$$

Applying the substitution $z=\frac{1}{x}$, it yields

$$z^{\Delta}(t)=\frac{-(\gamma-\lambda)+\nu}{1-\mu(t)(\lambda-\gamma)}z(t)+\frac{\gamma-\lambda}{1-\mu(t)(\lambda-\gamma)},$$

which has the solution

$$z(t)=e_{\alpha}(t,0)\,z(0)+\int_0^t e_{\alpha}(t,\sigma(s))\frac{\gamma-\lambda}{1-\mu(t)(\lambda-\gamma)}\Delta s$$

with $\alpha = \nu \ominus (\gamma - \lambda)$. The solution is equivalent to

$$\begin{aligned} z(t) &= e_{\alpha}(t,0)\, z(0) + \frac{\gamma - \lambda}{\nu - (\gamma - \lambda)} \int_0^t e_{\alpha}(t, \sigma(s))\, \alpha \Delta s \\ &= e_{\alpha}(t,0)\, z(0) + \frac{\gamma - \lambda}{\nu - \gamma + \lambda} \left(e_{\alpha}(t,0) - 1\right). \end{aligned}$$

Re-substituting $x = \frac{1}{z}$, one has

$$x(t) = \frac{x(0)}{e_{\alpha}(t,0)\left(1 + \frac{x(0)(\gamma-\lambda)}{\nu-\gamma+\lambda}\right) - \frac{x(0)(\gamma-\lambda)}{\nu-\gamma+\lambda}},$$

and

$$\begin{aligned} I^{\Delta}(t) &= \lambda I^{\sigma}(t)\, x(t) + \nu S^{\sigma}(t) - \gamma I^{\sigma}(t) \\ &= (\lambda x(t) - \gamma)\, I^{\sigma}(t) + \nu S^{\sigma}(t). \end{aligned}$$

From $x(t) = \frac{S(t)}{S(t)+I(t)}$ it follows that $S(t) = \frac{I(t)x(t)}{1-x(t)}$ and

$$\begin{aligned} I^{\Delta}(t) &= (\lambda x(t) - \gamma)\, I^{\sigma}(t) + \frac{\nu x^{\sigma}(t)}{1 - x^{\sigma}(t)} I^{\sigma}(t) \\ &= -\left(-\lambda x(t) + \gamma - \frac{\nu x^{\sigma}(t)}{1 - x^{\sigma}(t)}\right) I^{\sigma}(t) \\ &= -g(t)\, I^{\sigma}(t), \end{aligned}$$

where $g(t) = -\lambda x(t) + \gamma - \frac{\nu x^{\sigma}(t)}{1-x^{\sigma}(t)}$. The solution is equivalently expressed as in (13). □

As in Example 3.1, a simple application is obtained from our Theorem 4.1 when we restrict ourselves to the continuous case.

Example 4.1 If $\mathbb{T} = \mathbb{R}$, then system (12) reduces to

$$\begin{cases} S'(t) = -\lambda \frac{S(t)I(t)}{S(t)+I(t)} - \nu S(t), \\ I'(t) = \lambda \frac{S(t)I(t)}{S(t)+I(t)} + \nu S(t) - \gamma I(t), \\ R'(t) = \gamma I(t), \end{cases} \tag{15}$$

and, by Theorem 4.1, the solution to system (15) is

$$\begin{cases} S(t) = \frac{x(t)I(t)}{1-x(t)}, \\ I(t) = e^{\int_0^t g(s)ds} I(0), \end{cases}$$

where

$$g(t) = -\lambda x(t) + \gamma - \frac{\nu x(t)}{1 - x(t)}$$

and

$$x(t) = \frac{x(0)}{e^{\int_0^t \nu-(\gamma-\lambda)ds}\left(1 + \frac{x(0)(\gamma-\lambda)}{\nu-\gamma+\lambda}\right) - \frac{x(0)(\gamma-\lambda)}{\nu-\gamma+\lambda}}$$

with

$$x(0) = \frac{S(0)}{S(0) + I(0)}.$$

8.5 ACKNOWLEDGMENTS

Torres is supported by the R&D Unit CIDMA and funded by *Fundação para a Ciência e a Tecnologia*, I.P. (FCT) under grants UIDB/04106/2020 (https://doi.org/10.54499/UIDB/04106/2020) and UIDP/04106/2020 (https://doi.org/10.54499/UIDP/04106/2020), and within the project 2022.03091.PTDC, "Mathematical Modelling of Multi-scale Control Systems: applications to human diseases" (CoSysM3, https://doi.org/10.54499/2022.03091.PTDC), financially supported by national funds (OE) through FCT/MCTES.

REFERENCES

1. P. Agarwal, J. J. Nieto, M. Ruzhansky and D. F. M. Torres, *Analysis of infectious disease problems (Covid-19) and their global impact*, Springer, Singapore, 2021.
2. P. Agarwal, J. J. Nieto and D. F. M. Torres, *Mathematical analysis of infectious diseases*, Academic Press, London, UK, 2022.
3. I. Area, F. Ndaïrou, J. J. Nieto, C. J. Silva and D. F. M. Torres, Ebola model and optimal control with vaccination constraints, *J. Ind. Manag. Optim.* 14 (2018), no. 2, 427–446.
4. B. Aulbach and S. Hilger, A unified approach to continuous and discrete dynamics, *Colloq. Math. Soc. János Bolyai* 53 (1990), 37–56.

5. M. Bohner and A. Peterson, *Dynamic equations on time scales*, Birkhäuser Boston, Boston, MA, 2001.
6. M. Bohner and A. Peterson, *Advances in dynamic equations on time scales*, Birkhäuser Boston, Boston, MA, 2003.
7. M. Bohner and S. H. Streipert, The SIS-model on time scales, *Pliska Stud. Math.* 26 (2016), 11–28.
8. M. Bohner, S. Streipert and D. F. M. Torres, Exact solution to a dynamic SIR model, *Nonlinear Anal. Hybrid Syst.* 32 (2019), 228–238.
9. V. S. Borkar and D. Manjunath, Revisiting SIR in the age of COVID-19: explicit solutions and control problems, *SIAM J. Control Optim.* 60 (2022), no. 2, S370–S395.
10. W. Gleissner, The spread of epidemics, *Appl. Math. Comput.* 27 (1988), no. 2, 167–171.
11. H. W. Hethcote, The mathematics of infectious diseases, *SIAM Rev.* 42 (2000), no. 4, 599–653.
12. S. Hilger, Analysis on measure chains—a unified approach to continuous and discrete calculus, *Results Math.* 18 (1990), no. 1–2, 18–56.
13. W. O. Kermack and A. G. McKendrick, A contribution to the mathematical theory of epidemics, *Proc. Roy. Soc. Lond. A* 115 (1927), 700–721.
14. A. Rachah and D. F. M. Torres, Analysis, simulation and optimal control of a SEIR model for Ebola virus with demographic effects, *Commun. Fac. Sci. Univ. Ank. Sér. A1 Math. Stat.* 67 (2018), no. 1, 179–197.
15. C. J. Silva, G. Cantin, C. Cruz, R. Fonseca-Pinto, R. Passadouro, E. Soares dos Santos and D. F. M. Torres, Complex network model for COVID-19: human behavior, pseudo-periodic solutions and multiple epidemic waves, *J. Math. Anal. Appl.* 514 (2022), no. 2, Paper No. 125171, 25 pp.

CHAPTER 9

Stability Criteria of Nonlinear Generalized Proportional Fractional Delayed Systems

Hanaa Zitane and Delfim F. M. Torres

9.1 INTRODUCTION

With the rapid development and progress of fractional calculus, there has been an increasing interest in the investigation of time delay fractional systems since they allow the description of systems in which the rate of change depends not only on the present and delayed state but also on the whole past memory. These systems have diverse applications in science and engineering [10, 17]. During the past decades, the finite time stability of fractional delay systems has been widely studied [3, 8, 21, 23]. The main approaches to analyze this kind of stability include the fractional Halanay inequality [19], Hölder's inequality [7], Grönwall's inequality [16, 22], delayed Mittag-Leffler type matrix functions [18], and the weighted integral inequality [12]. Here, we consider the finite time stability problem of a class of nonlinear generalized proportional fractional systems (GPFSs) with time delays:

$$\begin{cases} {}^{C}D_0^{\alpha,\mu} y(t) = \exp\left(\frac{\mu-1}{\mu}t\right)(Ay(t) + By(t-\tau) \\ \qquad\qquad + f(t, y(t), y(t-\tau))), & t \in [0, T], \\ y(t) = \varphi(t), & t \in [-\tau, 0], \end{cases} \tag{1}$$

DOI: 10.1201/9781003530602-9

where A and B are constant $n \times n$ matrices, $T > 0$ is a real number, $\tau > 0$ is a time delay, $\varphi(\cdot)$ is a continuous function on $[-\tau, 0]$, $f : [0, T] \times \mathbb{R}^n \times \mathbb{R}^n \longrightarrow \mathbb{R}^n$ is a given nonlinear continuous function with $f(t, 0, 0) = 0$, and ${}^{C}D_0^{\alpha,\mu}$ is the Caputo generalized proportional fractional derivative (GPFD) as suggested by Jarad et al. [11]. This type of fractional operator has different interesting advantages: it preserves the semigroup property, which is essential for solving certain complicated fractional systems; it possesses a nonlocal character; and it tends to the original function and its derivative upon a limiting process [2]. Moreover, it provides an undeviating generalization to the existing Caputo fractional derivative. For further information about the GPFD, and differential systems evolving under GPFDs and their applications, we refer the reader to [2, 6, 9, 14] and references therein.

The stability of GPFSs, with and without delay, has been seldom investigated [1, 5, 15]. In [1], the second method of Lyapunov is used to analyze the exponential and the Mittag-Leffler stability of fractional order systems evolving GPFDs. Also, in [15], sufficient conditions that ensure the stability of nonlinear hybrid fractional integro-differential equations with Dirichlet boundary conditions are provided in the sense of Ulam–Hyers–Rassias. For the stability of time delay systems, it has been tackled only by means of the Lyapunov approach, where quadratic Lyapunov functions and their GPFD are applied to investigate the exponential and asymptotic stability of a scalar nonlinear integro-differential GPFS with bounded delays [5]. This serves as inspiration for our current work.

We study the finite time stability problem for a class of nonlinear GPFSs with time delays. The main contributions of this work can be summarized as follows:

(i) A simple sufficient condition is established for the stability of system (1) over a finite time interval in the homogeneous case by means of the Bellman–Grönwall approach.

(ii) An explicit delay-dependent criterion is obtained to ensure the finite-time stability of nonlinear time-delay GPFSs by using Hölder's and Jensen's inequalities.

(ii) Compared to the existing work [5], here we consider a more general and multi-dimensional class of GPFSs with delays. Furthermore, the stability is investigated based on non-Lyapunov approaches.

9.2 PRELIMINARIES

Throughout the text, $\|\cdot\|$ stands for the Euclidean norm and the spectral norm for a vector and a matrix, respectively. Also, the norm function with the initial condition is given by

$$\|\varphi\|_C = \sup_{t\in[-\tau,0]} \|\varphi(t)\|.$$

We begin by recalling the definitions of generalized proportional fractional operators that are employed throughout the manuscript. Further information about generalized proportional calculus can be found in [2, 11, 14].

Definition 1 (See [11]) Let $\alpha > 0$, $\mu \in (0,1]$ and $h \in C^1([0,T],\mathbb{R})$. The left generalized proportional fractional integral of function h is defined by

$${}^{C}I_0^{\alpha,\mu} h(t) = \frac{1}{\mu^{\alpha}\Gamma(\alpha)} \int_0^t \exp\left(\frac{\mu-1}{\mu}(t-s)\right)(t-s)^{\alpha-1} h(s)\, ds,$$

where $\Gamma(\cdot)$ is Euler's Gamma function [20].

Definition 2 (See [11]) Let $\alpha \in (0,1)$ and $\mu \in (0,1]$. The left generalized proportional fractional order derivative of function $v \in C^1([0,T],\mathbb{R})$ is given by

$${}^{C}D_0^{\alpha,\mu} h(t) = \frac{1}{\mu^{1-\alpha}\Gamma(1-\alpha)} \int_0^t \exp\left(\frac{\mu-1}{\mu}(t-s)\right)(t-s)^{-\alpha} D^{1,\mu} h(s)\, ds,$$

where $D^{1,\mu}h(t) = (1-\mu)h(t) + \mu h'(t)$.

Remark 1 When $\mu = 1$, the GPFDs reduce to the Caputo fractional derivatives [20].

The existence of a solution to system (1) is given by the following lemma.

Lemma 1 (See [2]) *A function $y\colon [-\tau,T] \longrightarrow \mathbb{R}^n$ is a mild solution of system (1) if and only if it satisfies*

$$\begin{cases} y(t) = \varphi(0)\exp\left(\frac{\mu-1}{\mu}t\right) + {}^{C}I_0^{\alpha,\mu} \exp\left(\frac{\mu-1}{\mu}s\right)[Ay(t) \\ \qquad + By(t-\tau) + f(t,y(t),y(t-\tau))], & t\in[0,T], \\ y(t) = \varphi(t), & t\in[-\tau,0]. \end{cases} \tag{2}$$

Moreover, if for any functions $y, z\colon [-\tau, T] \longrightarrow \mathbb{R}^n$ *there exists a constant* $L_f > 0$ *such that*

$$\begin{aligned}&\|f(t, y(t), y(t-\tau)) - f(t, z(t), z(t-\tau))\| \\ &\quad \le L_f(\|y(t) - z(t)\| + \|y(t-\tau) - z(t-\tau)\|), \quad t \in [0, T], \end{aligned} \tag{3}$$

then system (1) has a unique mild solution.

Lemma 2 (Generalized Proportional Fractional Grönwall Inequality [2]) *Suppose* $\beta > 0$, $\mu > 0$, g *and* J *are nonnegative and locally integrable functions on* $[0, t_f)$ $(t_f \le \infty)$ *and* h *is a nonnegative, nondecreasing, and continuous function on* $[0, t_f)$ *satisfying* $h(t) \le L$, *where* L *is a constant. In addition, if*

$$g(t) \le J(t) + h(t) \int_0^t \exp\left(\frac{\mu - 1}{\mu}(t-s)\right)(t-s)^{\alpha-1} g(s)\, ds,$$

then

$$g(t) \le J(t) + \int_0^t \left[\sum_{n=1}^{+\infty} \frac{(h(t)\Gamma(\alpha))^n}{\Gamma(n\alpha)} \exp\left(\frac{\mu-1}{\mu}(t-s)\right)(t-s)^{n\alpha-1} J(s)\right] ds,$$
$$t \in [0, t_f].$$

Moreover, if function J *is non-decreasing on* $[0, t_f)$, *then*

$$g(t) \le f(t) E_\alpha(h(t)\Gamma(\alpha)t^\alpha), \quad t \in [0, t_f], \tag{4}$$

where $E_\alpha(\cdot)$ *is the Mittag-Leffler function of one parameter [20] given by*

$$E_\alpha(x) = \sum_{k=0}^{+\infty} \frac{x^k}{\Gamma(\alpha k + 1)}, \quad x \in \mathbb{C}.$$

We conclude this section with the definition of finite-time stability of system (1).

Definition 3 For given positive numbers c_1 and c_2, $c_1 \le c_2$, system (1) is finite-time stable with respect to $\{c_1, c_2, T\}$ if

$$\|\varphi\|_C \le c_1 \Rightarrow \|y(t)\| < c_2, \quad t \in [0, T]. \tag{5}$$

Remark 2 If we let $\mu = 1$ in system (1), then one obtains the finite-time stability definition of Caputo fractional systems with delays [16].

9.3 FINITE TIME STABILITY OF DELAYED GPFS

In this section, we begin by studying the stability of the GPFS (1) over a finite time interval in the homogeneous case $f = 0$.

Theorem 1 *System (1) is finite-time stable with respect to* $\{c_1, c_2, T\}$, $c_1 \leq c_2$, *iff satisfies condition (3) and*

$$\left(1 + \frac{(\|A\| + \|B\|)\, t^{\alpha}}{\mu^{\alpha}\Gamma(\alpha+1)}\right) E_{\alpha}\left(\frac{(\|A\| + \|B\|)}{\mu^{\alpha}} t^{\alpha}\right) \leq \frac{c_2}{c_1}, \tag{6}$$

for all $t \in [0, T]$.

Proof. From Lemma 1, the solution of system (1) can be written as

$$\begin{aligned} y(t) =& \varphi(0) \exp\left(\frac{\mu-1}{\mu} t\right) + \frac{1}{\mu^{\alpha}\Gamma(\alpha)} \int_0^t \exp\left(\frac{\mu-1}{\mu}(t-s)\right) \\ &\times \exp\left(\frac{\mu-1}{\mu} s\right) (t-s)^{\alpha-1} \left[A(s)y(s) + B(s)y(s-\tau)\right] ds \end{aligned}$$

for all $t \in [0, T]$. Taking into account condition (3) and using $\exp\left(\frac{\mu-1}{\mu} s\right) \leq 1$ for all $s \in [0, t]$, we obtain

$$\begin{aligned} \|y(t)\| \leq& \|\varphi(0)\| + \frac{1}{\mu^{\alpha}\Gamma(\alpha)} \int_0^t \exp\left(\frac{\mu-1}{\mu}(t-s)\right)(t-s)^{\alpha-1} \\ & \left[\|A\|\|y(s)\| + \|B\|\|y(s-\tau)\|\right] ds. \end{aligned} \tag{7}$$

For all $t \in [0, T]$, let us consider $z(t) = \sup_{0 \leq \theta \leq t} \|y(\theta)\|$. Then, one has

$$\|y(s-\tau)\| \leq z(s) + \|\varphi\|_C, \quad s \in [0, t]. \tag{8}$$

Combining relation (8) and inequality (7) implies that

$$\begin{aligned} \|y(t)\| \leq& \|\varphi(0)\| + \frac{1}{\mu^{\alpha}\Gamma(\alpha)} \int_0^t \exp\left(\frac{\mu-1}{\mu}(t-s)\right)(t-s)^{\alpha-1} \\ & (\|A\| + \|B\|)\left(z(s) + \|\varphi\|_C\right) ds. \end{aligned} \tag{9}$$

It follows that

$$\|y(t)\| \leq \|\varphi\|_C + \frac{(\|A\| + \|B\|)\, t^\alpha}{\mu^\alpha \Gamma(\alpha+1)} \|\varphi\|_C + \frac{(\|A\| + \|B\|)}{\mu^\alpha \Gamma(\alpha)} \int_0^t \exp\left(\frac{\mu-1}{\mu}(t-s)\right)(t-s)^{\alpha-1} z(s)\, \mathrm{d}s \tag{10}$$

and, using the change of variable $x = t - s$, we get

$$\|y(t)\| \leq \|\varphi\|_C + \frac{(\|A\| + \|B\|)\, t^\alpha}{\mu^\alpha \Gamma(\alpha+1)} \|\varphi\|_C + \frac{(\|A\| + \|B\|)}{\mu^\alpha \Gamma(\alpha)} \int_0^t \exp\left(\frac{\mu-1}{\mu}x\right) x^{\alpha-1} z(t-x)\, \mathrm{d}x. \tag{11}$$

Moreover, by taking $t = \theta$ in (11) with $\theta \in [0, t]$ and using $\theta^\alpha \leq t^\alpha$, we obtain

$$\|y(\theta)\| \leq \left[1 + \frac{(\|A\| + \|B\|)\, t^\alpha}{\mu^\alpha \Gamma(\alpha+1)}\right] \|\varphi\|_C + \frac{(\|A\| + \|B\|)}{\mu^\alpha \Gamma(\alpha)} \int_0^\theta \exp\left(\frac{\mu-1}{\mu}x\right) x^{\alpha-1} z(\theta-x)\, \mathrm{d}x. \tag{12}$$

Since function z is nonnegative, it implies that

$$\int_0^t \exp\left(\frac{\mu-1}{\mu}x\right) x^{\alpha-1} z(t-x)\, dx$$

is an increasing function with respect to $t \geq 0$, which yields

$$\|y(\theta)\| \leq \left[1 + \frac{(\|A\| + \|B\|)\, t^\alpha}{\mu^\alpha \Gamma(\alpha+1)}\right] \|\varphi\|_C + \frac{(\|A\| + \|B\|)}{\mu^\alpha \Gamma(\alpha)} \int_0^t \exp\left(\frac{\mu-1}{\mu}x\right) x^{\alpha-1} z(t-x)\, \mathrm{d}x.$$

Hence,

$$z(t) \leq \left[1 + \frac{(\|A\| + \|B\|)\, t^\alpha}{\mu^\alpha \Gamma(\alpha+1)}\right] \|\varphi\|_C + \frac{(\|A\| + \|B\|)}{\mu^\alpha \Gamma(\alpha)} \int_0^t \exp\left(\frac{\mu-1}{\mu}(t-s)\right)(t-s)^{\alpha-1} z(s)\, ds.$$

Denote $J(t) = \left[1 + \frac{(\|A\|+\|B\|)t^{\alpha}}{\mu^{\alpha}\Gamma(\alpha+1)}\right] \|\varphi\|_C$, which is a nondecreasing function. By using Lemma 2 with

$$h(t) = \frac{\|A\| + \|B\|}{\mu^{\alpha}\Gamma(\alpha)},$$

one obtains

$$\|y(t)\| \leq \|\varphi\|_C \left[1 + \frac{(\|A\| + \|B\|)\, t^{\alpha}}{\mu^{\alpha}\Gamma(\alpha+1)}\right] E_{\alpha}\left(\frac{(\|A\| + \|B\|)}{\mu^{\alpha}} t^{\alpha}\right).$$

Therefore, by virtue of $\|\varphi\|_C \leq c_1$ and (6), it follows that $\|y(t)\| < c_2$ for all $t \in [0, T]$. □

Remark 3 If we let $\mu = 1$ in system (1) and Theorem 1, then one retrieves the condition

$$\left(1 + \frac{(\|A\| + \|B\|)\, t^{\alpha}}{\Gamma(\alpha+1)}\right) E_{\alpha}\left((\|A\| + \|B\|)\, t^{\alpha}\right) \leq \frac{c_2}{c_1}, \quad \forall t \in [0, T],$$

for the finite-time stability of the Caputo fractional-order time-delay system, established in [16].

Now, we shall characterize a sufficient condition to ensure the finite-time stability of the GPFS (1).

Theorem 2 *System (1) is finite-time stable with respect to* $\{c_1, c_2, T\}$, $c_1 \leq c_2$, *if f satisfies condition (3) and*

$$\sqrt[r]{\frac{3^{\frac{1}{\alpha}} r + \left(3^{\frac{1}{\alpha}}\psi + r\phi + \psi\phi\right)\exp\left((\psi + r)\, t\right)}{r + \psi}} \leq \frac{c_2}{c_1}, \quad t \in [0, T], \tag{13}$$

where

$$\psi = \frac{3^{\frac{1}{\alpha}}\left(\left(\|A\| + L_f\right)^r + \left(\|B\| + L_f\right)^r \exp(-r\tau)\right)\omega^r}{\mu^k\Gamma^r(\alpha)}, \tag{14}$$

$$\phi = \frac{3^{\frac{1}{\alpha}}\left(\|B\| + L_f\right)^r \left(1 - \exp(-\tau q)\right)}{r\mu^k\Gamma^r(\alpha)}\omega^r, \tag{15}$$

$k = 1 + \alpha$, $r = 1 + \frac{1}{\alpha}$, *and* $\omega = \left(\frac{\Gamma(\alpha^2)}{k^{\alpha^2}}\right)^{\frac{1}{k}}$.

Proof. According to Lemma 1, the mild solution of system (1) is given by

$$y(t) = \varphi(0)\exp\left(\frac{\mu-1}{\mu}t\right) + \frac{1}{\mu^{\alpha}\Gamma(\alpha)}\int_0^t \exp\left(\frac{\mu-1}{\mu}(t-s)\right)$$
$$\exp\left(\frac{\mu-1}{\mu}s\right)(t-s)^{\alpha-1}$$
$$\times\left[A(s)y(s) + B(s)y(s-\tau) + f(s, y(s), y(s-\tau))\right] ds.$$

By virtue of condition (3) with $f(s,0,0)=0$, it follows that

$$\|y(t)\| \le \|\varphi(0)\| + \frac{1}{\mu^{\alpha}\Gamma(\alpha)}\int_0^t (t-s)^{\alpha-1}$$
$$\left[\left(\|A\| + L_f\right)\|y(s)\| + \left(\|B\| + L_f\right)\|y(s-\tau)\|\right] ds,$$

which implies that

$$\|y(t)\| \le \|\varphi(0)\| + \frac{\|A\| + L_f}{\mu^{\alpha}\Gamma(\alpha)}\int_0^t (t-s)^{\alpha-1}\exp(s)\exp(-s)\|y(s)\|\, ds$$
$$+ \frac{\|B\| + L_f}{\mu^{\alpha}\Gamma(\alpha)}\int_0^t (t-s)^{\alpha-1}\exp(s)\exp(-s)\|y(s-\tau)\|\, ds.$$

By Hölder's inequality [4], we get

$$\|y(t)\| \le \|\varphi(0)\| + \frac{\|A\| + L_f}{\mu^{\alpha}\Gamma(\alpha)}\left(\int_0^t (t-s)^{k(\alpha-1)}\exp(ks)\, ds\right)^{\frac{1}{k}}$$
$$\times\left(\int_0^t \exp(-rs)\|y(s)\|^r ds\right)^{\frac{1}{r}}$$
$$+ \frac{\|B\| + L_f}{\mu^{\alpha}\Gamma(\alpha)}\left(\int_0^t (t-s)^{k(\alpha-1)}\exp(ks)\, ds\right)^{\frac{1}{k}}$$
$$\times\left(\int_0^t \exp(-rs)\|y(s-\tau)\|^r ds\right)^{\frac{1}{r}} \tag{16}$$

with $k = 1 + \alpha$ and $r = 1 + \frac{1}{\alpha}$. Also, one has

$$\int_0^t (t-s)^{k(\alpha-1)} \exp(ks)\, ds \leq \frac{\exp(kt)\,\Gamma(k(\alpha-1)+1)}{k^{k(\alpha-1)+1}} = \frac{\Gamma(\alpha^2)\exp(kt)}{k^{\alpha^2}}. \tag{17}$$

Combining inequalities (16) and (17) yields

$$\|y(t)\| \leq \|\varphi(0)\| + \frac{(\|A\| + L_f)\,\omega\,\exp(t)}{\mu^{\alpha}\Gamma(\alpha)}\left(\int_0^t \exp(-rs)\,\|y(s)\|^r ds\right)^{\frac{1}{r}} + \frac{(\|B\| + L_f)\,\omega\,\exp(t)}{\mu^{\alpha}\Gamma(\alpha)}\left(\int_0^t \exp(-rs)\,\|y(s-\tau)\|^r ds\right)^{\frac{1}{r}}$$

with $\omega = \left(\frac{\Gamma(\alpha^2)}{k^{\alpha^2}}\right)^{\frac{1}{k}}$. This implies that

$$\|y(t)\| \leq \|\varphi(0)\| + \frac{(\|A\| + L_f)\,\omega\,\exp(t)}{\mu^{\alpha}\Gamma(\alpha)}\left(\int_0^t \exp(-rs)\,\|y(s)\|^r ds\right)^{\frac{1}{r}} + \frac{(\|B\| + L_f)\,\omega\,\exp(t)}{\mu^{\alpha}\Gamma(\alpha)}\left(\int_{-\tau}^t \exp(-r(s+\tau))\,\|y(s)\|^r ds\right)^{\frac{1}{r}}. \tag{18}$$

Now, by applying Jensen's inequality [13] to inequality (18), we obtain

$$\|y(t)\|^r \leq 3^{\frac{1}{\alpha}}\left[\|\varphi(0)\|^r + \frac{(\|A\| + L_f)^r\omega^r\exp(rt)}{\mu^k\Gamma^r(\alpha)}\left(\int_0^t \exp(-rs)\,\|y(s)\|^r \mathrm{ds}\right) + \frac{(\|B\| + L_f)^r\omega^r\exp(rt)}{\mu^k\Gamma^r(\alpha)}\left(\int_{-\tau}^t \exp(-r(s+\tau))\,\|y(s)\|^r \mathrm{ds}\right)\right].$$

Then,

$$\|y(t)\|^r \leq 3^{\frac{1}{\alpha}}\|\varphi(0)\|^r + \psi\,\exp(rt)\int_0^t \exp(-rs)\,\|y(s)\|^r ds + \frac{3^{\frac{1}{\alpha}}(\|B\| + L_f)^r\omega^r e^{q(t-\tau)}}{\mu^k\Gamma^r(\alpha)}\int_{-\tau}^0 \exp(-rs)\,\|y(s)\|^r ds,$$

where ψ is defined in (14), which implies that

$$\|y(t)\|^r \leq 3^{\frac{1}{\alpha}} \|\varphi\|_C^r + \exp(rt) \|\varphi\|_C^r \phi + \psi \exp(rt) \int_0^t \exp(-rs) \|y(s)\|^r ds$$

with ϕ given by (15). It follows that

$$\exp(-rt) \|y(t)\|^r \leq \left(3^{\frac{1}{\alpha}} \exp(-rs) + \phi\right) \|\varphi\|_C^r + \psi \int_0^t \exp(-rs) \|y(s)\|^r ds.$$

Therefore, by applying Grönwall's inequality, one gets

$$\exp(-rt) \|y(t)\|^r \leq \left(3^{\frac{1}{\alpha}} \exp(-rs) + \phi\right) \|\varphi\|_C^r + \int_0^t \psi \left(3^{\frac{1}{\alpha}} \exp(-rs) + \phi\right) \|\varphi\|_C^r \exp(\psi(t-s))\, ds,$$

and

$$\|y(t)\|^r \leq \frac{3^{\frac{1}{\alpha}} r + \left(3^{\frac{1}{\alpha}} \psi + r\phi + \psi\phi\right) \exp((\psi + r)t)}{r + \psi} \|\varphi\|_C^r. \tag{19}$$

Hence, by combining condition (13) and inequality (19), we conclude with the finite-time stability of the GPFS (1). □

9.4 APPLICATIONS

In this section, two numerical examples are provided to illustrate the effectiveness of the proposed results.

Example 1 Let $\alpha = 0.2$, $\mu = 0.8$, $\tau = 0.2$, $T = 5$, $\omega(t) = (0.7 \quad 0.7)^T$, and consider the homogeneous GPFS with time delay given by

$$\begin{cases} {}^C D_0^{0.2,0.8} y(t) = \exp(0.25t) \left(\begin{pmatrix} 0.1 & 0 \\ 0 & -0.2 \end{pmatrix} \begin{pmatrix} y_1(t) \\ y_2(t) \end{pmatrix} \right. \\ \qquad \left. + \begin{pmatrix} -0.3 & 0 \\ 0 & -0.2 \end{pmatrix} \begin{pmatrix} y_1(t-0.2) \\ y_2(t-0.2) \end{pmatrix} \right), & t \in [0,5], \\ y(t) = (0.7 \quad 0.7)^T, & t \in [-0.2, 0]. \end{cases} \tag{20}$$

The aim is to verify condition (6) with respect to $\{c_1 = 1, c_2 = 8, T = 5\}$. We have

$$\|\omega\|_C = 0.9899 < c_1,$$

$$\|A\| = 0.2 \text{ and } \|B\| = 0.3.$$

Therefore, condition (6) holds over the interval $[0, 5]$. Then, from Theorem 1, we deduce that system (20) is finite time stable with respect to $\{c_1 = 1, c_2 = 8, T = 5\}$.

Example 2 Let $\alpha = 0.4$, $\mu = 0.5$, $\tau = 0.3$, $T = 3$, $\omega(t) = (0.5 \tanh(t) \quad 0)^T$, and consider the two-state fractional delayed nonlinear GPFS

$$\begin{cases} {}^{C}D_0^{0.4,0.5} y(t) = \exp(-t)\left(Ay(t) + By(t-0.3)\right. \\ \qquad\qquad \left.+0.02(\tanh(y(t)) + \tanh(y(t-0.3)))\right), & t \in [0, 3], \\ y(t) = (0.5 \tanh(t) \quad 0)^T, & t \in [-0.3, 0], \end{cases} \tag{21}$$

with

$$A = \begin{pmatrix} -0.2 & 0 \\ 0 & 0.3 \end{pmatrix}, \quad B = \begin{pmatrix} 0 & 0 \\ 0.3 & 0.4 \end{pmatrix}$$

and

$$f(t, y(t), y(t-\tau)) = 0.02\left(\tanh(y(t)) + \tanh(y(t-\tau))\right).$$

One needs to check condition (6) with respect to $\{c_1 = 0.4, c_2 = 4, T = 3\}$. The nonlinear term f satisfies the generalized Lipschitz condition (3) with $L_f = 0.02$. Also, one has

$$\|\omega\|_C = 0.2266 < c_1,$$

$$\|A\| = 0.3 \text{ and } \|B\| = 0.5.$$

Then, condition (6) holds over $[0, 3]$. Consequently, from Theorem 1, we deduce the finite-time stability of system (21) with respect to $\{c_1 = 0.4, c_2 = 4, T = 3\}$.

9.5 CONCLUSION

We have dealt with the problem of finite-time stability for generalized proportional fractional systems (GPFSs) with time delays. A sufficient condition, which ensures finite-time stability for homogeneous delayed GPFSs, based on the generalized Grönwall inequality, was obtained. Also, an explicit delay-dependent criterion that allows stability over a finite time interval for a class of nonlinear GPFSs is provided. The effectiveness of the proposed criteria has been illustrated by numerical examples.

9.6 FUNDING

Zitane and Torres are supported by The Center for Research and Development in Mathematics and Applications (CIDMA) through the Portuguese Foundation for Science and Technology (FCT – Fundação para a Ciência e a Tecnologia), projects UIDB/04106/2020 (https://doi.org/10.54499/UIDB/04106/2020) and UIDP/04106/2020 (https://doi.org/10.54499/UIDP/04106/2020).

REFERENCES

1. R. Almeida, R. Agarwal, S. Hristova and D. O'Regan, Quadratic Lyapunov functions for stability of the generalized proportional fractional differential equations with applications to neural networks, *Axioms* **10** (2022), no. 4, Art. 322, 14 pp.
2. J. Alzabut, T. Abdeljawad, F. Jarad and W. Sudsutad, A Gronwall inequality via the generalized proportional fractional derivative with applications, *J. Inequal. Appl.* **2019** (2019), Paper No. 101, 12 pp.
3. G. Arthi and N. Brindha, On finite-time stability of nonlinear fractional-order systems with impulses and multi-state time delays, *Results in Control Optim.* **2** (2021), Paper No. 100010, 7 pp.
4. E. F. Beckenbach and R. Bellman, *Inequalities*, Springer-Verlag, Inc., New York, 1983.
5. M. Bohner and S. Hristova, Stability for generalized Caputo proportional fractional delay integro-differential equations, *Bound. Value Probl.* **2022** (2022), Paper No. 14, 15 pp.

6. D. Boucenna, D. Baleanu, A. Ben Makhlouf and A. M. Nagyf, Analysis and numerical solution of the generalized proportional fractional Cauchy problem, *Appl. Numer. Math.* **167** (2021), 173–186.
7. F. Du and B. Jia , Finite-time stability of a class of nonlinear fractional delay difference systems, *Appl. Math. Lett.* **98** (2019), 233–239.
8. F. Du and J. G. Lu, New criterion for finite-time stability of fractional delay systems, *Appl. Math. Lett.* **104** (2020), Paper No. 106248, 7 pp.
9. M. Farman, A. Shehzad, A. Akgül, D. Baleanu and M. De la Sen, Modeling and analysis of a measles epidemic model with the constant proportional Caputo operator, *Symmetry* **15** (2023), no. 2, Paper No. 468, 22 pp.
10. Y. Gao and N. Li, Fractional order PD control of the Hopf bifurcation of HBV viral systems with multiple time delays, *Alex. Eng. J.* **83** (2023), 18 pp.
11. F. Jarad, T. Abdeljawad and J. Alzabut, Generalized fractional derivatives generated by a class of local proportional derivatives, *Eur. Phys. J. Spec. Top.* **226** (2017), no. 16, 3457–3471.
12. Y. Jia, C. Lin and B. Chen, Finite-time stability of singular time-delay systems based on a new weighted integral inequality, *J. Frank. Inst.* **360** (2023), no. 7, 5092–5103.
13. M. Kuczma, *An introduction to the theory of functional equations and inequalities*, Birkhäuser Verlag, Basel, 2009.
14. Z. Laadjal, T. Abdeljawad and F. Jarad, On existence-uniqueness results for proportional fractional differential equations and incomplete gamma functions, *Adv. Difference Equ.* **2020** (2020), Paper No. 641, 16 pp.
15. Z. Laadjal and F. Jarad, Existence, uniqueness and stability of solutions for generalized proportional fractional hybrid integro-differential equations with Dirichlet boundary conditions, *AIMS Math.* **8** (2023), no. 1, 1172–1194.
16. M. P. Lazarević and A. M. Spasić, Finite-time stability analysis of fractional order time-delay systems: Gronwall's approach, *Math. Comput. Modelling.* **49** (2009), no. 3-4, 475–481.
17. Q. Li, D. Sun, H. Liu and W. Zhao, Stability and bifurcation control of a delayed fractional eco-epidemiological system with saturated incidence, *Results in Phys.* **54** (2023), Paper No. 107019, 13 pp.
18. M. Li and J. Wang, Finite time stability of fractional delay differential equations, *Appl. Math. Lett.* **64** (2017), 170–176.

19. T. T. H. Nguyena, N. T. Nguyen and M. N. Tran, Global fractional Halanay inequalities approach to finite-time stability of nonlinear fractional order delay systems, *J. Math. Anal. Appl.* **525** (2023), no. 1, Paper No. 127145, 16 pp.
20. J. Sabatier, O. P. Agrawal and J. A. Machado, *Advances in Fractional Calculus*, Springer Netherlands, Dordrecht, 2007.
21. X. Yang, X. Wu and Q. Song, Caputo-Wirtinger integral inequality and its application to stability analysis of fractional-order systems with mixed time-varying delays, *Appl. Math. Comput.* **460** (2024), Paper No. 128303, 12 pp.
22. Z. Yang, J. Zhang, J. Hu and J. Mei, New results on finite-time stability for fractional-order neural networks with proportional delay, *Neurocomputing* **442** (2021), 327–336.
23. H. Zitane and D. F. M. Torres, Finite time stability of tempered fractional systems with time delays, *Chaos Solitons Fractals* **177** (2023), Art. 114265, 10 pp.

CHAPTER 10

On the Hamburger-Oberhettinger-Soni Modular Relations

K. Chakraborty, S. Kanemitsu, and L.-W. Yu

10.1 THE FOURIER-BESSEL EXPANSION AND ITS VARIANT

In this chapter, we shall elucidate the results—mainly summation formulas—of Hamburger [Hamburger (1922)], [Oberhettinger and Soni (1972)] in the light of a simpler version (Lemma 1) of the principle established in [Liu *et al.* (2023)]. Our main result on summation formula reads

Theorem 1 *The Poisson summation formula (1.34) à la Hamburger is a manifestation of the principle, Lemma 1, with the Fourier transform pair. The Soni-Oberhettinger formula of the form*

$$\sum_{n=1}^{\infty} a_n f(\lambda_n) = P(\cdot) + \sum_{n=1}^{\infty} b_n \hat{f}(\mu_n), \hat{f}(y) = \int_0^{\infty} f(x) K(xy)\, dx \qquad (1.1)$$

is a consequence of Lemma 1 with $F(z)$ replaced by the Mellin transform (2.1).

In the rest of this section, we provide basic material along with elucidating [Soni (1966)] as an example of the partial fraction expansion (a variant of the Fourier-Bessel expansion) in Theorem 2. We illustrate this with an example [Soni (1966), (5)]:

$$2\sum_{n=1}^{\infty} d(n)\left(K_0\left(4\pi e\sqrt{xn}\right) + K_0\left(4\pi \bar{e}\sqrt{xn}\right)\right) = \sigma(x) = \mathrm{P}_K(x) + \frac{x}{\pi}\sum_{n=1}^{\infty} \frac{d(n)}{n^2 + x^2}, \qquad (1.2)$$

DOI: 10.1201/9781003530602-10

where

$$\epsilon = e^{\frac{\pi}{4}i}, \quad \bar{\epsilon} = e^{\frac{\pi}{4}i}, \tag{1.3}$$

is a partial fraction expansion and is a special case of (1.28) with $Z(s) = \zeta(s)^2 = \sum_{n=1}^{\infty} \frac{d(n)}{n^s}$. Here $\zeta(s)$ denotes the Riemann zeta-function (for which we refer to [Titchmarsh (1951)]) and $d(n)$ the divisor function. Throughout, we write $s = \sigma + it$. In (1.2), $P_K(x)$ is a residual function given by (1.25), $\sigma(x)$ is defined by (1.19), and

$$2K_0(2z) = \frac{1}{2\pi i}\int_L \Gamma(s)^2 z^{-2s} ds \tag{1.4}$$

is the modified Bessel function of the third kind, where L is a (possibly indented) vertical Bromwich path $\sigma = c > 0$, often denoted (c). For Bessel functions, cf. e.g., [Watson (1944)].

[Soni (1964)], [Soni (1966)], and [Oberhettinger and Soni (1972)] are more or less concerned with [Koshlyakov (1934)], [Koshlyakov (1928/29a)], and [Koshlyakov (1928/29b)], but missing [Koshlyakov (1954)]. Here we adopt the standpoint of Koshlyakov [Koshlyakov (1954)] and let the Dedekind zeta-function represent the zeta-functions satisfying the functional equation (1.8). For more details, we refer to [Liu *et al.* (2023)]. For example, by $\zeta(s)$, we mean a zeta-function satisfying the Riemann functional equation (1.30) and write $\phi(s) = \sum_{n=1}^{\infty} \frac{a_n}{n^s}$, although it is a constant multiple of $\zeta(s)$.

We assemble data on the Dedekind zeta-function. We write $s = \sigma + it$ throughout. Let Ω be an algebraic number field of discriminant Δ and degree $\chi \le 2$

$$\chi = r_1 + 2r_2, \tag{1.5}$$

where r_1 resp. $2r_2$ indicates the real resp. imaginary conjugates, and let

$$\zeta_\Omega(s) = \sum_{n=1}^{\infty} \frac{a_n}{n^s}, \quad a_n = \sum_{N\mathfrak{a}=n} 1, \tag{1.6}$$

be the Dedekind zeta-function, which is absolutely convergent for $\sigma > 1$ on the grounds that

$$a_n = O(n^\eta), \tag{1.7}$$

for every $\eta > 0$.

The functional equation reads

$$A^{-s}\Gamma^{r_1}\left(\frac{s}{2}\right)\Gamma^{r_2}(s)\zeta_\Omega(s) = A^{-(1-s)}\Gamma^{r_1}\left(\frac{1-s}{2}\right)\Gamma^{r_2}(1-s)\zeta_\Omega(1-s), \tag{1.8}$$

where our A is the inverse of Koshlyakov's:

$$A = \frac{2^{r_2}\pi^{\frac{\chi}{2}}}{\sqrt{|\Delta|}}. \tag{1.9}$$

The Dedekind zeta-function has a simple pole at $s = 1$ with residue

$$\rho = \frac{2^{r+1}\pi^{r_2}Rh}{W\sqrt{|\Delta|}}, \tag{1.10}$$

where $r = r_1 + r_2 - 1$ is the rank of the unit group, h is the class number of Ω, and W is the number of roots of unity in Ω. This may be expressed in the case $\chi \le 2$ as

$$\rho = -\frac{2^{r+1}\pi^{r_2}\zeta_\Omega^{(r)}(0)}{\sqrt{|\Delta|}}. \tag{1.11}$$

Putting

$$G(s) = \frac{\Gamma^{r_1}\left(\frac{1-s}{2}\right)\Gamma^{r_2}(1-s)}{\Gamma^{r_1}\left(\frac{s}{2}\right)\Gamma^{r_2}(s)}, \tag{1.12}$$

we express (1.8) as

$$\zeta_\Omega(s) = A^{2s-1}G(s)\zeta_\Omega(1-s). \tag{1.13}$$

We use the estimate without notice

$$\zeta_\Omega(s) = O\left(|t|^{\chi(1-\sigma)}\right), \quad \sigma \le 0 \tag{1.14}$$

uniformly in the strip $\sigma_1 \le \sigma \le \sigma_2$.

Necessary information on algebraic numbers is available in many books, cf. e.g., [Narkiewicz (2004)]. We shall mainly work with the case of the rational field and a zeta-function satisfying the Riemann functional equation

(1.30) and the case of a real quadratic field $(r_1, r_2) = (2, 0)$ and the square of the Riemann zeta-function, which share the same functional equation with $G(s) = \frac{\Gamma^2\left(\frac{1-s}{2}\right)}{\Gamma^2\left(\frac{s}{2}\right)}$. For the Hecke functional equation case $(r_1, r_2) = (0, 1)$ with a more general line of reflection, we refer to [Liu *et al.* (2023)].

Following Koshlyakov, we define the $\mathfrak{K}$-functions, $\mathfrak{K} = X = X_{r_1,r_2}$, $\mathfrak{K} = K = K_{r_1,r_2}$, and $\mathfrak{K} = L = L_{r_1,r_2}$:

$$X_{r_1,r_2}(x) = \frac{1}{2\pi i}\int_{(c)} \Gamma^{r_1}\left(\frac{s}{2}\right)\Gamma^{r_2}(s)\, x^{-s} ds, \quad c > 0, \tag{1.15}$$

for $x > 0$ ($Re x > 0$), where and in what follows (c) indicates the Bromwich path $\sigma = c, -\infty < t < \infty$.

$$X_{1,0}(x) = 2e^{-x^2}, \quad X_{0,1}(x) = e^{-x}, \quad X_{2,0}(x) = 4K_0(2x). \tag{1.16}$$

As in [Koshlyakov (1954), (2.8)], let

$$K_{r_1,r_2}(x) = \frac{1}{2\pi i}\int_{(c)} \frac{\pi}{2\cos\frac{\pi s}{2}} G(1-s)\, x^{-s} ds. \tag{1.17}$$

As in [Koshlyakov (1954), p. 114, (2.5), (2.6)] or as can be readily checked by (1.4), we have

$$\begin{aligned} K_{1,0}(x) &= \sqrt{\pi}\, e^{-2x}, \\ K_{0,1}(x) &= -2\,\mathrm{kei}_0\left(4\sqrt{\frac{x}{4}}\right) = \frac{1}{i}\left(K_0\left(2\epsilon\sqrt{x}\right) - K_0\left(2\bar{\epsilon}\sqrt{x}\right)\right), \\ K_{2,0}(x) &= 4\,\mathrm{ker}_0\left(4\sqrt{x}\right) = 2\left(K_0\left(4\epsilon\sqrt{x}\right) + K_0\left(4\bar{\epsilon}\sqrt{x}\right)\right), \end{aligned} \tag{1.18}$$

where kei_0 and ker_0 are modified Kelvin functions ([Erdélyi *et al.* (1953), p. 6], [Prudnikov *et al.* (1986)]).

After [Koshlyakov (1954), I, (2.9), (2.13)] the σ-function is defined by $(c > 1)$

$$\sigma(x) = \frac{A}{\pi}\sum_{n=1}^{\infty} a_n K_{r_1,r_2}\left(A^2 xn\right) = \frac{1}{2\pi i}\int_{(c)} \frac{1}{2\cos\frac{\pi s}{2}} A^{1-2s} G(1-s)\,\zeta_\Omega(s)\, x^{-s} ds, \tag{1.19}$$

where $K_{r_1,r_2}(x)$ is defined by (1.17), and (r_1, r_2) is defined in (1.5).

The Koshlyakov L-function is a relative of K-functions and is related through

$$L_{r_1,r_2}(x) = \frac{1}{2}\left(K_{r_1,r_2}(-ix) + K_{r_1,r_2}(ix)\right)\text{([Koshlyakov (1954), (7.11)]).} \tag{1.20}$$

It is defined by ([Koshlyakov (1954), (8.9)]) as

$$\begin{aligned} L_{r_1,r_2}(x) &= \frac{1}{2\pi i}\int_{(c)} \frac{\pi}{2} G(1-s)\, x^{-s} ds \\ &= \frac{1}{2\pi i}\int_{(c)} \frac{\pi}{2} \frac{\Gamma^{r_1}\left(\frac{s}{2}\right)\Gamma^{r_2}(s)}{\Gamma^{r_1}\left(\frac{1-s}{2}\right)\Gamma^{r_2}(1-s)} x^{-s} ds, c > 0, x > 0 \end{aligned} \tag{1.21}$$

We have

$$\begin{aligned} L_{1,0}(x) &= \sqrt{\pi}\ \cos 2x, L_{0,1}(x) = \frac{\pi}{2} J_0\left(2\sqrt{x}\right), \\ L_{2,0}(x) &= \pi\left(\frac{2}{\pi}K_0\left(4\sqrt{x}\right) - Y_0\left(4\sqrt{x}\right)\right). \end{aligned} \tag{1.22}$$

Proof is given in [Kanemitsu and Tsukada (2014), pp. 105–132, Example 3.8].

We let $\mathrm{P}_{\mathfrak{K}}(x) = R_0 + R_1$ be the residual function, where

$$R_j = Res\left(\chi_{\mathfrak{K}}(s) z^{-s}, s = j\right), \quad j = 0, 1, \tag{1.23}$$

where

$$\begin{aligned} \chi_{\mathfrak{K}}(s) &= \frac{\pi}{2\cos\frac{\pi s}{2}} G(1-s) x^{-s} \quad \text{for} \quad \mathfrak{K} = K_{r_1,r_2} \\ \chi_{\mathfrak{K}}(s) &= \Gamma^{r_1}\left(\frac{s}{2}\right)\Gamma^{r_2}(s) x^{-s} \quad \text{for} \quad \mathfrak{K} = X_{r_1,r_2} \\ \chi_{\mathfrak{K}}(s) &= \frac{\pi}{2}\frac{\Gamma^{r_1}\left(\frac{s}{2}\right)\Gamma^{r_2}(s)}{\Gamma^{r_1}\left(\frac{1-s}{2}\right)\Gamma^{r_2}(1-s)} x^{-s} \quad \text{for} \quad \mathfrak{K} = L_{r_1,r_2} \end{aligned} \tag{1.24}$$

For $\mathfrak{K} = K = K_{r_1,r_2}$ we have

$$\begin{aligned} \mathrm{P}_K(x) &= -\frac{1}{2}\rho - \frac{\zeta_\Omega(0)}{\pi}\frac{1}{x} \quad \text{for} \quad \zeta_\Omega(s) \\ \mathrm{P}_K(x) &= -\frac{1}{4\pi x} - \frac{1}{2}\log x - \gamma \quad \text{for} \quad \zeta(s)^2 \end{aligned} \tag{1.25}$$

For $\mathfrak{K} = X = X_{r_1,r_2}$ we have

$$\mathrm{P}_X(x) = 2^{r_1}\zeta_\Omega^{(r)}(0) - 2^{r_1}\zeta_\Omega^{(r)}(0)\,x^{-1} \quad \text{for} \quad \zeta_\Omega(s)$$
$$\mathrm{P}_X(x) = \frac{1}{4}\left(\gamma - \log\frac{4\pi}{x} - \frac{1}{x}(\gamma - \log 4\pi x)\right) \quad \text{for} \quad \zeta(s)^2 \tag{1.26}$$

Theorem 2 (Koshlyakov) *The functional equation (1.8) (which we apply in the form (1.13)) is equivalent to each of the following. The Bochner modular relation*

$$\sum_{n=1}^{\infty} a_n X_{r_1,r_2}(Axn) = \mathrm{P}_X(x) + x^{-1}\sum_{n=1}^{\infty} a_n X_{r_1,r_2}\left(Ax^{-1}n\right), \quad \operatorname{Re} x > 0. \tag{1.27}$$

The partial fraction expansion

$$\frac{A}{\pi}\sum_{n=1}^{\infty} a_n K_{r_1,r_2}\left(A^2xn\right) = \sigma(x) = \mathrm{P}_K(x) + \frac{x}{\pi}\sum_{n=1}^{\infty}\frac{a_n}{n^2+x^2}. \tag{1.28}$$

Deduction of (1.8) from (1.27) is made in [Koshlyakov (1954), I, p. 120, (3.4)]. For (1.28), cf. [Koshlyakov (1954), I, p. 117, (3.4)] as modified in [Kanemitsu and Tsukada (2014), p. 129, Theorem 4.7]. Deduction of (1.8) from (1.28) is done in [Koshlyakov (1954), II, pp. 225–226] based on the formula [Koshlyakov (1954), II, p. 225, (16.1)], which should read (supplementing the missing second integral on the right) as

$$\zeta_\Omega(s) = \rho\frac{\alpha^{1-s}}{s-1} + \int_\alpha^{\alpha+i\infty}\sigma(-iz)\,z^{-s}dz + \int_\alpha^{\alpha-i\infty}\sigma(iz)\,z^{-s}dz, \ 0 < \alpha < 1. \tag{1.29}$$

Formulas similar to this appear in [Hamburger (1922), p. 137] and [Soni (1966), p. 547], and Hamburger and Soni used them to deduce (1.34) from (1.32) and (1.28) from (1.27), respectively. We shall prove in §2 that (1.29) is a special case of a version of the Plana summation formula, Lemma 3, and clarify the works of Hamburger and Soni.

[Kanemitsu and Tsukada (2014), pp. 124–127], without concretizing, state (a general case of) the Fourier-Bessel expansion for the Dedekind zeta-function of the real quadratic field. Since the partial fraction expansion may be viewed as a special case of the Fourier-Bessel expansion, which is more easily seen to be equivalent to the functional equation, this will establish Theorem 2 more easily.

Corollary 1 ([Hamburger (1922)]) *The functional equation for the Riemann zeta-function*

$$\pi^{-\frac{s}{2}}\Gamma\left(\frac{s}{2}\right)\zeta(s) = \pi^{-\frac{1-s}{2}}\Gamma\left(\frac{1-s}{2}\right)\zeta(1-s) \tag{1.30}$$

is equivalent to the following equalities:

The theta-transformation formula (the Bochner modular relation)

$$\sum_{n=1}^{\infty} e^{-\pi n^2 x} = x^{-\frac{1}{2}}\sum_{n=1}^{\infty} e^{-\frac{\pi n^2}{x}} + \frac{1}{2}\left(x^{-\frac{1}{2}} - 1\right), \quad \operatorname{Re} x > 0. \tag{1.31}$$

The partial fraction expansion for the cotangent function $i\cot \pi i z$

$$1 + 2\sum_{n=1}^{\infty} e^{-2\pi n z} = \frac{1}{\pi z} + \frac{2z}{\pi}\sum_{n=1}^{\infty}\frac{1}{z^2+n^2} \tag{1.32}$$

The Fourier expansion

$$\sum_{n=1}^{[x]}(x-n) = \frac{x(x-1)}{2} - \frac{1}{2\pi^2}\sum_{n=1}^{\infty} n^{-2}(\cos 2\pi n x - 1) \tag{1.33}$$

The Poisson summation formula $(T > 0)$

$$T\sum_{n=-\infty}^{\infty}\phi(Tn) = \sum_{n=-\infty}^{\infty}\hat{\phi}\left(\frac{2\pi}{T}n\right) = \sum_{n=-\infty}^{\infty}\int_{-\infty}^{\infty} e^{-\frac{2\pi}{T}in}\phi(u)\,du \tag{1.34}$$

is valid for a function $\phi(u)$ *which is of bounded variation in any finite interval satisfying convergence conditions and such that*

$$f(z) := \int_{-\infty}^{\infty} e^{izu}\phi(u)\,du \tag{1.35}$$

is analytic in a certain (vertical) strip $-d < \sigma < c$*, say, where* $c, d > 0$.

The convergence conditions are as stated in ([Hamburger (1922), p. 136]): (2.14) and

$$\int_{-\infty}^{\infty}|\phi(\gamma+iy)|\,dy < \infty, \int_{-\infty}^{\infty}|\phi(-\gamma+iy)|dy < \infty \tag{1.36}$$

for some γ, $-d < \gamma < c$.

(1.34) is proved for the analytic function $f(z)$ in (1.35) and then inverted by the Fourier integral theorem. (1.34) in the form

$$T \sum_{n=-\infty}^{\infty} \phi(Tn) = \sum_{n=-\infty}^{\infty} \int_{\infty}^{\infty} e^{\frac{2\pi}{T} in} \phi(u)\, du \tag{1.37}$$

with $T = 2\pi$ leads to [Hamburger (1922), (V*)] while (1.34) with $T = 2\pi\lambda$ is a familiar form of [Katznelson (2004), pp. 141–142].

The first two are contained in Theorem 2. Proof of equivalence of (1.33) and (1.30) is given in [Liu *et al.* (2023)]. (1.34) is proved in Theorem 3.

(1.33) may be proved as follows. We integrate

$$\ell_0(x) = \frac{e^{2\pi ix}}{1 - e^{2\pi ix}} = \frac{1}{2}(-1 + i \cot \pi x) \tag{1.38}$$

from $\frac{1}{2}$ to x, $0 < x < 1$ to deduce that

$$\frac{1}{2\pi i}(\ell_1(x) + \log 2) = \frac{1}{2}\left(-\left(x - \frac{1}{2}\right) + \frac{i}{\pi} \log \sin \pi x\right).$$

Here we take the expansion of the Lerch zeta-function for granted

$$\ell_1(x) = -\log\left(1 - e^{2\pi ix}\right) = \sum_{n=1}^{\infty} \frac{e^{2\pi inx}}{n}, \qquad 0 < x < 1. \tag{1.39}$$

Comparing the real and imaginary parts, we arrive at the special case of $\varkappa = 1$ of the Fourier series formulas, cf. e.g., [Yamamoto (1977)]:

The Fourier series for the Clausen function $A_\varkappa(x)$ of order $\varkappa$ is

$$A_\varkappa(x) = \frac{\varkappa!}{2(2\pi i)^{\varkappa-1}} \sum_{n=-\infty}^{\infty}{}' \frac{\operatorname{sgn}(n)\, e^{2\pi inx}}{n^\varkappa} \tag{1.40}$$

while the Fourier series for the periodic Bernoulli polynomial $\overline{B}_\varkappa(x)$ of order is $\varkappa$

$$\overline{B}_\varkappa(x) = -\frac{\varkappa!}{(2\pi i)^\varkappa} \sum_{n=-\infty}^{\infty}{}' \frac{e^{2\pi inx}}{n^\varkappa} \tag{1.41}$$

where the prime on the summation sign means that the term $n = 0$ is omitted and $sgn(n)$ is the sign of $n \neq 0$, i.e., 1 resp. -1 for $n > 0$ resp. $n > 0$.

10.2 SOME SUMMATION FORMULAS ACCORDING TO THE PRINCIPLE

Let

$$f(x) = \frac{1}{2\pi i}\int_{(c)} x^{-s}F(s)\,ds, \quad \operatorname{Re} x > 0,\ c > 0,$$
$$F(s) = \int_0^\infty \xi^s f(\xi)\,\frac{d\xi}{\xi} \tag{2.1}$$

be the Mellin transform pair that satisfies the conditions of convergence necessitated in our discussion. (f, F) will always be used as the Mellin transform pair. For Mellin transforms, cf. [Oberhettinger (1974)], [Paris and Kaminski (2001)], etc.

Let $\{\lambda_n\}$ be a strictly increasing sequence of real numbers with $\lambda_1 > 0$. Let

$$Z(s) = \sum_{n=1}^{\infty} \frac{a_n}{\lambda_n^s} \tag{2.2}$$

be absolutely convergent for $\operatorname{Re} s = \sigma > 1$. The abscissa of absolute convergence can be σ_a^*, but we assume it to be 1 for the sake of simplicity. Suppose it satisfies the functional equation of the form (where we understand the right-hand side member may be a different Dirichlet series)

$$Z(s) = G(s)Z(1-s), \tag{2.3}$$

where $G(s)$ is a certain gamma factor to be specified in each occasion, and the line of reflection is chosen to be $\frac{1}{2}$ instead of a more general $\frac{r}{2} \in \mathbb{R}$. We may interpret the right-hand side as another Dirichlet series $W(s) = \sum_{n=1}^{\infty} \frac{b_n}{\mu_n^s}$, say. There is another theory due to Landan and Walfisz which deals with different zeta-functions with gamma factors that are different in number [Kanemitsu and Tsukada (2014), p. 35, Table 2.2].

We have for $c > 1$

$$Z_f(s) = \sum_{n=1}^{\infty} a_n f(\lambda_n) = \frac{1}{2\pi i}\int_{(c)} F(z)Z(z)\,dz \tag{2.4}$$

in the first instance. Here and in what follows, we use this type of suggestive notation to mean that the processing factors—f and F in this case—may depend also on the extraneous parameter s (which may be x with $Rex > 0$, say).

The following lemma is Corollary 1 to Theorem 1 [Liu *et al.* (2023)] and is an extract of the case of unprocessed sum (2.4). This is a result of common knowledge (á la Shannon) but is the underlying principle.

Lemma 1 *Suppose the integration path (c) in (2.4) may be shifted to $(-d)$, $\frac{1}{2} < d < 1$, say with the resulting residual function* $\mathrm{P}(s)$ *(sum of residues in the vertical strip $-d < \sigma < c$):*

$$Z_f(s) = \frac{1}{2\pi i}\int_{(-d)} F(z)Z(z)\,dz + \mathrm{P}(s) = J(s) + \mathrm{P}(s), \tag{2.5}$$

say. Suppose that

$$F(z) = \int \mathfrak{G}(w,z)\,\mathfrak{F}(w)\,dw, \tag{2.6}$$

where the integral for $F(z)$ may be the infinite integral over $(0,\infty)$, the contour integral, or may indicate the integrand itself, and that the order of integration is interchangeable:

$$\frac{1}{2\pi i}\int_{(-d)} dz \int dw = \int dw \frac{1}{2\pi i}\int_{(-d)} dz. \tag{2.7}$$

Then

$$J(s) = \int I(w)\,\mathfrak{F}(w)\,dw, \tag{2.8}$$

where

$$I(w) := \frac{1}{2\pi i}\int_{(-d)} \mathfrak{G}(w,z)\,G(z)\,Z(1-z)\,\gamma(z)\,dz = \sum_{n=1}^{\infty} a_n K(w,n). \tag{2.9}$$

Here $\gamma(z) = \gamma(z,s)$ is a simple function specified at each occasion and

$$K(w,n) = \frac{1}{2\pi i}\int_{(-d)} \mathfrak{G}(w,z)\,G(z)\,\lambda nz - 1\gamma(z)\,dz. \tag{2.10}$$

Then, if (2.10) is expressed in a closed form, it gives a closed form for $Z_f(z)$.

Lemma 1 is a simplified version of the Principle. For example, to prove Lerch's transformation formula, we need its full force with all processing gamma factors incorporated, cf. [Liu *et al.* (2023), Proof of Theorem 2].

In what follows, we consider $\zeta_\Omega(s)$ as $Z(s)$ and the gamma factor $G(s)$ as given by (1.12). Therefore the basic sequences $\{\lambda_n\}$, $\{\mu_n\}$ are $\{n\}$.

Lemma 2 *We have*

$$2\pi i\sigma(iz) = -\pi i\rho - \frac{2\zeta_\Omega(0)}{z} + \sum_{n=1}^{\infty} a_n\left(\frac{1}{z-n} + \frac{1}{z+n}\right), \tag{2.11}$$

i.e., $\sigma(iz)$ is a meromorphic function with simple poles at every integer, except at poles,

$$\sigma(iz) = -\sigma(-iz) - \rho. \tag{2.12}$$

For $\mathrm{Re}z \to \infty$,

$$\sigma(z) = o\left(|z|^{-K}\right) \tag{2.13}$$

for any $K > 0$.

The following lemma reveals the procedures adopted by Hamburger and Soni. Cf. [Koshlyakov (1954), II, pp. 222–226].

Lemma 3 (A version of the Plana summation formula) *Let* $0 < \alpha < \min\{\lambda_1, 1\} < \beta$, *let* C *denote the rectangle with vertices at* $(\alpha, \alpha - iT)$, $(\alpha, \alpha + iT)$, $(\beta, \beta + iT)$, $(\beta, \beta - iT)$, $T > 0$, *and let* $f(s)$ *be an arbitrary function analytic along the contour* C *and satisfy the following limit conditions.*

$$\int_\alpha^\beta f(x+iT)\sigma(T-ix)\,dx \to 0, \int_\alpha^\beta f(x-iT)\sigma(T+ix)\,dx \to 0 \text{ as } T \to \infty \tag{2.14}$$

$$J(\beta, \infty) \to 0 \text{ as } \beta \to \infty, \tag{2.15}$$

where

$$J(x,T) = \int_x^{x+iT} f(z)\sigma(-iz)\,dz + \int_x^{x-iT} f(z)\sigma(iz)\,dz. \tag{2.16}$$

Then

$$\Sigma_f(s) := \sum_{n=1}^{\infty} a_n f(n) = -\int_{(\alpha)} f(z)\sigma(iz)\,dz \tag{2.17}$$

in the first instance. If (α) *is changed by the path* $(0+)$ *with indentation at the origin, then (2.17) becomes*

$$\Sigma_f(s) = \sum_{n=1}^{\infty} a_n f(n) = \zeta_\Omega(0) f(0) + \int_{(0+)} f(z)\sigma(iz)\,dz. \tag{2.18}$$

$\Sigma_f(s)$ *has another expression ([Koshlyakov (1954), II, p. 223, IV])*

$$\Sigma_f(s) = \rho \int_\alpha^\infty f(x)\,dx + J(\alpha, \infty), \tag{2.19}$$

which amounts to a version of the Plana summation formula

$$\Sigma_f(s) = \sum_{n=1}^{\infty} a_n f(n) = \zeta_\Omega(0) f(0) + \rho \int_0^\infty f(x)\,dx + J(0, \infty) \tag{2.20}$$

where

$$J(0, \infty) = \int_0^{i\infty} f(z)\sigma(-iz)\,dz + \int_0^{-i\infty} f(z)\sigma(iz)\,dz. \tag{2.21}$$

Proof. By Lemma 2 and the Cauchy residue theorem, we have

$$\sum_{n=1}^{[\beta]} a_n f(n) = \int_C f(z)\sigma(iz)\,dz \tag{2.22}$$

which leads to (2.17) on letting $T \to \infty$, $\beta \to \infty$ in view of (2.14) and (2.15). Using Lemma 2, (2.17) becomes:

$$\sum_{n=1}^{[\beta]} a_n f(n) = \rho \int_\alpha^\infty f(x)\,dx + J(\alpha, \infty) - J(\beta, \infty), \tag{2.23}$$

which leads to (2.19) by (2.15). By Lemma 2, the limiting case $\alpha \to 0$ of (2.19) needs the extra term $\zeta_\Omega(0) f(0)$ to arrive at (2.20). □

Theorem 3 *With the aid of a version of the Plana summation formula, Lemma 3, the Poisson summation formula (1.34) of Hamburger follows from the partial fraction expansion (1.28), which in turn follows from (1.27) and leads to the functional equation (1.8).*

Proof. (1.27)$\Longrightarrow$ (1.28). It turns out that by coincidence, Soni's function $\sigma(z)$ is Koshlyakov's $\sigma(iz)$. If the upper part, resp. lower part, of the path (0+) in (2.18) is rotated by $-\frac{\pi}{2}$ resp. $\frac{\pi}{2}$, then it becomes the limit of the path of Soni. This leads to [Soni (1966), (IV)], which leads to (1.28) in the case of the divisor problem. Other cases similarly follow: (1.28) $\Longrightarrow$ (1.34). Corresponding to Hamburger, we use (1.28) in the form

$$\sigma_H(iz) := \sigma(iz) + \frac{1}{2}\rho = -\frac{\zeta_\Omega(0)}{\pi}\frac{1}{iz} + \frac{1}{2\pi i}\sum_{n=1}^{\infty} a_n\left(\frac{1}{z-n} + \frac{1}{z+n}\right) \tag{2.24}$$

in place of $\sigma(iz)$, so that $\sigma_H(iz)$ is an odd function. Arguing as before, we arrive at

$$\Sigma_f(s) = \sum_{n=1}^{\infty} a_n f(n) = -\int_{(\alpha)} f(z)\sigma_H(iz)\,dz = \int_{-\infty}^{\infty} f(\alpha+iy)\sigma_H(\alpha+iy)\,dy$$

corresponding to (2.17), which leads to

$$\Sigma_f(s) = -\frac{1}{2}a_0 + \int_{(0+)} f(z)\sigma_H(iz)\,dz \tag{2.25}$$

where $a_0 = -\frac{\zeta_\Omega(0)}{\pi}$. Similarly,

$$\Sigma_{f-}(s) := \sum_{n=1}^{\infty} a_{-n} f(-n) = \int_{(-\alpha)} f(z)\sigma_H(iz)\,dz,$$

where $a_{-n} = a_n$ and the limit case corresponding to (2.25). It follows that

$$\begin{aligned}\sum_{n=-\infty}^{\infty} a_n f(n) &= \int_{(0+)} \sigma_H(iz)\left(f(z)+f(-z)\right)\,dz \\ &= \int_{-\infty}^{\infty} \sigma_H(ix)\left(f(x)+f(-x)\right)\,dx.\end{aligned} \tag{2.26}$$

From (1.28) and (2.24), we have

$$\sigma_H(ix) = b_0 + \frac{A}{\pi}\sum_{n=1}^{\infty} a_n K_{r_1,r_2}\left(iA^2xn\right), \tag{2.27}$$

where $b_0 = \frac{1}{2}\rho$. Substituting this in (2.26), we deduce that

$$\sum_{n=-\infty}^{\infty} a_n f(n) = \frac{A}{\pi} \sum_{n=-\infty}^{\infty} b_n \int_{-\infty}^{\infty} K_{r_1,r_2}\left(iA^2 xn\right) f(x)\, dx, \tag{2.28}$$

where $b_n = a_n$ for $n \neq 0$. Substituting (1.18), in the rational case, (2.28) reads

$$\sum_{n=-\infty}^{\infty} f(n) = \sum_{n=-\infty}^{\infty} \int_{-\infty}^{\infty} e^{-2\pi ixn} f(x)\, dx. \tag{2.29}$$

Here, Hamburger applies the Fourier integral theorem:

$$\phi(x) = \frac{1}{2\pi} \int_{-\infty}^{\infty} e^{ixu} \int_{-\infty}^{\infty} e^{-iuy} \phi(y)\, dy du. \tag{2.30}$$

To make (2.29) consistent with (1.34) with ϕ defined by (1.35), we put

$$f_\lambda(x) = \lambda f(\lambda x), \quad \lambda = \frac{2\pi}{T}. \tag{2.31}$$

Then the right-hand side becomes $2\pi\phi\left(2\pi\lambda^{-1}n\right) = 2\pi\phi(Tn)$, while the left-hand side is $\lambda\sum_{n=-\infty}^{\infty}\hat{\phi}\left(\frac{2\pi}{T}n\right)$. Hence (1.34) follows. In Hamburger's case, changing the sequence $\{l_n\}$ by $\{-l_n\}$ leads to the same form, which is the most familiar one.

(2.29) translates into (1.34) with $T = 2\pi$.

In the notation of (1.35), in the rational case

$$\int_{-\infty}^{\infty} K_{r_1,r_2}(izx)\,\phi(x)\, dx = \sqrt{\pi} f(-2z). \tag{2.32}$$

In this case, it is simpler if we work with the rectangle C' with vertices at $(\alpha, \alpha - iT)$, $(\alpha, \alpha + iT)$, $(-\alpha, -\alpha + iT)$, $(-\alpha, -\alpha - iT)$, $T > 0$, and assuming convergence conditions, we count all the residues at in, $n \in \mathbb{Z}$. (1.34) $\Longrightarrow$ (1.8). (2.19) with $f(z) = z^{-s}$ leads to (1.29), in which we let $\alpha \to 0+$ for $\sigma < 1$ to arrive at a simplified form [Koshlyakov (1954), II, (16.2)] of (2.20)

$$\zeta_\Omega(s) = J(0, \infty) = 2 \sin\frac{\pi}{2}s \int_0^{\infty} x^{-s}\sigma(x)\, dx \tag{2.33}$$

valid for $\sigma < r$. On the other hand, with the Mellin inversion of (1.19), we have

$$\zeta_\Omega(s) = 2A^{1-2s} \cos\frac{\pi s}{2} G(s) \int_0^{\infty} \sigma(x) x^{-s} ds, \sigma > 1 - r. \tag{2.34}$$

Comparing (2.33) and (2.34) implies (1.8). □

10.3 SUMMATION FORMULAS

Theorem 4 *Suppose the Mellin inversion $F(s)$ of a continuous function $f(x)$ in (2.1) satisfies the following.*

$$F(s) = O\left(|t|^{m\sigma - n}\right), \tag{3.1}$$

uniformly in $-d \leq \sigma \leq c$, where $-1 < -d \leq -\delta < 0, 1 < c$, and $m > 0, n > 0$ are subject to the condition

$$m\delta + n - \chi(d+1) \geq 1 \tag{3.2}$$

and that $F(s)$ is regular in the strip $-d \leq \sigma \leq c$ except for a simple pole at $s = 0$ with residue β.

Suppose the only singularity of $Z(s)$ in the strip is $s = 1$, where it has a (possibly) double pole

$$Z(s) = \frac{A_1}{(s-1)^2} + \frac{A_0}{s-1} + O(1), \quad s \to 1, \quad F(s) = \frac{\beta}{s} + O(1), \quad s \to 0. \tag{3.3}$$

Then

$$\sum_{n=1}^{\infty} a_n f(n) = \mathrm{P}(\cdot) + J(\cdot), \tag{3.4}$$

where

$$J(\cdot) = \frac{2A}{\pi} \sum_{n=1}^{\infty} a_n \int_0^{\infty} f(x)\, L_{r_1.r_2}\left(A^2 xn\right) dx. \tag{3.5}$$

and, writing $f(0) = \beta$, we have

$$\mathrm{P}(\cdot) = Z(0) f(0) + \int_0^{\infty} (A_0 + A_1 \log x) f(x)\, dx. \tag{3.6}$$

Further, (3.4) is, via the Bochner modular relation (1.27), equivalent to the functional equation.

TABLE 10.1 Factors of the Integrand

Function	$\mathfrak{G}(w,z)$	$\mathfrak{F}(w)$	$\gamma(z)$
-	w^{z-1}	$f(w)$	$A\left(A^2\right)^{z-1}$

10.3.1 Proof

The exponent of $|t|$ of $F(s)\,Z(s)$ is

$$m\sigma - n + \chi(1-\sigma) \le -m\delta - n + \chi d + \chi < -1.$$

Hence shifting the integration path is possible. For the interchange of integration and summation, we apply the Stirling formula [Erdélyi *et al.* (1953), p. 47, (6)]:

$$\Gamma(s) = \sqrt{2\pi}\, e^{-\frac{\pi}{2}|t|}|t|^{\sigma-\frac{1}{2}}, \quad |t| \to \infty. \tag{3.7}$$

The auxiliary functions shown are in Table 10.1 and (2.10) reads

$$K(w,n) = A\frac{1}{2\pi i}\int_{(-d)} G(z)\left(A^2wn\right)^{z-1}dz = \frac{2A}{\pi}L_{r_1.r_2}\left(A^2wn\right). \tag{3.8}$$

□

Corollary 2 *Under the same conditions as in Theorem 4, the Poisson summation formula*

$$\frac{1}{2}f(0) + \sum_{n=1}^{\infty} f(n) = F(1) + \sum_{n=1}^{\infty}\int_0^{\infty} f(x)\,\cos 2\pi xn\,\mathrm{d}x. \tag{3.9}$$

is equivalent to the functional equation (1.30).

The Voronoĭ summation formula

$$-\frac{1}{4}f(0) + \sum_{n=1}^{\infty} d(n)f(n) = \int_0^{\infty}(2\gamma + \log x)\,dx + \sum_{n=1}^{\infty} d(n)\,G(n),$$

$$G(y) = 2\int_0^{\infty} f(x)\,\pi\left(\frac{2}{\pi}K_0\left(4\sqrt{xy}\right) - Y_0\left(4\sqrt{xy}\right)\right)dx \tag{3.10}$$

is equivalent to the functional equation for $\zeta^2(s)$.

Proof. In the rational case, (3.4) leads to (3.9).□

For the Voronoĭ summation formula, cf. e.g., [Laurinčikas (1998)].

10.4 VARIANT OF THE PRINCIPLE

Hitherto, we have treated the case where the line of reflection is the same for all zeta-functions, i.e., (2.2). But when we consider the product of zeta-functions $Z_j(s)$, the line of reflection may be different $Z_j(r_j - s) = G_l(s) Z_j(s)$. The Principle still applies to such a case with a slight variation. We illustrate this with the product of two Riemann zeta-functions. [Gupta and Maji (2021)] is the case where we apply the functional equation to both of them and shift the line of integration in (2.4) to the left so that the product $Z_j(r_j - s)$ is absolutely convergent, i.e., $\max r_j - \sigma > c$. On the other hand, [Krätzel (1981)] is the case where we apply the functional equation to only one factor and transform the product into the form which gives rise to symmetry. We restrict to the Hecke gamma transform case, giving rise to the Bochner modular relation or the transformation formula for the Lambert series.

10.4.1 The Case of Gupta and Maji

Let $k \geq 2$ and r be both even integers. Let

$$D_{k,r}(n) = \sum_{d^k|n} \left(\frac{n}{d^k}\right)^r.$$

Then the generating Dirichlet series $Z(s)$ in (2.2) is

$$Z(s) := \zeta(ks)\zeta(s-r) = \sum_{n=1}^{\infty} \frac{D_{k,r}(n)}{n^s}. \quad \sigma > \sigma_a^* := \max\left\{\frac{1}{k}, r+1\right\} \tag{4.1}$$

and (2.3) reads

$$Z(s) = G(s)\, W(1-s), \tag{4.2}$$

where

$$W(1-s) = \frac{1}{(2\pi)^r \pi^2}\left((2\pi)^{k+1}\right)^s \zeta(1-ks)\zeta(1-s+r) \tag{4.3}$$

and

$$\begin{aligned} G(s) &= \Gamma(1-ks)\Gamma(1-s+r)\sin\frac{\pi ks}{2}\sin\frac{\pi(s-r)}{2} \\ &= \frac{(-1)^{\frac{r}{2}}}{2}\sin\pi s\frac{\sin\frac{\pi ks}{2}}{\cos\frac{\pi}{2}s}\Gamma(1-ks)\Gamma(1-s+r). \end{aligned} \tag{4.4}$$

We also introduce the counterpart of (4.1):

$$W(s) := \zeta(ks+1-k)\zeta(s+r) = \sum_{n=1}^{\infty} \frac{S_{k,r}(n)}{n^s}. \quad \sigma > \sigma_b^* := \max\{1, 1-r\} \tag{4.5}$$

We consider the most basic case of $f(x) = e^{-x}$, $F(s) = \Gamma(s)$ in (2.1). Then (2.4) reads ($Rex > 0$)

$$Z_f(x) = \sum_{n=1}^{\infty} D_{k,r}(n) e^{-nx} = \frac{1}{2\pi i} \int_{(c)} \Gamma(s) Z(s) x^{-s} ds, \quad c > \sigma_a^*. \tag{4.6}$$

The situation is the same as in Theorem 4, and all the processing factors are trivial, so that (2.9) amounts to ($1 + d > \sigma_a^*$)

$$\begin{aligned} J(x) = I(x) &= \frac{1}{2\pi i} \int_{(-d)} \mathfrak{G}(s) G(s) W(1-s) \gamma(s) x^{-s} ds \\ &= \frac{1}{(2\pi)^r \pi^2} \frac{1}{2\pi i} \int_{(-d)} \frac{(-1)^{\frac{r}{2}}}{2} \Gamma(s) \sin \pi s \frac{\sin \frac{\pi k s}{2}}{\cos \frac{\pi}{2} s} \Gamma(1-ks) \Gamma(1-s+r) \\ &\quad \times \zeta(1-ks) \zeta(1-s-r) \left(\frac{(2\pi)^{k+1}}{x} \right)^s ds \\ &= \frac{(-1)^{\frac{r}{2}}}{(2\pi)^{r+1}} \frac{1}{2\pi i} \int_{(-d)} \frac{\sin \frac{\pi k s}{2}}{\Gamma(1-s) \cos \frac{\pi}{2} s} \Gamma(1-ks) \Gamma(1-s+r) \\ &\quad \times \zeta(1-ks) \zeta(1-s-r) \left(\frac{(2\pi)^{k+1}}{x} \right)^s ds \\ &= \sum_{n=1}^{\infty} S_{k,r}(n) K(x,n), \end{aligned} \tag{4.7}$$

where

$$K(x,n) = \frac{(-1)^{\frac{r}{2}}}{(2\pi)^{r+1}} \frac{1}{2\pi i} \int_{(-d)} \frac{\sin \frac{\pi k s}{2}}{\Gamma(1-s) \cos \frac{\pi}{2} s} \Gamma(1-ks) \Gamma \\ (1-s+r) n^{s-1} \left(\frac{(2\pi)^{k+1}}{x} \right)^s ds$$

$$= \frac{(-1)^{\frac{k+r}{2}-1}(2\pi)^{k-r}}{x} \frac{1}{2\pi i} \int_{(d+1)} \frac{\sin \frac{\pi k s}{2}}{\sin \frac{\pi}{2} s}$$

$$\frac{\Gamma(ks-k+1)\Gamma(s+r)}{\Gamma(s)} \left(\frac{(2\pi)^{k+1} n}{x} \right)^{-s} ds, \tag{4.8}$$

where we replaced s by $1-s$ in the second equality. Thus, as Lemma 1 asserts, it remains to find a closed expression for (the integral in) $K(x, n)$ and to calculate residues.

To find a closed form, Gupta and Maji express $\frac{\sin \frac{\pi k s}{2}}{\sin \frac{\pi}{2} s}$ as a finite sum of complex exponential functions, and applying the multiplication formula for $\frac{\Gamma(ks-k+1)\Gamma(s+r)}{\Gamma(s)}$, they arrive at Steen's function. Cf. e.g., [Gupta and Maji (2021), Theorem 3.1].

[Dixit *et al.* (2023)] considered the situation similar to the above and deduced Voronoĭ summation formulas. In place of (4.1) and (4.5), they put $Z(s) = \zeta(s)\zeta(ks-r)$ and $W(s) = \zeta(ks)\zeta\left(s+1-\frac{r+1}{k}\right)$, respectively. Their method is based on Koshlyakov's idea and depends on the Plana summation formula with $K_{r_1,r_2}(x)$ and $\sigma(x)$ replaced by the more involved entities:

$$K_r^{(k)}(x) = \frac{1}{2\pi i} \int_{(c)} \Gamma(s) \cos \frac{\pi}{2} s \Gamma\left(\frac{s-1-r}{k}+1\right) \frac{x^{-s}}{k} ds \tag{4.9}$$

and

$$\Phi_k(x; r) = C \sum_{n=1}^{\infty} S_r^{(k)}(n) K_r^{(k)}\left((2\pi)^{\frac{1}{k}+1}(nx)^{\frac{1}{k}}\right)$$

$$= \frac{C}{2\pi i k} \int_{(c)} \Gamma(s) \cos \frac{\pi}{2} s \Gamma\left(\frac{s-1-r}{k}+1\right) \zeta(s)$$

$$\zeta\left(\frac{s-r-1}{k}+1\right)\left((2\pi)^{\frac{1}{k}+1} x^{\frac{1}{k}}\right)^{-s} ds. \tag{4.10}$$

$\Phi_k(x; r)$ has properties corresponding to Lemma 2, especially the partial fraction expansion. The final result contains the H-function corresponding to Koshlyakov's L-function, which is connected to the K-function by (1.20). To prove a generalization of this relation is the hardest part of [Dixit *et al.* (2023)], which they attain by the uniqueness of the solution to a differential equation.

10.4.2 The Case of Krätzel

[Krätzel (1981)] deals with a generalization of the eta-function, which depends on the Hecke gamma transform of the zeta-function.

$$Z_{a.b}(s) := \frac{\pi}{\Gamma(s+1)\sin\frac{\pi}{2}s}\zeta(-as)\zeta(bs). \tag{4.11}$$

where a, b are natural numbers, $(a,b)=1$. We give a brief account. $Z_{a.b}(s)$ satisfies the Hecke functional equation

$$\Gamma(s)Z_{a.b}(s) = \Gamma(-s)Z_{b,a}(-s). \tag{4.12}$$

This is because the factor $\frac{\Gamma(s)}{\Gamma(s+1)}$ being $\frac{1}{s}$ remains invariant under the change of variable $s \to -s$, For the moment, we work with $\left(\operatorname{Re} x > 0 \text{ and } |\arg z| < \frac{\pi}{2ab}\right)$

$$\eta_{a,b}(x) := \prod_{m=1}^{\infty}\prod_{\nu=1}^{a-1}\left(1 - e^{2\pi i\epsilon_{2\mu+1}(4a)n^{\frac{b}{a}}x^b}\right), \tag{4.13}$$

where $\epsilon_{2\mu+1}(4a) = e^{2\pi i\frac{2\mu+1}{4a}}$. Then for $\varkappa > \frac{a}{b}$, we have by the Hecke gamma transform

$$\log\eta_{a,b}(x) = -\frac{1}{2\pi i}\int_{(\varkappa)}\Gamma(s)\zeta(s+1)\zeta\left(\frac{b}{a}s\right)\sum_{\nu=1}^{a-1}\left(2\pi ie^{2\pi i\frac{2\nu+1}{4a}}x^b\right)^{-s}ds. \tag{4.14}$$

Now the sum becomes

$$\sum_{\nu=1}^{a-1}\left(ie^{2\pi i\frac{2\nu+1}{4a}}\right)^{-s} = \frac{\sin\frac{\pi}{2}s}{\sin\frac{\pi}{2a}s}.$$

Hence (4.14) becomes

$$\log\eta_{a,b}(x) = -\frac{1}{2\pi i}\int_{(\varkappa)}\Gamma(s)\frac{\sin\frac{\pi}{2}s}{\sin\frac{\pi}{2a}s}\zeta(s+1)\zeta\left(\frac{b}{a}s\right)\left(2\pi x^b\right)^{-s}ds. \tag{4.15}$$

Comparing this with (4.6), we find that Krätzel's case with $a=1, b=k, r-1$ corresponds to a special case of Gupta and Maji, i.e., r being the odd case.

It is more or less known that the r odd case leads to Hecke's functional equation and the transformation formula for the Lambert series.

Now we apply the functional equation only to one factor $\zeta(s+1)$:

$$\zeta(s+1) = -(2\pi)^s \frac{\pi}{\Gamma(s+1)\sin\frac{\pi}{2}s}\zeta(-s). \tag{4.16}$$

Substituting (4.16) in (4.15), we obtain

$$\log \eta_{a,b}(x) = \frac{1}{2\pi i}\int_{(\varkappa)} \frac{\Gamma(s)}{\Gamma(s+1)\sin\frac{\pi}{2a}s}\pi\zeta(-s)\zeta\left(\frac{b}{a}s\right)\left(x^b\right)^{-s} ds. \tag{4.17}$$

By the change of variable $s \to -a$ (4.17) becomes as in Krätzel,

$$\begin{aligned}\log \eta_{a,b}(x) &= \frac{1}{2\pi i}\int_{(\varkappa_1)} \frac{\Gamma(s)}{\Gamma(s+1)\sin\frac{\pi}{2}s}\pi\zeta(-as)\zeta(bs)\left(x^{ab}\right)^{-s} ds \\ &= \frac{1}{2\pi i}\int_{(\varkappa_2)} \Gamma(s) Z_{a,b}(s) x^{-abs} ds \end{aligned} \tag{4.18}$$

where $\varkappa_1 > \frac{1}{b}$, i.e., the Hecke gamma transform of $Z_{a,b}(s)$. As usual, shifting the integration path to $-\varkappa_2 < -\frac{1}{a}$, we encounter poles, and we are to find the sum of residues:

$$-\gamma_{a,b}(x) + \gamma_{b,a}\left(\frac{1}{x}\right) + \frac{1}{2}(b-a)\log 2\pi - \frac{ab}{2}\log x, \tag{4.19}$$

where

$$\gamma_{a,b}(x) = \frac{\pi\zeta\left(-\frac{b}{a}\right)}{\sin\frac{\pi}{2a}}x^b. \tag{4.20}$$

We note that by (4.12) the resulting integral is the same as (4.18) with x replaced by $\frac{1}{x}$.

Hence defining

$$\eta_{a,b}(x) = (2\pi)^{\frac{1-b}{2}} e^{\gamma_{a,b}(x)}\eta_{a,b}(x), \tag{4.21}$$

we conclude that

Theorem 5 *The transformation formula*

$$\eta_{a,b}(x) = x^{-\frac{ab}{2}}\eta_{b,a}\left(\frac{1}{x}\right) \tag{4.22}$$

is a consequence of the pseudo-modular relation (4.12) and the functional equation (4.16).

This type of pseudo-modular relation and a functional equation for one of the factors often appear in the literature, especially in the theory of Dedekind sums [Wang *et al.* (2023a)] and other areas [Wang *et al.* (2023b)].

10.5 EQUIVALENTS TO THE RIEMANN FUNCTIONAL EQUATION

Integrating (1.28), we obtain

$$\frac{A}{\pi}\sum_{n=1}^{\infty} a_n \int K_{r_1,r_2}\left(A^2xn\right)\, dx$$
$$= \int \mathrm{P}_K(x)\, dx + \frac{1}{2\pi}\sum_{n=1}^{\infty} a_n \log\left(n^2+x^2\right) + C, \tag{5.1}$$

where C is an integration constant. Here $\int \mathrm{P}_K(x)\, dx = -\frac{1}{2}\rho x - \frac{\zeta_\Omega(0)}{\pi}\log x$ and in the rational case, the left-hand side is $-\sum_{n=1}^{\infty}\frac{a_n}{2\pi n}e^{-2\pi xn}$. Hence

$$\frac{1}{2\pi}\log\left(1-e^{-2\pi x}\right) = -\sum_{n=1}^{\infty}\frac{1}{2\pi n}e^{-2\pi xn} = -\frac{1}{2}x + \frac{1}{2\pi}\log x$$
$$+\frac{1}{2\pi}\sum_{n=1}^{\infty}\log\left(n^2+x^2\right) + C. \tag{5.2}$$

Assuming all the terms are real, Hamburger put $z = x+iy$ with $y > 0$ for x, and let $x \to 0+$. Then the series on the right-hand side is imaginary only for $n < [y]$, so that $\frac{1}{2\pi}\sum_{n<[y]}\left(\log\left(y^2-n^2\right)+\pi i\right)$. Hence the imaginary part is $\frac{1}{2}\sum_{n<[y]}1 - \frac{1}{2}y + \frac{1}{2}$, which must be $-\sum_{n=1}^{\infty}\frac{1}{2\pi n}\sin(-2\pi yn)$. Hence

$$y - [y] - \frac{1}{2} = -\sum_{n=1}^{\infty}\frac{1}{\pi n}\sin 2\pi yn, \tag{5.3}$$

which is the special case $\varkappa = 1$ of (1.41). Further integrating, Hamburger arrived at (1.33). Hamburger did not consider the real part, but it is equally interesting. We understand (5.2) as the definite integral from 0 to x

$$\left[\frac{1}{2\pi}\log\frac{1-e^{-2\pi x}}{x}\right]_0^x = \left[-\frac{1}{2}x + \frac{1}{2\pi}\sum_{n=1}^{\infty}\log\left(n^2+x^2\right)\right]_0^x, \tag{5.4}$$

i.e.

$$-\sum_{n=1}^{\infty}\frac{1}{n}e^{-2\pi xn} = \log\left(1-e^{-2\pi x}\right)$$
$$= -\pi x + \log x + \log(2\pi) + \sum_{n=1}^{\infty}\log\frac{n^2+x^2}{n^2}, \tag{5.5}$$

Suppose $0 < y < 1$ and proceed as above. Then the real part of (5.5) reads

$$-\sum_{n=1}^{\infty}\frac{1}{n}\cos 2\pi yn = \log 2\pi y + \sum_{n=1}^{\infty}\log\frac{n^2-y^2}{n^2}. \tag{5.6}$$

By the product representation for the sine function

$$\frac{\sin\pi z}{\pi z} = \prod_{n=1}^{\infty}\left(1-\frac{z^2}{n^2}\right), \tag{5.7}$$

the right-hand side of (5.6) is $\log 2\ \sin 2\pi y$, and (5.6) amounts to the special case $\varkappa = 1$ of (1.41). By the reciprocal relation for the gamma function

$$\Gamma(s)\,\Gamma(1-s) = \frac{\pi}{\sin\pi s}, \tag{5.8}$$

we see that $\log\sin\pi s$ is the even part of $\log\Gamma(s)$.

(5.2) does not hold true in the ordinary sense, and to incorporate the convergence factor $-\frac{1}{2\pi}\sum_{n=1}^{\infty}\log n^2 = -\frac{1}{2\pi}\log\prod_{n=1}^{\infty}n^2$, we introduce the regularized product

$$\prod_{n=1}^{\infty}{}^{\text{reg}}\, n = e^{-\zeta'(0)} = \log\sqrt{2\pi} \tag{5.9}$$

and the constant is to be $-\frac{1}{2\pi}\log 2\pi$.

Theorem 6 *In addition to the identities in Corollary 1, the following are equivalent to the Riemann functional equation (1.30).* Γ **type**: *the product representation for the sine function (5.7), the reciprocal relation for the gamma function (5.8), and the Kummer Fourier series*

$$\log\frac{\Gamma(x)}{\sqrt{2\pi}} = -\frac{1}{2}\log(2\ \sin\pi x) + \frac{1}{2}(\gamma + \log 2\pi)(1-2x)$$
$$+\frac{1}{\pi}\sum_{n=1}^{\infty}\frac{\log n}{n}\sin 2\pi nx, \tag{5.10}$$

The Fourier series for the periodic Bernoulli polynomial (1.41) lies in the Fourier series type along *with the Kummer Fourier series. The* **Fourier-Bessel expansion type**: *partial fraction expansion for the cotangent function is a special case of the Fourier-Bessel expansion (Watson's formula) and the Hurwitz formula* $(0 < a < 1)$

$$\zeta(s,a) = -i(2\pi)^{s-1}\Gamma(1-s)\left(e^{\frac{\pi}{2}is}l_{1-s}(a) - e^{-\frac{\pi}{2}is}l_{1-s}(1-a)\right), \quad (5.11)$$

where $\zeta(s,a)$ *is the Hurwitz zeta-function.*

The integral expression

$$\zeta(1-s,x) = \frac{1}{2\pi i}\int_{\alpha} x^{-z}F(s,z)\,dz, \frac{1}{2} < \sigma, \max\{0, 1-\sigma\} < \alpha < \frac{1}{2}, \quad (5.12)$$

is a form of the functional equation (5.11), where

$$F(s,z) = 2\frac{\Gamma(s)}{(2\pi)^s}(2\pi)^{-z}\cos\frac{\pi}{2}(s-z)\Gamma(z)\zeta(s+z). \quad (5.13)$$

*[*Mikolás (1956)*], [Li et al. (2012)].*

Summation formula type: *Poisson, Plana.*

The proof of Γ type can be found in [Kanemitsu and Tsukada (2007), §5.4]. Summation formula types are proved in [Liu *et al.* (2023)] and in this note.

Acknowledgment: The third author wishes to express his thanks to Professor Dixit for sending his preprint [Dixit *et al.* (2023)] and for creative discussions.

REFERENCES

1. Dixit, A., Maji, B., and Vatwani, A. (2021). Voronoĭ summation formula for the generalized divisor function $\sigma_z^k(n)$. Preprint.
2. Erdélyi, A, Magnus, W., Oberhettinger, F. and Tricomi, F. G. (1953). *Higher transcendental functions*, I-III McGraw-Hill, New York, Toronto and London.
3. Gupta, A. and Maji, B. (2021). On Ramanujan's formula for $\zeta(1/2)$ and $\zeta(2m+1)$, *J. Math. Anal. Appl.* **507** (1), 1 March 2022, 125738.

4. Hamburger, H. (1922). Über einige Beziehungen, die mit der Funktionalgleichung der Riemannschen ζ-Funktion äquivalent sind, *Math. Ann.* **85**, 129–140.
5. Kanemitsu, S. and Tsukada, H. (2007). *Vistas of special functions*, World Scientific, Singapore.
6. Kanemitsu, S. and Tsukada, H. (2014). *Contributions to the theory of zeta-functions: The modular relation supremacy*, World Scientific, Singapore.
7. Katznelson, Y. (2004). *An introduction to harmonic analysis*, 3rd ed., Cambridge University Press, Cambridge.
8. Koshlyakov, N. S. (1928/29a). Application of the theory of sum-formulae to the investigation of a class of one-valued analytical functions in the theory of numbers, *Mess. Math.* **58**, 1–23.
9. Koshlyakov, N. S. (1928/29b). On Voronoĭ's sum formula, *Mess. Math.* **58**, 30–32.
10. Koshlyakov, N. S. (1934). On a general summation formula and its applications, *Dokl. Akad. Nauk SSSR.* **4**, 187–190.
11. Koshlyakov, N. S. (1954). Investigation of some questions of analytic theory of the rational and quadratic fields I-III (Russian), *Izv. Akad. Nauk SSSR, Ser. Mat.* **18**, 113–144, 213–260, 307–326; errata **19** (1955), 277.
12. Krätzel, E. (1981). Dedekindsche Funktionen und Summen, I, II *Period. Math. Hungar.* **12**, 113–123, 163–179.
13. Laurinčikas, A. (1998). On Voronoĭ summation formulae, Voronoĭ's impact on modern science. Book I., *Proc. Inst. Math. Natl. Acad. Sci. Ukr., Math. Appl.* **21** (1), P. Engel and H. Syta (Eds.), Inst, of Math., Kviv, pp. 117–136.
14. Li, H. -L., Kanemitu, S. and Wang, X. -H. (2012). Plana's summation formula as a modular relation and applications, *Int. Transf. Special Functions.* **23**, 183–190.
15. Liu, G. -D., Chakraborty, K. and Kanemitsu, S. (2023). *A unifying principle in the theory of modular relations*, Mathematics 11030535, MDPI, Switzerland.
16. Mikolás, M. (1956). Mellinsche Transformation und Orthogonalität bei $\zeta(s, u)$; Verallgemeinerung der Riemannschen Funktionalgleichung von $\zeta(s)$, *Acta Sci. Math. (Szeged)* **17**, 143–164.
17. Narkiewicz, W. (2004). *Elementary and algebraic theory of algebraic numbers*, 3rd ed., Springer Verlag, Berlin-Heidelberg.

18. Oberhettinger, F. (1974). *Tables of Mellin transforms*, Springer Verlag, Berlin-Heidelberg-New York.
19. Oberhettinger, F. and Soni, K. L. (1972). On some relations which are equivalent to the functional equations involving the Riemann zeta function, Math. Z. **127**, 17-34.
20. Paris, R. B. and Kaminski, D. (2001). *Asymptotics and Mellin-Barnes integrals*, Cambridge University Press, Cambridge.
21. Prudnikov, A, P. Bychkov, Yu. A. and Marichev, O. I. (1986). *Integrals and Series, Supplementary Chapters*, Izd. Nauka, Moscow.
22. Soni, K. (1964). *On some relations which are equivalent to the functional equations of type of Riemann's zeta function and tables of divisor transforms*, Ph.D. Thesis, Oregon State University, Corvallis.
23. Soni, K. (1966). Some relations associated with an extension of Koshliakov's formula, *Proc. Amer. Math. Soc.* **17**, 543–551.
24. Titchmarsh, E. C. (1951). *The Theory of the Riemann Zeta-Function*, 2nd edition revised by D. R. Heath-Brown, Oxford University Press, Oxford, 1986.
25. Wang, N. -L., Kuzumaki, T. and Kanemitsu, S. (2023a). *Generalized eta transformation formulas and Dedekind sums viewed as modular relations, Mathematics, MDPI, Switzerland, to appear.*
26. Wang, N. -L., Kuzumaki, T. and Kanemitsu, S. (2023b). *On Koshlyakov's kernels and Fourier-Bessel expansion, in preparation.*
27. Watson, G. N. (1944). *A treatise on the theory of Bessel functions*, 2nd ed., Cambridge University Press, Cambridge.
28. Yamamoto, Y. (1977). Dirichlet series with periodic coefficients, Proceedings of International Symposium "Algebraic Number Theory", Kyoto 1976, 275–289. JSPS, Tokyo.

CHAPTER 11

Extended and Efficient Secant-Type Methods Based on Generalized Schmidt-Schwetlick Conditions

Ioannis K. Argyros, Jinny Ann John, and Jayakumar Jayaraman

11.1 INTRODUCTION

In the realm of applied science and technology, a multitude of complex challenges can be effectively approached by framing them as nonlinear equations of the following form:

$$F(x) = 0. \tag{1.1}$$

In this context, $F : \Omega \subset B_1 \to B_2$ represents a differentiable function according to the Fréchet sense, where B_1, B_2 signifies a complete normed linear space, and Ω denotes a non-empty, open, and convex set.

The pursuit of closed-form solutions for these nonlinear equations typically presents formidable difficulties. Consequently, iterative methods have emerged as the go-to approach for seeking solutions to such problems. Among these methods, Newton's method [19, 26, 18, 21, 1, 11, 28, 9, 2, 3, 10, 7, 6] stands out as a widely adopted choice, primarily due to its remarkable quadratic convergence rate when tackling equations like (1.1).

DOI: 10.1201/9781003530602-11

In recent years, the fields of science and mathematics have witnessed significant advancements, leading to the discovery and application of various higher-order iterative techniques for solving nonlinear equations [12, 19, 26, 18, 21, 1, 11, 15, 28, 24, 9, 22, 23, 27, 16, 13, 17, 14, 20, 25, 2, 3, 4, 10, 8, 5, 7, 6]. However, these advanced methods often grapple with the drawback of demanding the computation of second- and higher-order derivatives, which can significantly hinder their practicality in real-world applications. The computational overhead associated with evaluating F during each iteration makes classical cubic convergent schemes less suitable. Furthermore, it's worth noting that many of these methods rely on Taylor expansions, necessitating derivatives of higher order not inherently present within the method itself.

An in-depth analysis of the local and semi-local behavior of iterative methods offers invaluable insights into their convergence properties, error bounds, and the region where solutions are unique. A multitude of studies have dedicated their focus to investigating the local and semi-local convergence aspects of efficient iterative techniques, yielding substantial outcomes in the form of convergence radii, error estimations, and expanded applicability of these methods [19, 26, 18, 21, 1, 14, 2, 3, 5, 7, 6]. Such findings carry significant weight, especially in guiding the selection of appropriate initial points for the iterative process.

In this chapter, a few variants of the secant method for solving nonlinear equations are scrutinized and investigated for their convergence properties. The methods are defined for $\alpha \geq 0$, $x_{-1}, x_0 \in \Omega$, $F : \Omega \subseteq B_1 \to B_2$ and each $n = 0, 1, 2, \ldots$ by

$$x_{n+1} = x_n - [x_{n-1}, x_n; F]^{-1} F(x_n), \tag{1.2}$$

$$\begin{aligned} y_n &= x_n - [x_{n-1}, x_n; F]^{-1} F(x_n), \\ x_{n+1} &= y_n - [x_{n-1}, x_n; F]^{-1} F(y_n), \end{aligned} \tag{1.3}$$

$$\begin{aligned} y_n &= x_n - [x_{n-1}, x_n; F]^{-1} F(x_n), \\ x_{n+1} &= x_n - [y_n, x_n; F]^{-1} F(x_n), \end{aligned} \tag{1.4}$$

$$\begin{aligned} y_n &= x_n - [x_{n-1}, x_n; F]^{-1} F(x_n), \\ x_{n+1} &= y_n - [y_n, x_n; F]^{-1} F(y_n), \end{aligned} \tag{1.5}$$

and

$$\begin{aligned} y_n &= x_n - [x_{n-1}, x_n; F]^{-1} F(x_n), \\ z_n &= (1-\alpha) x_n + \alpha y_n, \\ x_{n+1} &= y_n - [z_n, x_n; F]^{-1} F(y_n). \end{aligned} \tag{1.6}$$

The local convergence orders $\frac{1+\sqrt{5}}{2}$, 2, $1+\sqrt{2}$, $1+\sqrt{2}$, $1+\sqrt{3}$, respectively are established in [**12**] under the Taylor expansion series approach for $B_1 = B_2 = \mathbb{R}^k$, and by assuming the existence of derivatives not on these methods, making them unsuitable for solving non-differentiable equations. In this chapter, generalized Schmidt-Schwetlick [**13**] conditions are employed to show convergence that do not require derivatives, making them suitable for solving non-differentiable equations. The more interesting semi-local convergence analysis not studied in [**12**] for the last four methods is developed in this chapter using majorizing sequences. It is worth noticing that our approach extends the applicability of the secant method (1.2) given in [**12**] in the local and semi-local cases. Moreover, the analysis is developed in the more general setting of a Banach space. Our approach can be used to extend the applicability of other methods using divided differences such as Kurchatov's or other methods along the same lines [1, 11, 15, 28, 24, 9, 16, 13, 17, 25, 2, 4, 10].

Let us consider the academic and motivational example. Consider Ω to be an interval containing 0 and 1. Define $f(t) = \gamma_1 t^4 \log t + \gamma_2 t^5 + \gamma_3 t^4$ for $t \neq 0$ with $\gamma_1 \neq 0$, $\gamma_2 + \gamma_3 = 0$, and $f(t) = 0$ for $t = 0$. Then, the results in [**12**], see e.g. Section 2.3, cannot assure the convergence of these methods to the solution $t^* = 1$ of the equation $f(t) = 0$, since, the functions $f^{(j)}$, $j = 4, 5, ...$ are not continuous at $t = 0$. But these methods converge to the solution, say for $t_{-1} = 0.9$ and $t_0 = 0.95$. Hence, the convergence results in [**12**] can be weakened.

The convergence conditions for the local case are:

- $\left\|\mathcal{L}^{-1}\left([x,y;F] - \mathcal{L}\right)\right\| \leq \phi_0\left(\|x-y\|, \|x-x^*\|, \|y-x^*\|\right)$,
- $\left\|\mathcal{L}^{-1}\left([x,y;F] - [y,x^*;F]\right)\right\| \leq \phi\left(\|y-x\|, \|x-x^*\|, \|y-x^*\|\right)$,
- $\left\|\mathcal{L}^{-1}\left([x,x^*;F] - \mathcal{L}\right)\right\| \leq \bar{\phi}_0\left(\|x-x^*\|\right)$.

The convergence conditions for the semi-local analysis case are:

- $\left\|\mathcal{L}^{-1}\left([x, y; F]-\mathcal{L}\right)\right\| \leq \psi_0\left(\|x-x_0\|, \|y-x_0\|, \|y-x\|\right)$,
- $\left\|\mathcal{L}^{-1}\left([x, y; F]-[z, y; F]\right)\right\| \leq \psi\left(\|x-z\|, \|x-y\|, \|z-y\|\right)$,
- $\left\|\mathcal{L}^{-1}\left([x, x^*; F]-\mathcal{L}\right)\right\| \leq \overline{\psi}_0\left(\|x^*-x_0\|, \|x-x_0\|\right)$

where $\mathcal{L}$ stands for an invertible operator, and the functions $\phi_0, \phi, \overline{\phi}_0, \psi_0, \psi$, and $\overline{\psi}_0$ are non-decreasing and continuous on the non-negative axis or a subset of it.

11.2 LOCAL ANALYSIS OF CONVERGENCE

Method (1.2):

The main result for method (1.2) is:

Theorem 2.1 *Suppose the conditions (C_1)-(C_3) hold;*

The equation $\phi_0(t, t, 2t)-1=0$ has a smallest positive solution denoted by ρ;

The equation $h_1(t)-1=0$ has a smallest positive solution denoted by $r_1 \in (0, \rho)$, where

$$h_1(t)=\frac{\phi(t, t, 2t)}{1-\phi_0(t, t, 2t)}$$

and

$$S[x*, r1] \subset \Omega.$$

Then, the sequence $\{x_n\}$ generated by the method (1.2) is convergent to x^* provided that $x_{-1}, x_0 \in S_0=S(x^*, r_1)-\{x^*\}$ such that

$$\|x_{n+1}-x^*\| \leq h_1(r)\|x_n-x^*\|. \tag{2.1}$$

Proof. By the choice of x_{-1}, $x_0 \in S_0$, and (C_1), we have in turn

$$\begin{aligned}\|\mathcal{L}^{-1}([x_{-1}, x_0; F]-\mathcal{L})\| &\leq \phi_0(\|x_{-1}-x^*\|, \|x_0-x^*\|, \|x_{-1}-x_0\|)\\ &\leq \phi_0(r_1, r_1, 2r_1)<1.\end{aligned} \tag{2.2}$$

It follows by (2) and the Banach Lemma on invertible operators [19, 26, 18, 21, 1] that $[x_{-1}, x_0; F]$ is invertible, and

$$\left\|[x_{-1}, x_0; F]^{-1}\mathcal{L}\right\| \leq \frac{1}{1-\phi_0(\|x_{-1}-x^*\|, \|x_0-x^*\|, \|x_{-1}-x_0\|)}. \tag{2.3}$$

Moreover, the iterate x_1 is well defined by the first step of method (1.2) for $n = 0$, and we can write

$$x_1 - x^* = x_0 - x^* - [x_{-1}, x_0; F]^{-1} F(x_0)$$
$$= [x_{-1}, x_0; F]^{-1} ([x_{-1}, x_0; F] - [x_0, x^*; F]) (x_0 - x^*). \quad (2.4)$$

By applying the condition (C_2), the definition of the function h_1, the radius r_1, and (2), we get in turn from (2)

$$\|x_1 - x^*\| \leq \frac{\phi (\|x_{-1} - x^*\|, \|x_0 - x^*\|, \|x_{-1} - x_0\|) \|x_0 - x^*\|}{1 - \phi_0 (\|x_{-1} - x^*\|, \|x_0 - x^*\|, \|x_{-1} - x_0\|)} \quad (2.5)$$

$$\leq h_1 (r_1) \|x_0 - x^*\| < r_1. \quad (2.6)$$

Thus, the iterate $x_1 \in S(x^*, r_1) - \{x^*\}$, and the estimate (2.1) holds if $n = 0$. Simply, replace x_{-1}, x_0 by x_m, x_{m+1}, respectively in the preceding estimations to complete the induction for (2.1). Then, from the estimation

$$\|x_{m+1} - x^*\| \leq c\|x_m - x^*\| < r_1, \quad (2.7)$$

where, $c = h_1 (\max \{\|x_{-1} - x^*\|, \|x_0 - x^*\|\}) \in [0, 1)$, we deduce that $x_{m+1} \in S(x^*, r_1)$, and $\lim_{m \to +\infty} = x^*$. □

A domain is specified in the next result that contains only x^* as a solution of the equation $F(x) = 0$.

Proposition 2.2 *There exists a solution $\bar{x} \in S(x^*, \bar{r})$ of the equation $F(x) = 0$ for some $\bar{r} > 0$;*

The condition (C_3) holds in the ball $S(x^*, \bar{r})$, and there exists $\bar{\bar{r}} \geq \bar{r}$ so that

$$\bar{\phi}_0 (\bar{\bar{r}}) < 1. \quad (2.8)$$

Define the domain $\Omega_1 = \Omega \cap S[x^*, \bar{\bar{r}}]$. Then, the only solution of the equation $F(x) = 0$ in the domain Ω_1 is x^*.

Proof. Suppose that $\bar{x} \neq x^*$ and $F(\bar{x}) = 0$. Then, the divided difference $T = [x^*, \bar{x}; F]$ is well defined. It follows by the condition (C_3) and (2.8) that

$$\left\|\mathcal{L}^{-1} (T - \mathcal{L})\right\| \leq \bar{\phi}_0 (\|\bar{x} - x^*\|) \leq \bar{\phi}_0 (\bar{\bar{r}}) < 1.$$

Hence, the linear operator $\mathcal{L}$ is invertible. Therefore, from the identity

$$x^* - \bar{x} = T^{-1} (F(x^*) - F(\bar{x})) = T^{-1} (0) = 0,$$

we conclude that $\bar{x} = x^*$. □

Remark 2.3

- In Proposition 2.2, we can certainly pick $\bar{x} = x^*$ and $\bar{r} = r_1$.
- A popular but not the most flexible choice for $\mathcal{L} = [x_{-1}, x_0; F]$ [**12**] or $\mathcal{L} = F'(x^*)$ or $\mathcal{L} = I$. Note that the local convergence is shown using the existence of the Fréchet derivative $F'(x)$, which makes the Secant method not applicable to solve non-differentiable equations, although the derivative does not appear in method (1.2). In our Theorem 2.1, the derivative is not assumed or implied by the conditions (C_1)–(C_3). Therefore, Theorem 2.1 can be relied upon to show the local convergence of the Secant method (1.2) to a solution of a non-differentiable equation.
- The conditions (C_1)–(C_3) generalize the ones given in [24, 22, 23]. In fact, they can be reduced to the ones in [24, 22, 23] if we take $\phi_0 = \phi$ and $\phi(t_1, t_2, t_3) = \alpha t_1 + \beta t_2 + \gamma t_3$ for some non-negative constants $\alpha, \beta,$ *and* γ.
- The condition (C_2) can be weakened as follows: Define the domain $D = \Omega \cap S(x^*, \rho)$. Then, notice that we can replace (C_2) by $\|\mathcal{L}^{-1}([x, y; F] - [y, x^*; F])\| \leq \bar{\phi}(\|x - x^*\|, \|y - x^*\|, \|y - x\|)$. But then $\bar{\phi} \leq \phi$, since $D \subset \Omega$. This way, the at least as tight function $\bar{\phi}$ can replace ϕ in the previous results.
- It is worth emphasizing that the intent of this chapter is to obtain the largest possible convergence, which allows a wider choice of initial points as well as the largest possible domain containing only one solution. The convergence order $\frac{1+\sqrt{5}}{2}$ is well established with the choice of the function ϕ_0 and ϕ as in (iii) [**24**].

The rest of the methods defined in the Introduction are studied analogously. But the role of the function ϕ is simply extended to include all combinations of the iterates involved, and the convergence radius uses r_1 as well as the solution of a second h function as follows.

Method (1.3):

The second function is

$$h_2(t) = \frac{\phi(t, t, h_1(t)t, 2t, t + h_1(t)t, t + h_1(t)t)h_1(t)}{1 - \phi_0(t, t, 2t)}.$$

Suppose that the equation $h_2(t) - 1 = 0$ has a smallest solution $r_2 \in (0, \rho)$. Define

$$r = \min\{r_1, r_2\}. \tag{2.9}$$

The corresponding local result to Theorem 2.1 is:

Theorem 2.4 *Suppose that the conditions* (C_1)*–*(C_3) *hold, and r given by (2.9) replaces* r_1*. Then, the sequence* $\{x_n\}$ *generated by (1) is convergent to* x^**.*
Proof. Notice the estimate

$$x_{n+1} - x^* = [x_{n-1}, x_n; F]^{-1}([x_{n-1}, x_n; F] - [y_n, x^*; F])(y_n - x^*),$$

leading to

$$\begin{aligned}&\|x_{n+1} - x^*\|\\ &\leq \frac{\phi(\|x_n - x^*\|, \|x_n - x^*\|, \|y_n - x^*\|, \|x_{n-1} - x_n\|, \|x_{n-1} - y_n\|, \|x_n - y_n\|)\|y_n - x^*\|}{1 - \phi_0(\|x_{n-1} - x^*\|, \|x_n - x^*\|, \|x_n - x_{n-1}\|)}\\ &\leq h_2(r)\|x_n - x^*\| \leq \|x_n - x^*\| < r.\end{aligned}$$

□

The uniqueness of the solution x^* is given in Proposition 2.2. Clearly, the remarks for this method are omitted as already given in Remark 2.3.

Method (1.4):

The function h_2 is defined by

$$h_2(t) = \frac{\phi(t, h_1(t)t, t + h_1(t)t)}{1 - \phi_0(t, h_1(t)t, t + h_1(t)t)}.$$

Suppose that the equation $h_2(t) - 1 = 0$ has a smallest solution $r_3 \in (0, \rho)$. Set

$$r = \min\{r_1, r_3\}. \tag{2.10}$$

Theorem 2.5 *Suppose that the conditions* (C_1)*–*(C_3) *hold, and r is given by (2.10). Then, the sequence* $\{x_n\}$ *generated by method (1) is convergent to* x^**.*
Proof. The motivation for the function h_2 follows from the calculation

$$\begin{aligned}x_{n+1} - x^* &= x_n - x^* - [y_n, x_n; F]^{-1}F(x_n)\\ &= [y_n, x_n; F]^{-1}([y_n, x_n; F] - [x_n, x^*; F])(x_n - x^*),\end{aligned}$$

so

$$\|x_{n+1} - x^*\| \leq \frac{\phi(\|x_n - x^*\|, \|y_n - x^*\|, \|y_n - x_n\|)\|x_n - x^*\|}{1 - \phi_0(\|x_n - x^*\|, \|y_n - x^*\|, \|y_n - x_n\|)}$$
$$\leq h_2(r)\|x_n - x^*\| \leq \|x_n - x^*\| < r.$$

□

Method (1.5):

The function h_2 is defined by

$$h_2(t) = \frac{\phi(t, h_1(t)t, t + h_1(t)t)h_1(t)}{1 - \phi_0(t, h_1(t)t, t + h_1(t)t)}.$$

Suppose that the equation $h_2(t) - 1 - 0$ has a smallest solution $r_4 \in (0, \rho)$. Set

$$r = \min\{r_1, r_4\}. \tag{2.11}$$

Theorem 2.6 *Suppose that the conditions (C_1)-(C_3) hold. Then, the sequence $\{x_n\}$ generated by method (1) is convergent to x^*.*

Proof. This time we rely on the estimation

$$x_{n+1} - x^* = [y_n, x_n; F]^{-1}([y_n, x_n; F] - [y_n, x^*; F])(y_n - x^*)$$

leading to

$$\|x_{n+1} - x^*\| \leq \frac{\phi(\|y_n - x^*\|, \|x_n - x^*\|, \|y_n - x_n\|)\|y_n - x^*\|}{1 - \phi_0(\|x_n - x^*\|, \|y_n - x^*\|, \|y_n - x_n\|)}$$
$$\leq h_2(r)\|x_n - x^*\| \leq \|x_n - x^*\| < r.$$

□

Method (1.6):

There are three functions$h$ in this case since there are three sub-steps instead of two. The functions h_1, h_3 are h_1, h_2, respectively, and are given in method (1). But in the second function,

$$\overline{h}_2(t) = |1 - \alpha|t + |\alpha|h_1(t)t.$$

Suppose that in the equation, $\overline{h}(t) - 1 = 0$ has a smallest solution $\tilde{r} \in (0, \rho)$. Define

$$r = \min\{r_1, \tilde{r}, r_3\}. \tag{2.12}$$

Theorem 2.7 *Suppose that the conditions* (C_1)*–*(C_3) *hold. Then, the sequence generated by method (1) is convergent to* x^*.

Proof. Notice that

$$\begin{aligned} z_n - x^* &= (1-\alpha)x_n + \alpha y_n - x^* \\ &= (1-\alpha)(x_n - x^*) + \alpha(y_n - x^*), \\ \|z_n - x^*\| &\leq |1-\alpha|\,\|x_n - x^*\| + |\alpha|\,\|y_n - x^*\| \\ &\leq \bar{h}_2(r)\,\|x_n - x^*\| \leq \|x_n - x^*\| < r. \end{aligned}$$

□

The semi-local analysis of convergence follows by replacing the conditions (C_1)–(C_2), x^*, the functions φ by (H_1)–(H_3), x_0, the functions ψ, respectively, and using majorizing sequences [1–7, 21] as given in the next section.

11.3 SEMI-LOCAL ANALYSIS OF CONVERGENCE

A scalar sequence is constructed that plays the majorizing role for the method (1.2) (see Theorem **3.1**).

Suppose:

- The equation $\psi_0(t,t,2t) - 1 = 0$ has a smallest positive solution denoted by q_0. Let $x_0, x_{-1} \in \Omega$. Define the parameters a_{-1} and a_0 by $a_{-1} = 0$ and $\|x_0 - x_{-1}\| \leq a_0$ such that $a_0 < q_0$. Then, the condition (H_1) gives for $x = x_{-1}$ and $y = x_0$

$$\begin{aligned} \left\|\mathcal{L}^{-1}([x_{-1}, x_0; F] - \mathcal{L})\right\| &\leq \psi_0(\|x_{-1} - x_0\|, \|x_0 - x_0\|, \|x_0 - x_{-1}\|) \\ &\leq \psi_0(a_0, 0, a_0) < 1. \end{aligned}$$

Thus, the linear operator $[x_{-1}, x_0; F]$ is invertible, and the iterate x_1 is well defined by method (1.2) for $n = 0$. Define the parameter a_1 by $a_1 \geq a_0 + \left\|[x_{-1}, x_0; F]^{-1} F(x_0)\right\|$. Then, we have $\|x_1 - x_0\| \leq a_1 - a_0$. Define the sequence $\{a_n\}$ for each $n = 2, 3, \ldots$ by

$$b_{n+1} = \psi(a_{n+1} - a_{n-1}, a_{n+1} - a_n, a_n - a_{n-1})(a_{n+1} - a_n)\, and$$

$$a_{n+2} = a_{n+1} + \frac{b_{n+1}}{1 - \psi 0\,(an - 1 - a0, an - a0, an - an - 1)}.$$

Next, a convergence criterion is presented for the sequence $\{a_n\}$.

- There exists $q \in [0, q_0)$ such that

$$\psi_0 (a_{n-1} - a_0, a_n - a_0, a_n - a_{n-1}) < 1, \text{and } a_n \leq q.$$

It follows from this condition and (H_4) that

$$0 \leq a_n \leq a_{n+1} \leq q$$

and there exists $a^* \in [0, q]$ such that

$$\lim_{n \to +\infty} a_n = a^*.$$

- $S[x_0, a^* - a_0] \subset \Omega$.

The main semi-local convergence result follows for method (1.2).

Theorem 3.1 *Suppose that the conditions* (H_1)*–*(H_6) *hold. Then, the sequence* $\{x_n\}$ *generated by method (1.2) is well defined in* $S[x_0, a^* - a_0]$*, remains in* $S[x_0, a^* - a_0]$ *for each* $n = 0, 1, 2, \ldots$ *, and is convergent to some solution* $x^* \in S[x_0, a^* - a_0]$ *of the equation* $F(x) = 0$*. Moreover, the following assertion holds*

$$\|x^* - x_n\| \leq a^* - a_n. \tag{3.2}$$

Proof. Mathematical induction shall establish the estimate

$$\|x_{n+1} - x_n\| \leq a_{n+1} - a_n \quad \text{for each} \quad n = -1, 0, 1, 2, \ldots. \tag{3.3}$$

Estimate (3.3) holds if $n = -1, 0$. We also have

$$\|x_1 - x_0\| \leq \| - [x_{-1}, x_0; F]^{-1} F(x_0) \| \leq a_1 - a_0 \leq a^* - a_0,$$

so the iterate $x_1 \in S[x_0, a^* - a_0]$. Let $u, v \in S[x_0, a^* - a_0]$. It follows by the condition (H_1) that

$$\left\| \mathcal{L}^{-1} ([u, v; F] - \mathcal{L}) \right\| \leq \psi_0 (\|u - x_0\|, \|v - x_0\|, \|v - u\|) < 1.$$

Hence, the linear operator $[u, v; F]$ is invertible and

$$\left\| [u, v; F]^{-1} \mathcal{L} \right\| \leq \frac{1}{1 - \psi_0 (\|u - x_0\|, \|v - x_0\|, \|u - v\|)} \tag{3.4}$$

We can write by method (1.2)

$$\begin{aligned} F(x_{m+1}) &= F(x_{m+1}) - F(x_m) - [x_{m-1}, x_m; F](x_{m+1} - x_m) \\ &= ([x_{m+1}, x_m; F] - [x_{m-1}, x_m; F])(x_{m+1} - x_m) \end{aligned}$$

leading to condition (H_2)

$$\begin{aligned}\left\|\mathcal{L}^{-1}F(x_{m+1})\right\| \leq & \psi\left(\|x_{m+1}-x_{m-1}\|,\|x_{m+1}-x_m\|,\|x_m-x_{m-1}\|\right)\\ & \|x_{m+1}-x_m\|\\ \leq & \psi\left(a_{m+1}-a_{m-1},a_{m+1}-a_m,a_m-a_{m-1}\right)=b_{m+1} \quad (3.5)\end{aligned}$$

Thus, for $u = x_m$, $v = x_{m+1}$, (1.2), (3.4), (3), and the induction hypothesis give in turn

$$\begin{aligned}\|x_{m+2}-x_{m+1}\| &\leq \|[x_m,x_{m+1};F]^{-1}\mathcal{L}\|\|\mathcal{L}^{-1}F(x_{m+1})\|\\ &\leq \frac{b_{m+1}}{1-\psi_0\left(\|x_{m-1}-x_0\|,\|x_m-x_0\|,\|x_m-x_{m-1}\|\right)}\\ &\leq \frac{b_{m+1}}{1-\psi_0\left(a_{m-1}-a_0,a_m-a_0,a_m-a_{m-1}\right)}=a_{m+2}-a_{m+1},\end{aligned}$$

and

$$\begin{aligned}\|x_{m+2}-x_0\| &\leq \|x_{m+2}-x_{m+1}\|+\|x_{m+1}-x_0\|\\ &\leq a_{m+2}-a_{m+1}+a_{m+1}-a_0=a_{m+2}-a_0<a^*.\end{aligned}$$

Hence, the induction for (3.3) is completed and $\{x_m\} \subset S[x_0, a^* - a_0]$. Notice that the sequence $\{a_m\}$ is fundamental by (H_5) as convergent. So, by (3.3), the sequence $\{x_m\}$ is also convergent in the Banach space B_1, and as such it is convergent to some $x^* \in S[x_0, a^* - a_0]$. By letting $m \to +\infty$ in (3) and using the continuity of the operator F, we deduce that $F(x^*) = 0$. Moreover, it follows by (3.3) and the triangle inequality that

$$\|x_{m+j}-x_m\| \leq a_{m+j}-a_m. \quad (3.6)$$

Finally, by letting $j \to +\infty$ in (3.6), we obtain (3.2). □

The uniqueness of a solution of the equation $F(x) = 0$ in a certain domain is established in the next result.

Proposition 3.2 *Suppose:*

The equation $F(x) = 0$ has a solution $w \in S(x_0, p)$, for some $p > 0$;

The condition (H_3) holds in the ball $S(x_0, p)$, and there exists $p_1 \geq p$ such that

$$\psi_0\left(p,p_1,p+p_1\right)<1. \quad (3.7)$$

Define the domain $D_1 = \Omega \cap S[x_0, p_1]$. Then, the only solution of the equation $F(x) = 0$ in the domain D_1 is w.

Proof. Let $w_1 \in D_1$ be such that $F(w_1) = 0$ and $w_1 \neq w$. Then, the linear operator $T_1 = [w, w_1; F]$ is well defined. It follows by the condition (H_1) and (3.7) that

$$\left\|\mathcal{L}^{-1}(T_1 - \mathcal{L})\right\| \leq \bar{\psi}_0(\|w - x_0\|, \|w_1 - x_0\|, \|w_1 - w\|)$$
$$\leq \psi_0(p, p_1, p + p_1) < 1.$$

Therefore, the operator T_1 is invertible. Then, from the identity

$$w_1 - w = T^{-1}(F(w_1) - F(w)) = T^{-1}(0) = 0,$$

we conclude that $w_1 = w$. □

Remark 3.3

(i) If all the conditions of Theorem 3.1 hold in Proposition 3.2, then take $w = x^*$ and $p = a^*$.

(ii) The limit point a^* in the condition (H_6) can be replaced by q_0 defined in (H_4).

(iii) Other comments can be found in Remark 2.3.

The conclusions of Theorem 3.1 hold for the rest of the methods, but the majorizing sequences differ.

Method (1.3):

The majorant sequence $\{a_n\}$ is defined by

$$\lambda_n = \psi(b_n - a_{n-1}, b_n - a_n, a_n - a_{n-1}),$$
$$a_{n+1} = b_n + \frac{\lambda_n}{1 - \psi_0(a_{n-1}, a_n, a_n - a_{n-1})},$$
$$c_{n+1} = (1 + \psi_0(a_{n+1}, b_n))(a_{n+1} - b_n)$$
$$+ \psi(b_n - a_{n-1}, b_n - a_n, a_n - a_{n-1})(b_n - a_n)$$

and

$$b_{n+1} = a_{n+1} + \frac{c_{n+1}}{1 - \psi_0(a_n, a_{n+1}, a_{n+1} - a_n)}.$$

The motivational calculations are

$$
\begin{aligned}
F(y_n) &= F(y_n) - F(x_n) - [x_{n-1}, x_n; F](y_n - x_n) \\
&= ([y_n, x_n; F] - [x_{n-1}, x_n; F])(y_n - x_n), \\
\|\mathcal{L}^{-1}F(y_n)\| &\le \psi(\|y_n - x_{n-1}\|, \|y_n - x_n\|, \|x_n - x_{n-1}\|) \\
&\le \psi(b_n - a_{n-1}, b_n - a_n, a_n - a_{n-1}) = \lambda_n, \\
\|x_{n+1} - y_n\| &\le \|[x_{n-1}, x_n; F]^{-1}\mathcal{L}\| \|\mathcal{L}^{-1}F(y_n)\| \\
&\le \frac{\lambda_n}{1 - \psi_0(a_{n-1}, a_n, a_n - a_{n-1})} = a_{n+1} - b_n, \\
F(x_{n+1}) &= F(x_{n+1}) - F(x_n) - [x_{n-1}, x_n; F](y_n - x_n) \\
&= F(x_{n+1}) - F(y_n) + (F(y_n) - F(x_n) \\
&\quad - [x_{n-1}, x_n; F](y_n - x_n)), \\
\|\mathcal{L}^{-1}F(x_{n+1})\| &\le (1 + \psi_0(a_{n+1}, b_n))(a_{n+1} - b_n) \\
&\quad + \psi(b_n - a_{n-1}, b_n - a_n, a_n - a_{n-1})(b_n - a_n) = d_{n+1}
\end{aligned}
$$

and

$$
\begin{aligned}
\|y_{n+1} - x_{n+1}\| &\le \|[x_n, x_{n+1}; F]^{-1} \mathcal{L}\| \, \|\mathcal{L}^{-1} F(x_{n+1})\| \\
&\le \frac{d_{n+1}}{1 - \psi_0(a_n, a_{n+1}, a_{n+1} - a_n)} = b_{n+1} - a_n + 1.
\end{aligned}
$$

Method (1.4):

The iterate b_{n+1} is as defined in method (1). But

$$
a_{n+1} = b_n + \frac{\psi(b_n - a_{n-1}, a_n - a_{n-1}, b_n - a_n)(b_n - a_n)}{1 - \psi_0(a_n, b_n, b_n - a_n)}.
$$

The motivational calculations are

$$
\begin{aligned}
x_{n+1} - y_n &= \left([x_{n-1}, x_n; F]^{-1} - [y_n, x_n; F]^{-1}\right) F(x_n) \\
&= -\left([y_n, x_n; F]^{-1} - [x_{n-1}, x_n; F]^{-1}\right) F(x_n) \\
&= -[y_n, x_n; F]^{-1} ([x_{n-1}, x_n; F] - [y_n, x_n; F]) [x_{n-1}, x_n; F]^{-1} F(x_n) \\
&= [y_n, x_n; F]^{-1} ([x_{n-1}, x_n; F] - [y_n, x_n; F]) (y_n - x_n),
\end{aligned}
$$

and

$$
\begin{aligned}
\|x_{n+1} - y_n\| &\le \frac{\psi(\|x_{n-1} - y_n\|, \|x_{n-1} - x_n\|, \|x_n - y_n\|) \|y_n - x_n\|}{1 - \psi_0(a_n, b_n, b_n - a_n)} \\
&\le \frac{\psi(b_n - a_{n-1}, a_n - a_{n-1}, b_n - a_n)(b_n - a_n)}{1 - \psi_0(a_n, b_n, b_n - a_n)} = a_{n+1} - b_n.
\end{aligned}
$$

Method (1.5):

The iterate b_{n+1} is again as defined in method (1.5). But

$$a_{n+1} = b_n + \frac{\lambda_n}{1 - \psi_0(a_n, b_n, b_n - a_n)},$$

where λ_n is as given in method (1). The motivational calculation is

$$\begin{aligned} x_{n+1} - y_n &= -[y_n x_n; F]^{-1} F(y_n), \\ \|x_{n+1} - y_n\| &\leq \left\| [y_n, x_n; F]^{-1} \mathcal{L} \right\| \left\| \mathcal{L}^{-1} F(y_n) \right\| \\ &\leq \frac{\lambda_n}{1 - \psi_0(a_n, b_n, b_n - a_n)} = a_{n+1} - b_n. \end{aligned}$$

Method (1.6):

The majorant sequence$\{a_n\}$ is defined by

$$\begin{aligned} c_n &= b_n + |1 - \alpha|(b_n - a_n), \\ a_{n+1} &= b_n + \frac{\lambda_n}{1 - \psi_0(a_n, c_n, c_n - a_n)}, \end{aligned}$$

where the iterate b_{n+1} is as in method (1). The motivational calculations are

$$\begin{aligned} z_n - y_n &= (1 - \alpha) x_n + \alpha y_n - y_n = (1 - \alpha)(x_n - y_n), \\ \|z_n - y_n\| &\leq |1 - \alpha| \, \|y_n - x_n\| \leq |1 - \alpha|(b_n - a_n) = c_n - b_n, \\ \|x_{n+1} - y_n\| &\leq \left\| -[z_n, x_n; F]^{-1} \mathcal{L} \right\| \left\| \mathcal{L}^{-1} F(y_n) \right\| \\ &\frac{\lambda_n}{1 - \psi_0(a_n, c_n, c_n - a_n)} = a_{n+1} - b_n. \end{aligned}$$

Remark 3.4

- The denominators in the definition of the functions h can be dropped together with (C_1) and (C_3) if Myshovskit-type conditions are assumed [**24**]. As an example

  $$\|[x, y; F]^{-1}([x, y; F] - [y, x^*; F])\| \leq \tilde{\phi}(\|x - x^*\|, \|y - x^*\|, \|y - x\|)$$

  for each $x, y \in \Omega$ and some continuous and non-decreasing function. Clearly, this function is such that

  $$\tilde{\phi}(t, t, 2t) \leq \frac{\phi(t, t, 2t)}{1 - \phi_0(t, t, 2t)}.$$

Consequently, the convergent domain is enlarged even further.
The same remarks follow for the semi-local convergence analysis of these methods.

- These ideas can be used to extend the applicability of other methods such as Kurchatov's, Steffensen's, and others [1, 11, 15, 28, 24, 9, 16, 13, 17, 25, 2, 4, 10] along the same lines.

11.4 CONCLUSION

Through a comprehensive examination, this study sheds light on the convergence properties of various secant method variants for solving nonlinearnonlinear equations. The findings underscore the importance of considering generalized convergence conditions to extend the applicability of these methods to non-differentiable equations. Additionally, the introduction of semi-local convergence analysis enriches our understanding of the convergence behavior, further enhancing the practical utility of these methods.

REFERENCES

1. Argyros, I. *Computational Theory of Iterative Methods.* Elsevier, 2007.
2. Argyros, I., and Gutiérrez, J. A unified approach for enlarging the radius of convergence for Newton's method and applications. *Nonlinear Functional Analysis and ApplicationsNonlinear Funct. Anal. Appl* 10 (2005), 555–563.
3. Argyros, I., and Gutierrez, J. A unifying local and semilocal convergence analysis of Newton-like methods. *Advances In Nonlinear Variational Inequalities* 10, 1 (2007), 01.
4. Argyros, I. K. A Kantorovich-type analysis for a fast iterative method for solving nonlinear equations. *Journal of Mathematical Analysis and Applications* 332, 1 (2007), 97–108.
5. Argyros, I. K., Argyros, C. I., John, J. A., and Jayaraman, J. On Twotwo Competingcompeting Methodsmethods with Optimaloptimal Eightheighth Orderorder Convergenceconvergence. *International Journal of Applied and Computational Mathematics* 9, 5 (2023), 72.

6. Argyros, I. K., Jayaraman, J., John, J. A., and Regmi, S. Extended Zabrejko Theorytheory Ofof Thethe Newton-Kantorovich iIteration aAnd tThe PtÁK-Potra eError eEstimates. In *Annales Universitatis Scientiarum Budapestinensis de Rolando Eötvös Nominatae.* Sectio Computatorica, (2023), vol. 55.
7. Argyros, I. K., John, J. A., Jayaraman, J., and Regmi, S. Extended Newton'S Methodmethod Onon Banachbanach Spacespace With-with Aa Convergenceconvergence Structurestructure. *Advances and Applications in Mathematical Sciences* 22, 9 (2023), 1977–1989.
8. Brezinski, C. *Accélération de la convergence en analyse numérique*, vol. 584. Springer, 2006, vol. 584.
9. Chen, K.-W. Generalization of Steffensen's method for operator equations in Banach space. *Commentationes Mathematicae Universitatis Carolinae* 5, 2 (1964), 47–77.
10. Ezquerro, J., Grau-Sánchez, M., Grau, A., Hernández, M., Noguera, M., and Romero, N. On iterative methods with accelerated convergence for solving systems of nonlinear equations. *Journal of Optimization Theory and Applications* 151, 1 (2011), 163–174.
11. Grau-Sánchez, M., Grau, À., and Noguera, M. Frozen divided difference scheme for solving systems of nonlinear equations. *Journal of Computational and Applied Mathematics* 235, 6 (2011), 1739–1743.
12. Grau-Sánchez, M., and Noguera, M.A technique to choose the most efficient method between secant method and some variants. *Applied Mathematics and Computation* 218, 11 (2012), 6415–6426.
13. Grau-Sánchez, M., Noguera, M., and Gutiérrez, J. M. Frozen iterative methods using divided differences "à la Schmidt–Schwetlick". *Journal of Optimization Theory and Applications* 160 (2014), 931–948.
14. Hernández, M., and Rubio, M. Semilocal convergence of the secant method under mild convergence conditions of differentiability. *Computers Mathematics with Applications* 44, 3--4 (2002), 277–285.
15. Hernández, M., Rubio, M., and Ezquerro, J. Secant-like methods for solving nonlinear integral equations of the Hammerstein type. *Journal of Computational and Applied Mathematics* 115, 1--2 (2000), 245–254.
16. Hernández, M., Rubio, M., and Ezquerro, J. Secant-like methods for solving nonlinear integral equations of the Hammerstein type. *Journal of Computational and Applied Mathematics* 115, 1--2 (2000), 245–254.

17. Kurchatov, V. On a method of linear interpolation for the solution of functional equations. In *Doklady Akademii Nauk* (1971), vol. 198,. Russian Academy of Sciences, 1971, vol. 198, pp. 524–526.
18. Ortega, J., and Rheinboldt, W. *Iterative Solution of Nonlinear Equations in Several Variables*, vol. 30. SIAM, 1970, vol. 30.
19. Ostowski, A. *Solution of Equations and System of Equations*, Academic Press, Cambridge, 1960.
20. Potra, F. A., and Pták, V. A generalization of Regula Falsi. *Numerische Mathematik* 36 (1980), 333–346.
21. Potra, F.-A., and Pták, V. *Nondiscrete Iinduction and Iiterative Pprocesses.* Research Notes in Mathematics, 1994, vol. 103.
22. Schmidt, J. W. Eine Übertragung der Regula Falsi auf Gleichungen in Banachräumen I. *ZAMM-Journal of Applied Mathematics and Mechanics/Zeitschrift für Angewandte Mathematik und Mechanik* 43, 1--2 (1963), 1–8.
23. Schmidt, J. W. Konvergenzgeschwindigkeit der Regula falsi und des Steffensen-Verfahrens im Banachraum. *ZAMM-Journal of Applied Mathematics and Mechanics/Zeitschrift für Angewandte Mathematik und Mechanik* 46, 2 (1966), 146–148.
24. Schmidt, J. W., and Schwetlick, H. Ableitungsfreie verfahren mit höherer konvergenzgeschwindigkeit. Computing 3 (1968), 215–226.
25. Shakhno, S. On an iterative algorithm with superquadratic convergence for solving nonlinear operator equations. *Journal of Computational and Applied Mathematics* 231, 1 (2009), 222–235.
26. Traub, J. F. *Iterative Methods for the Solution of Equations*, vol. 312. American Mathematical Soc., 1982, vol. 312.
27. Ulm, S. Die Algorithmen des verallgemeinerten Verfahrens vonSteffensen. *Ž. Vyčisl. Mat. i Mat. Fiz.* 6 (1964), 1093–1097.
28. Weerakoon, S., and Fernando, T. A variant of Newton's method with accelerated third-order convergence. *Applied Mathematics Letters* 13, 8 (2000), 87–93.

CHAPTER 12

Summation of Schlömilch-Type Series

Slobodan B. Tričković and Miomir S. Stanković

12.1 INTRODUCTION

Schlömilch series were originally defined as Fourier-type series expansion of a twice continuously differentiable function $f(x)$ in the interval $(0, \pi)$ in terms of the Bessel function of the first kind, named after the German mathematician Oskar Schlömilch [26]:

$$f(x) = \sum_{n=1}^{\infty} a_n J_\nu (nx) . \tag{1.1}$$

Using the Erdélyi-Kober operator of fractional order, Rusev [25] gives a necessary and sufficient condition for a holomorphic function $f(x)$ to be represented by the series (1.1) where $\nu = 0$. Rayleigh [24] showed that this series plays a significant role in physics, and it is applicable when one investigates periodic transverse vibrations uniformly distributed in direction through the two dimensions of the membrane.

Instead of Bessel functions, some functions related to them may appear in (1.1). That is why the series we deal with here with one can be regarded as belonging to the class of Schlömilch-type series. Apart from Bessel and related functions, it is of interest to also consider the series

$$\sum_{n=1}^{\infty} a_n D_\nu (nx) , \tag{1.2}$$

DOI: 10.1201/9781003530602-12

where $D_\nu(nx)$ is an integral of the Bessel, Struve, and related functions, and the series

$$\sum_{n=1}^{\infty} a_n g(ny) D_\nu(nx) \tag{1.3}$$

with g presenting a trigonometric function. In place of Bessel or related functions, we can take a product of these functions multiplied by a trigonometric function. All such series we call the Schlömilch-type series. Obtaining a summation formula for these series is based on finding its trigonometric series representation. So one can regard trigonometric series in a broader sense, as a fundamental goal of the investigation.

So we consider the Schlömilch series as a series of the type (see [41])

$$\sum_{n=1}^{\infty} a_n \phi_\nu(nx),$$

where ϕ_ν denotes the Bessel J_ν, Neumann Y_ν, Struve $\mathbf{H}_\nu$ functions, or the modified Bessel function of the second kind K_ν. First, we deal with finding a sum of the series in terms of Bessel or Struve functions, i.e., [39]

$$S_\alpha^\phi = \sum_{n=1}^{\infty} \frac{(s)^{n-1} \phi_\nu((an-b)x)}{(an-b)^\alpha},$$

where $\alpha \in \mathbb{R}$, $a = \left\{ \begin{array}{c} 1 \\ 2 \end{array} \right\} b = \left\{ \begin{array}{c} 0 \\ 1 \end{array} \right\}$, $s = 1$ or -1, ϕ_ν denotes Bessel $J_\nu(x)$ or Struve $\mathbf{H}_\nu(x)$ functions of the first kind of order ν. Here, the comma between a and b is deliberately omitted, whereby we emphasize a rule we make use of in the sequel, meaning that $b = 0$ corresponds to $a = 1$, and if $a = 2$ then $b = 1$.

We evaluate and represent these series in terms of the Riemann zeta function and related functions of reciprocal powers. The obtained sums can, in certain cases, be brought into closed form.

Also, in a Schlömilch series, the general term contains one or more Bessel functions or associated functions of the form $J_\nu(nx)$, if the summation is over n. So we will sum the series involving the product of two Bessel functions

$$S_\alpha^{J,J} = \sum_{n=1}^{\infty} \frac{(s)^{n-1} J_\mu((an-b)x) J_\nu((an-b)x)}{(an-b)^\alpha},$$

as well as the series involving a product of Bessel or Struve functions and a trigonometric function

$$S_{\alpha}^{\phi,f} = \sum_{n=1}^{\infty} \frac{(s)^{n-1} \phi_{\nu}((an-b)x) f((an-b)y)}{(an-b)^{\alpha}},$$

where $f =$ sin or cos, and the series involving the product of two Bessel functions and one trigonometric function

$$S_{\alpha}^{J,J,f} = \sum_{n=1}^{\infty} \frac{(s)^{n-1} J_{\mu}((an-b)x) J_{\nu}((an-b)x)}{(an-b)^{\alpha}} f((an-b)y).$$

12.2 BESSEL AND RELATED FUNCTIONS

Bessel functions $J_{\nu}(x)$ defined by the Swiss mathematician Daniel Bernoulli and then generalized and developed by Friedrich Bessel while studying the dynamics of gravitational systems in the second decade of the 19th century, are canonical solutions of the homogeneous Bessel differential equation

$$x^2 \frac{\mathrm{d}^2 y}{\mathrm{d}x^2} + x \frac{\mathrm{d}y}{\mathrm{d}x} + (x^2 - \nu^2) y = 0 \tag{2.1}$$

for an arbitrary complex number ν. To determine its solution, one defines the function by its series expansion around $x = 0$

$$y(x) = \sum_{m=0}^{\infty} a_n x^{m+\alpha}, a_0 \neq 0,$$

and find first that $\alpha = \pm\nu$. Further it can be found by applying the Frobenius method to Bessel's equation (see 9.1.10 in [1, p. 360]). The particular solution

$$J_{\nu}(x) = \sum_{m=0}^{\infty} \frac{(-1)^m \left(\frac{x}{2}\right)^{2m+\nu}}{m! \Gamma(m+\nu+1)}, \tag{2.2}$$

is called Bessel's function of the first kind order ν [23, 18]. For integer or positive ν, Bessel functions of the first kind are finite at the origin ($x = 0$), while for negative non-integer ν, Bessel functions of the first kind diverge as x approaches zero.

For $\nu = \frac{1}{2}$ in the denominator of (2.2), by using Legendre's duplication formula, we have

$$\Gamma\left(m+1+\frac{1}{2}\right) = \frac{\sqrt{\pi}\,\Gamma(2m+2)}{2^{2m+1}\Gamma(m+1)} = \frac{(2m+1)!}{2^{2m+1}m!}\sqrt{\pi}. \tag{2.3}$$

Thus, we ascertain that the Bessel function can be expressed through the sine function

$$J_{1/2}(x) = \sqrt{\frac{x}{2}} \sum_{m=0}^{\infty} \frac{(-1)^m x^{2m}}{2^{2m} m! \frac{(2m+1)!}{2^{2m+1}m!}\sqrt{\pi}} = \sqrt{\frac{2}{x\pi}} \sum_{m=0}^{\infty} \frac{(-1)^m x^{2m+1}}{(2m+1)!}$$
$$= \sqrt{\frac{2}{x\pi}} \sin x.$$

From (2.2), the Bessel function can be expressed using the regularized hypergeometric function as well. So, by using the Pochhammer symbol

$$(\alpha)_n = \alpha(\alpha+1)(\alpha+2)\cdots(\alpha+n-1) = \frac{\Gamma(\alpha+n)}{\Gamma(\alpha)},$$

and taking $p = 0, q = 1$ in

$${}_pF_q(a_1, a_2, ..., a_p; b_1, b_2, ..., b_q; z) = \sum_{n=0}^{\infty} \frac{(a_1)_n \cdots (a_p)_n}{(b_1)_n \cdots (b_q)_n} \cdot \quad \frac{z^n}{n!}, |z| < 1. \tag{2.4}$$

we have

$$J_\nu(x) = \frac{\left(\frac{x}{2}\right)^\nu}{\Gamma(\nu+1)} \sum_{m=0}^{\infty} \frac{1}{(\nu+1)_m} \frac{\left(-\frac{x^2}{4}\right)^m}{m!} = \frac{\left(\frac{x}{2}\right)^\nu}{\Gamma(\nu+1)} {}_0F_1\left(\nu+1; -\frac{x^2}{4}\right).$$

Equation (2.1) is unchanged when ν is replaced by $-\nu$. This means that $J_{-\nu}(x)$ is also a solution of (2.1), and one can check directly that for ν being a non-integer, solutions $J_\nu(x)$ and $J_{-\nu}(x)$ are linearly independent. A general solution of (2.1) is then a linear combination of these two particular solutions

$$y(z) = C_1 J_\nu(z) + C_2 J_{-\nu}(z), \quad \nu \neq n, n \in \mathbb{N},$$

where series $J_\nu(z)$ and $J_{-\nu}(z)$ converge in the whole complex plane except for $z = 0$. Conversely, for ν as an integer, say $\nu = n$, there holds

$$J_{-n}(x) = (-1)^n J_n(x),$$

meaning that $J_{-n}(x)$ is linearly dependent on $J_n(x)$.

Bessel himself originally proved that for nonnegative integers n, the equation $J_n(x) = 0$ has an infinite number of solutions in x (see [3]). When the functions $J_n(x)$ are plotted on the same graph, however, none of the zeros seem to coincide for different values of n except for the zero at $x = 0$. This phenomenon is known as Bourget's hypothesis, after the 19th-century French mathematician who studied Bessel functions. Specifically, it states that for any integers $n \geqslant 0$ and $m \geqslant 1$, the functions $J_n(x)$ and $J_{n+m}(x)$ have no common zeros other than the one at $x = 0$. The hypothesis was proved by Carl Ludwig Siegel in 1929 (see [41, pp. 484–485]).

There is another solution $Y_\nu(z)$ of the Bessel differential equation (2.1), called a Bessel function of the second kind or Neumann's function. It is singular at the origin and related to $J_\nu(z)$ as follows

$$Y_\nu(x) = \frac{J_\nu(x)\cos\nu\pi - J_{-\nu}(x)}{\sin\nu\pi}, \quad \nu \neq \pm n, n \in \mathbb{N}. \tag{2.5}$$

So, a general solution can be expressed like this:

$$y(x) = AJ_\nu(x) + BY_\nu(x). \tag{2.6}$$

Taking $\nu - 1$ instead of ν in (2.2), we find

$$\begin{aligned}
J_{\nu-1}(x) &= \sum_{m=0}^{\infty} \frac{(-1)^m \left(\frac{x}{2}\right)^{2m+\nu-1}}{m!\Gamma(m+\nu)} = \frac{2}{x}\sum_{m=0}^{\infty} \frac{(-1)^m \left(\frac{x}{2}\right)^{2m+\nu}(\nu+m)}{m!\Gamma(m+\nu+1)} \\
&= \frac{2\nu}{x}J_\nu(x) + \frac{2}{x}\sum_{m=1}^{\infty} \frac{(-1)^m \left(\frac{x}{2}\right)^{2m+\nu}}{(m-1)!\Gamma(m+\nu+1)} \\
&= \frac{2\nu}{x}J_\nu(x) - \sum_{k=0}^{\infty} \frac{(-1)^k \left(\frac{x}{2}\right)^{2k+\nu+1}}{k!\Gamma(k+\nu+2)} = \frac{2\nu}{x}J_\nu(x) - J_{\nu+1}(x).
\end{aligned}$$

The same holds for $Y_\nu(x)$, so this recurrence relation demonstrates that for $\nu = n \in \mathbb{N}$ and $x = 1$, both $J_n(1)$ and $Y_n(1)$ satisfy the difference equation [21]

$$a_{n+1} - 2na_n + a_{n-1} = 0$$

having as a general solution

$$a_n = C_1 J_n(1) + C_2 Y_n(1).$$

However, there are two linearly independent solutions to Bessel's equation as well, defined as linear combinations of the Bessel functions $J_\nu(x)$ and $Y_\nu(x)$

$$H_\nu^{(1)}(x) = J_\nu(x) + iY_\nu(x), \quad H_\nu^{(2)}(x) = J_\nu(x) - iY_\nu(x),$$

known also as Bessel functions of the third kind or the Hankel functions of the first and second kind, respectively, named after the German mathematician Hermann Hankel. Thus, a general solution can be presented in the form of

$$y(x) = C_1{}'H\nu(1)(x) + C_2{}'H\nu(1)(x).$$

If one considers a modified Bessel differential equation (see [41])

$$x^2\frac{\mathrm{d}^2y}{\mathrm{d}x^2} + x\frac{\mathrm{d}y}{\mathrm{d}x} + \left(x^2 + \nu^2\right)y = 0,$$

one finds that its general solution can be expressed in two ways

$$y(x) = C_1I_\nu(x) + C_2I_{-\nu}(x), \quad y(x) = C_1'I_\nu(x) + C_2'K_\nu(x),$$

where $\nu \neq n, n \in \mathbb{Z}, C_1, C_2, C_1', C_2'$ are constants. $I_\nu(z)$ is a modified Bessel function of the first kind (see [23, p. 729])

$$I_\nu(x) = \sum_{m=0}^{\infty}\frac{\left(\frac{x}{2}\right)^{\nu+2m}}{m!\Gamma(\nu+m+1)},$$

and $K_\nu(x)$ is a modified Bessel function of the second kind or MacDonald's function [15, p. 19]

$$K_\nu(x) = \frac{\pi}{2\sin\nu\pi}\left(I_{-\nu}(x) + I_\nu(x)\right).$$

The Struve function $\mathbf{H}_\nu(z)$, introduced by Hermann Struve (1882), a German astronomer, is a solution of the nonhomogeneous Bessel differential equation

$$x^2\frac{\mathrm{d}^2y}{\mathrm{d}x^2} + x\frac{\mathrm{d}y}{\mathrm{d}x} + \left(x^2 - \nu^2\right)y = \frac{4\left(\frac{x}{2}\right)^{\nu+1}}{\sqrt{\pi}\,\Gamma\left(\nu+\frac{1}{2}\right)},$$

and can be expressed by the power series

$$\mathbf{H}_\nu(x) = \sum_{m=0}^{\infty} \frac{(-1)^m \left(\frac{x}{2}\right)^{\nu+2m+1}}{\Gamma\left(m+\frac{3}{2}\right)\Gamma\left(\nu+m+\frac{3}{2}\right)}. \tag{2.8}$$

The complex number ν is the order of the Struve function and is often an integer of the first kind order ν. As regards a recurrence relation similar to (2.7), we derive

$$\mathbf{H}_{\nu-1}(x) = \frac{\left(\frac{x}{2}\right)^\nu}{\sqrt{\pi}\,\Gamma\left(\nu+\frac{3}{2}\right)} + \frac{2\nu}{x}\mathbf{H}_\nu(x) - \mathbf{H}_{\nu+1}(x). \tag{2.9}$$

There is an approach to the Bessel functions for integer values of n in the integral form, called Bessel's integrals [32]:

$$J_n(x) = \frac{1}{\pi}\int_0^\pi \cos(nt - x\sin t)\,dt. \tag{2.10}$$

This way, Bessel defined these functions [3, p. 28], from which he derived several of their properties.

Another integral representation of Bessel and Struve functions, because of their similar forms, enables us to give an integral representation comprising both of them. It is called Poisson's integral

$$\phi_\nu(z) = \frac{2\left(\frac{z}{2}\right)^\nu}{\Gamma\left(\frac{1}{2}\right)\Gamma\left(\nu+\frac{1}{2}\right)}\int_0^{\pi/2} \sin^{2\nu}\theta\, f(z\cos\theta)\,d\theta, \quad \operatorname{Re}\nu > -\frac{1}{2}, \tag{2.11}$$

where $\phi_\nu = \left\{\begin{array}{c} J_\nu \\ \mathbf{H}_\nu \end{array}\right\} f = \left\{\begin{array}{c} \cos \\ \sin \end{array}\right\}$. Poisson [22] and Lommel [13] proved that for $2\nu \in \mathbb{N}_0$, it is a solution of the homogeneous Bessel's differential equation.

Apart from (2.11), we deal with the integral representation (see [1, p. 496])

$$\phi_\nu(x) = \frac{2\left(\frac{x}{2}\right)^\nu}{\sqrt{\pi}\,\Gamma\left(\nu+\frac{1}{2}\right)}\int_0^1 \left(1-t^2\right)^{\nu-\frac{1}{2}} f(xt)\,dt, \quad \operatorname{Re}\nu > -\frac{1}{2}. \tag{2.12}$$

where $\phi_\nu = \left\{\begin{array}{c} J_\nu \\ \mathbf{H}_\nu \end{array}\right\} f = \left\{\begin{array}{c} \cos \\ \sin \end{array}\right\}$.

12.3 ANGER AND WEBER FUNCTIONS

Anger and Weber functions [14] were introduced by C. T. Anger (1855) and H. F. Weber (1879), respectively. The Anger functions satisfy the equation [20]

$$z^2\frac{d^2y}{dz^2}+z\frac{dy}{dz}+\left(z^2-\nu^2\right)y=\frac{(z-\nu)\,\sin\pi\nu}{\pi},$$

so they are closely related to Bessel functions because for $\nu \in \mathbb{N}_0$, it becomes homogeneous, meaning that $\mathbf{J}_\nu(z) = J_\nu(z)$.

Since the Anger and Weber functions have similar integral representations, we can write only one formula encompassing both of them

$$\phi_\nu(x)=\frac{1}{\pi}\int_0^\pi f(\nu\theta-x\,\sin\theta)\,d\theta, \tag{3.1}$$

where $\phi_\nu = \left\{\begin{matrix}\mathbf{J}_\nu\\ \mathbf{E}_\nu\end{matrix}\right\} f = \left\{\begin{matrix}\cos\\ \sin\end{matrix}\right\}$. To assign a more suitable form to (3.1), we make use of additional trigonometric formulas including all four cases

$$f(\alpha-\beta)=f(\alpha)\,\cos\beta-\tilde{f}\,(\alpha)\,\sin\beta, \tag{3.2}$$

where $f = \left\{\begin{matrix}\cos\\ \sin\end{matrix}\right\} \tilde{f} = \left\{\begin{matrix}-\sin\\ \cos\end{matrix}\right\}$. From (3.1), there follows

$$\phi_\nu(x)=\frac{1}{\pi}\int_0^\pi f(\nu\theta)\,\cos(x\,\cos\theta)\,d\theta-\frac{1}{\pi}\int_0^\pi \tilde{f}\,(\nu\theta)\,\sin(x\,\cos\theta)\,d\theta.$$

Let $f = \cos$. Then $\phi = \mathbf{J}$ and $\tilde{f} = -\sin$. Introducing the substitution $\theta = \frac{\pi}{2} - t$ in the second integral, it takes the following form

$$\frac{1}{\pi}\int_0^\pi \sin\nu\theta\,\sin(z\,\sin\theta)\,d\theta=\frac{1}{\pi}\int_{-\pi/2}^{\pi/2}\sin\left(\frac{\nu\pi}{2}-\nu t\right)\theta\,\sin(z\,\cos t)\,dt.$$

Now, we suppose that $\nu \in \mathbb{N}_0$, and distinguish between even and odd ν, i.e., $\nu = 2p$ and $\nu = 2p-1$, $p \in \mathbb{N}$. The case $\nu = 2p$ yields

$$\frac{(-1)^{p-1}}{\pi}\int_{-\pi/2}^{\pi/2}\sin 2pt\,\sin(z\,\cos t)\,dt=0,$$

Thus, there remains only the first integral

$$\mathbf{J}_{2p}(z) = \frac{1}{\pi}\int_0^{\pi} \cos(2p\theta - z\sin\theta)\, d\theta = \frac{1}{\pi}\int_0^{\pi} \cos 2p\theta \cos(z\sin\theta)\, d\theta,$$

and in the case $\nu = 2p - 1$, it is the other way around, i.e.,

$$\begin{aligned}\mathbf{J}_{2p-1}(z) &= \frac{1}{\pi}\int_0^{\pi} \cos((2p-1)\theta - z\sin\theta)\, d\theta \\ &= \frac{1}{\pi}\int_0^{\pi} \sin(2p-1)\theta \sin(z\sin\theta)\, d\theta.\end{aligned}$$

Assume now $f = \sin$. Then $\phi = \mathbf{E}$ and $\tilde{f} = \cos$. Repeating the preceding procedure, we easily conclude that there holds

$$\mathbf{E}_{2p}(z) = \frac{1}{\pi}\int_0^{\pi} \sin(2p\theta - z\sin\theta)\, d\theta = -\frac{1}{\pi}\int_0^{\pi} \cos 2p\theta \sin(z\sin\theta)\, d\theta,$$

as well as

$$\begin{aligned}\mathbf{E}_{2p-1}(z) &= \frac{1}{\pi}\int_0^{\pi} \sin(2p-1)\theta - z\sin\theta) d\theta \\ &= \frac{1}{\pi}\int_0^{\pi} \sin(2p-1)\theta \cos(z\sin\theta)\, d\theta.\end{aligned}$$

Taking account of both cases, $\nu = 2p$ and $\nu = 2p-1$ ($p \in \mathbb{N}_0$), we can write

$$\phi_{2p-r}(x) = \frac{1}{\pi}\int_0^{\pi} f((2p-r)\,\theta)\, g(x\sin\theta)\, d\theta, \tag{3.3}$$

with $\phi_{2p-r} = \left\{\begin{matrix}\mathbf{J}_{2p-r} \\ \mathbf{E}_{2p-r}\end{matrix}\right\} g = \left\{\begin{matrix} f \\ \tilde{f}\end{matrix}\right\}$, where $r = \left\{\begin{matrix}0 \\ 1\end{matrix}\right\} f = \left\{\begin{matrix}\cos \\ \sin\end{matrix}\right\} \tilde{f} = \left\{\begin{matrix}-\sin \\ \cos\end{matrix}\right\}$.

If ν is not an integer, the Anger and Weber functions can be expressed as linear combinations of each other

$$\begin{aligned}\mathbf{J}_\nu(z) &= \operatorname{ctg}\pi\nu\, \mathbf{E}_\nu(z) - \csc\pi\nu\, \mathbf{E}_{-\nu}(z), \\ \mathbf{E}_\nu(z) &= -ctg\,\pi\nu\, \mathbf{J}_\nu(z) + \csc\pi\nu\, \mathbf{J}_{-\nu}(z).\end{aligned}$$

On the other hand, Weber functions are closely related to Struve functions in the case of ν being a non-negative integer, i.e., [16, 20]:

$$\mathbf{H}_{\nu}(z) = -\mathbf{E}_{\nu}(z) + \frac{1}{\pi}\sum_{k=0}^{\left[\frac{\nu-1}{2}\right]} \frac{\Gamma\left(k+\frac{1}{2}\right)\left(\frac{z}{2}\right)^{\nu-2k-1}}{\Gamma\left(\nu-k+\frac{1}{2}\right)},$$

$$\mathbf{H}_{-\nu}(z) = -\mathbf{E}_{-\nu}(z) + \frac{(-1)^{\nu+1}}{\pi}\sum_{k=0}^{\left[\frac{\nu-3}{2}\right]} \frac{\Gamma\left(n-k-\frac{1}{2}\right)\left(\frac{z}{2}\right)^{-\nu+2k+1}}{\Gamma\left(k+\frac{3}{2}\right)}, \quad (3.4)$$

where $\nu \in \mathbb{N}_0$.

12.4 SERIES OVER BESSEL OR STRUVE FUNCTIONS

In the sequel, we derive summation formulas for the Schlömilch-type series

$$S_{\alpha}^{\phi} = \sum_{n=1}^{\infty} \frac{(s)^{n-1}\phi_{\nu}((an-b)x)}{(an-b)^{\alpha}}, \quad (4.1)$$

where $\phi_{\nu}(z)$ denotes $J_{\nu}(z)$, $\mathbf{H}_{\nu}(z)$, $\mathbf{J}_{\nu}(z)$, $\mathbf{E}_{\nu}(z)$, i.e., Bessel, Struve, Anger, Weber functions.

First of all, we need the following lemma.

Lemma 4.1 *Let $\alpha > \nu$. The series*

$$\sum_{n=1}^{\infty} \frac{\cos(nx\cos\theta)}{n^{\alpha-\nu}}, \sum_{n=1}^{\infty} \frac{\sin(nx\cos\theta)}{n^{\alpha-\nu}}, \quad (4.2)$$

are uniformly convergent with respect to x on $(0, 2\pi)$ and θ on each segment $\left[\epsilon, \frac{\pi}{2}-\epsilon\right] \subset \left(0, \frac{\pi}{2}\right)$, $\epsilon > 0$.

Proof. We are referring to Dirichlet's test [12] stating that the series

$$\sum_{n=0}^{\infty} a_n(x)\, b_n(x)$$

is uniformly convergent in D, if partial sums $\sum_{k=0}^{n} a_k(x)$ are uniformly bounded in D, and the sequence $b_n(x)$, being monotonic for every fixed x, uniformly converges to 0, and first note that $0 < \cos\theta < 1$, and if

$0 < x < 2\pi$, we have $0 < \frac{x\cos\theta}{2} < \pi$. For each $\epsilon > 0$ satisfying $\epsilon \le \frac{x\cos\theta}{2} \le \pi - \epsilon$, we know that $\sin\frac{x\cos\theta}{2} \ge \sin\epsilon > 0$, so there follows

$$\left|1 + \sum_{k=1}^{n} \cos(kx\cos\theta)\right| = \left|\frac{\sin\frac{(n+1)x\cos\theta}{2}\cos\frac{nx\cos\theta}{2}}{\sin\frac{x\cos\theta}{2}}\right| \leqslant \frac{1}{\sin\frac{x\cos\theta}{2}} \le \frac{1}{\sin\epsilon},$$

meaning that partial sums $\sum_{k=1}^{n}\cos(kx\cos\theta)$ are uniformly bounded with respect to θ on each segment $\left[\epsilon, \frac{\pi}{2} - \epsilon\right] \subset \left(0, \frac{\pi}{2}\right), 0 < \epsilon < \frac{\pi}{2}$, and for $\nu > \mu$, the monotonically decreasing sequence $\frac{1}{n^{\alpha-\nu}}$ tends to 0. □

We rely on the series (4.2) when dealing with the series over Bessel and Struve functions, respectively.

Theorem 4.1 *The summation formula for the series over the Bessel functions (4.4) is as follows:*

$$\sum_{n=1}^{\infty} \frac{J_\nu(nx)}{n^\alpha} = \frac{\pi\left(\frac{x}{2}\right)^{\alpha-1}}{2\Gamma\left(\frac{\alpha-\nu+1}{2}\right)\Gamma\left(\frac{\alpha+\nu+1}{2}\right)\cos\frac{\pi(\alpha-\nu)}{2}} + \sum_{k=0}^{\infty} \frac{(-1)^k\left(\frac{x}{2}\right)^{\nu+2k}\zeta(\alpha-\nu-2k)}{k!\,\Gamma(\nu+k+1)}, \tag{4.3}$$

where the convergence region is $0 < x < 2\pi$ *and* $\alpha > 0, \nu > -\frac{1}{2}, \alpha > \nu$.

Proof. We use representations of Bessel or Struve functions by means of Poisson's integral (2.11). To demonstrate the procedure in a simpler way, we choose $a = 1, b = 0, s = 1, \phi_\nu = J_\nu$ in Poisson's integral (2.11), then we must take $f =$ cos, and place it in (4.1), so that we have

$$\sum_{n=1}^{\infty} \frac{J_\nu(nx)}{n^\alpha} = \frac{2\left(\frac{x}{2}\right)^\nu}{\sqrt{\pi}\,\Gamma\left(\nu+\frac{1}{2}\right)} \sum_{n=1}^{\infty} \int_0^{\pi/2} \frac{\sin^{2\nu}\theta\cos(nx\cos\theta)}{n^{\alpha-\nu}}\, d\theta. \tag{4.4}$$

According to the result of the preceding lemma, we are allowed to interchange summation and integration in (4.4)

$$\sum_{n=1}^{\infty} \frac{J_\nu(nx)}{n^\alpha} = \frac{2\left(\frac{x}{2}\right)^\nu}{\sqrt{\pi}\,\Gamma\left(\nu+\frac{1}{2}\right)} \int_0^{\pi/2} \sin^{2\nu}\theta \sum_{n=1}^{\infty} \frac{\cos(nx\cos\theta)}{n^{\alpha-\nu}}\, d\theta, \tag{4.5}$$

and to the right-hand series, we apply the formula [37]

$$\sum_{n=1}^{\infty} \frac{(s)^{n-1} f((an-b)x)}{(an-b)^{\alpha}} = \frac{c\pi x^{\alpha-1}}{2\Gamma(\alpha) f\left(\frac{\pi\alpha}{2}\right)} + \sum_{k=0}^{\infty} \frac{(-1)^k F(\alpha-2k-\delta)}{(2k+\delta)!} x^{2k+\delta}, \tag{4.6}$$

where we first replace x with $x \cos\theta$ and α with $\alpha-\nu$, taking $c=1, f=\cos$, $\delta=0$, $F=\zeta$, see Table 12.1, p. 272. Thus, we obtain

$$\sum_{n=1}^{\infty} \frac{J_\nu(nx)}{n^\alpha} = \frac{2\left(\frac{x}{2}\right)^\nu}{\sqrt{\pi}\,\Gamma\left(\nu+\frac{1}{2}\right)} \int_0^{\pi/2} \sin^{2\nu}\theta \left(\frac{\pi(x\cos\theta)^{\alpha-\nu-1}}{2\Gamma(\alpha-\nu)\cos\frac{\pi(\alpha-\nu)}{2}} \right.$$
$$\left. + \sum_{k=0}^{\infty} \frac{(-1)^k \zeta(\alpha-\nu-2k)(x\cos\theta)^{2k}}{(2k)!} \right) d\theta. \tag{4.7}$$

After integrating, we interchange the integral and the sum again on the right-hand side

$$\sum_{n=1}^{\infty} \frac{J_\nu(nx)}{n^\alpha} = \frac{\sqrt{\pi}\,x^{\alpha-1}}{2^\nu \Gamma\left(\nu+\frac{1}{2}\right)\Gamma(\alpha-\nu)\cos\frac{\pi(\alpha-\nu)}{2}} \int_0^{\pi/2} \sin^{2\nu}\theta \cos^{\alpha-\nu-1}\theta \, d\theta$$
$$+ \frac{2\left(\frac{x}{2}\right)^\nu}{\sqrt{\pi}\,\Gamma\left(\nu+\frac{1}{2}\right)} \sum_{k=0}^{\infty} \frac{(-1)^k \zeta(\alpha-\nu-2k)\, x^{2k}}{\Gamma\left(2\left(k+\frac{1}{2}\right)\right)}$$
$$\int_0^{\pi/2} \sin^{2\nu}\theta \cos^{2k}\theta \, d\theta, \tag{4.8}$$

where, by virtue of the beta function

$$\mathrm{B}(x,y) = 2\int_0^{\pi/2} \sin^{2x-1}\phi \cos^{2y-1}\phi \, d\phi = \int_0^1 t^{x-1}(1-t)^{y-1} dt,$$

we solve the first integral

$$\int_0^{\frac{\pi}{2}} \sin^{2\nu}\theta \cos^{\alpha-\nu-1}\theta \, d\theta = \int_0^{\frac{\pi}{2}} \sin^{2\left(\nu+\frac{1}{2}\right)-1}\theta \cos^{2\frac{\alpha-\nu}{2}-1}\theta \, d\theta$$
$$= \frac{1}{2}\mathrm{B}\left(\nu+\frac{1}{2}, \frac{\alpha-\nu}{2}\right) = \frac{\Gamma\left(\nu+\frac{1}{2}\right)\Gamma\left(\frac{\alpha-\nu}{2}\right)}{2\Gamma\left(\frac{\alpha+\nu+1}{2}\right)},$$

and the second one

$$\int_0^{\frac{\pi}{2}} \sin^{2\nu}\theta \cos^{2k}\theta \, d\theta = \int_0^{\frac{\pi}{2}} \sin^{2\left(\nu+\frac{1}{2}\right)-1}\theta \cos^{2\left(k+\frac{1}{2}\right)-1}\theta \, d\theta$$

$$= \frac{\Gamma\left(\nu+\frac{1}{2}\right)\Gamma\left(k+\frac{1}{2}\right)}{2\Gamma(\nu+k+1)}.$$

If we set $z = \frac{\alpha-\nu}{2}$ in Legendre's duplication formula

$$\sqrt{\pi}\,\Gamma(2z) = 2^{2z-1}\Gamma(z)\Gamma\left(z+\frac{1}{2}\right), \tag{4.9}$$

we get

$$\sqrt{\pi}\,\Gamma(\alpha-\nu) = \sqrt{\pi}\,\Gamma\left(2\frac{\alpha-\nu}{2}\right) = 2^{2\frac{\alpha-\nu}{2}-1}\Gamma\left(\frac{\alpha-\nu}{2}\right)\Gamma\left(\frac{\alpha-\nu}{2}+\frac{1}{2}\right)$$

$$= 2^{\alpha-\nu-1}\Gamma\left(\frac{\alpha-\nu}{2}\right)\Gamma\left(\frac{\alpha-\nu+1}{2}\right),$$

and similarly, by setting $z = k + \frac{1}{2}$ in Legendre's duplication formula, we have

$$\sqrt{\pi}\,\Gamma\left(2\left(k+\frac{1}{2}\right)\right) = 2^{2\left(k+\frac{1}{2}\right)-1}\Gamma\left(k+\frac{1}{2}\right)\Gamma\left(k+\frac{1}{2}+\frac{1}{2}\right)$$

$$= 2^{2k}\Gamma\left(k+\frac{1}{2}\right)\Gamma(k+1).$$

Replacing corresponding entries in (4.8), we obtain the summation formula (4.3) (see [31]). □

Providing $\alpha - \nu = 2m$, $m \in \mathbb{N}$, the right-hand side series in (4.3) truncates because Riemann's zeta function equals zero if the argument is a negative even integer. So, setting $\alpha = \nu + 2m$ in (4.3) brings the series in closed form

$$\sum_{n=1}^{\infty} \frac{J_\nu(nx)}{n^{\nu+2m}} = \frac{m!(-1)^m x^{\nu+2m-1}\sqrt{\pi}}{2^\nu (2m)!\Gamma\left(\nu+m+\frac{1}{2}\right)}$$

$$+ \sum_{k=0}^{m} \frac{(-1)^k \zeta(2m-2k)\left(\frac{x}{2}\right)^{\nu+2k}}{k!\,\Gamma(\nu+k+1)}, 0 < x < 2\pi, \tag{4.10}$$

where $m = 1, 2, 3, \ldots$; $\operatorname{Re}\nu > -2m - \frac{1}{2}$.

Proposition 1 *The formula (4.10) is equivalent to entry* 13 *in [23, p. 681].*

$$\sum_{n=1}^{\infty}\frac{J_{\nu}(nx)}{n^{2k+\nu}}=\frac{(-1)^{k+1}x^{2k+\nu}}{(2k)!2^{\nu+1}\sqrt{\pi}}\sum_{n=0}^{2k}\frac{\Gamma\left(k-\frac{n-1}{2}\right)}{\Gamma\left(k+\nu+1-\frac{n}{2}\right)}\binom{2k}{n}\left(\frac{2\pi}{x}\right)^{n}B_{n},\tag{4.11}$$

with $0<x<2\pi$*, and* B_n *are Bernoulli numbers [2]. Even though this entry looks different in comparison with (4.10), it presents the same sum in a different form.*

Proof. Knowing that $B_0=1, B_1=-\frac{1}{2}$, we expand the first two terms of the right-hand side sum of entry 13, rewriting it as follows

$$\begin{aligned}&\frac{(-1)^{k+1}x^{2k+\nu}}{(2k)!2^{\nu+1}\sqrt{\pi}}\sum_{n=0}^{2k}\frac{\Gamma\left(k-\frac{n-1}{2}\right)}{\Gamma\left(k+\nu+1-\frac{n}{2}\right)}\binom{2k}{n}\left(\frac{2\pi}{x}\right)^{n}B_{n}\\&=\frac{(-1)^{k+1}x^{2k+\nu}}{(2k)!2^{\nu+1}\sqrt{\pi}}\frac{\Gamma\left(k+\frac{1}{2}\right)}{\Gamma(k+\nu+1)}+\frac{k!(-1)^{k}x^{2k+\nu-1}\sqrt{\pi}}{(2k)!2^{\nu}\Gamma\left(k+\nu+\frac{1}{2}\right)}\\&+\frac{(-1)^{k+1}x^{2k+\nu}}{(2k)!2^{\nu+1}\sqrt{\pi}}\sum_{n=2}^{2k}\frac{\Gamma\left(k-\frac{n-1}{2}\right)}{\Gamma\left(k+\nu+1-\frac{n}{2}\right)}\binom{2k}{n}\left(\frac{2\pi}{x}\right)^{n}B_{n}.\end{aligned}\tag{4.12}$$

For the sake of brevity and conciseness, we are dealing only with the last sum. Since $B_{2n+1}=0$ for $n\geqslant 1$, there remain only even indices, so after the substitution $n=2j$ in the last sum, we obtain

$$\frac{(-1)^{k+1}x^{2k+\nu}}{2^{\nu+1}\sqrt{\pi}}\sum_{j=1}^{k}\frac{(2\pi)^{2j}\Gamma\left(k-j+\frac{1}{2}\right)}{x^{2j}(2j)!(2k-2j)!\Gamma(k-j+\nu+1)}B_{2j}.$$

Applying the relation between Bernoulli numbers and the ζ function [2]

$$B_{2n}=\frac{(-1)^{n-1}2(2n)!}{(2\pi)^{2n}}\zeta(2n),\quad n\geqslant 1,$$

this expression further reduces to

$$\frac{x^{2k+\nu}}{2^{\nu}\sqrt{\pi}}\sum_{j=1}^{k}\frac{(-1)^{k-j}\Gamma\left(k-j+\frac{1}{2}\right)\zeta(2j)}{x^{2j}(2k-2j)!\Gamma(k-j+\nu+1)}$$
$$=\sum_{j=1}^{k}\frac{(-1)^{k-j}x^{\nu+2(k-j)}(2k-2j-1)!!\zeta(2j)}{2^{\nu+k-j}(2k-2j)!\Gamma(k-j+\nu+1)}.$$

After introducing a new index $i = k - j$, it becomes

$$\frac{x^{2k+\nu}}{2^{\nu}\sqrt{\pi}}\sum_{j=1}^{k}\frac{(-1)^{k-j}\Gamma\left(k-j+\frac{1}{2}\right)\zeta(2j)}{x^{2j}(2k-2j)!\Gamma(k-j+\nu+1)}=\sum_{i=0}^{k-1}\frac{(-1)^{i}\left(\frac{x}{2}\right)^{\nu+2i}\zeta(2k-2i)}{i!\Gamma(i+\nu+1)}.$$

However, the first term on the right-hand side of (4.12) after rearrangement becomes

$$\frac{(-1)^{k+1}x^{2k+\nu}}{(2k)!2^{\nu+1}\sqrt{\pi}}\frac{\Gamma\left(k+\frac{1}{2}\right)}{\Gamma(k+\nu+1)}=-\frac{1}{2}\frac{(-1)^{k}\left(\frac{x}{2}\right)^{\nu+2k}}{(2k)!}\frac{2^{k}(2k-1)!!}{\Gamma(k+\nu+1)}$$
$$=\frac{(-1)^{k}\left(\frac{x}{2}\right)^{\nu+2k}}{k!\Gamma(k+\nu+1)}\zeta(0),$$

whereby we obtain the kth term of the last sum. In addition, the second term on the right-hand side of (4.12) is exactly the same as the first term on the right-hand side of (4.10), meaning that we have proved that (4.11) and (4.10) are equal. □

We note that from (4.3), there immediately follows

$$\sum_{n=1}^{\infty}\frac{J_{\nu}(2nx)}{(2n)^{\alpha}}=\frac{1}{2^{\alpha}}\sum_{n=1}^{\infty}\frac{J_{\nu}(n\cdot 2x)}{n^{\alpha}}=\frac{\pi x^{\alpha-1}}{2^{\alpha+1}\Gamma\left(\frac{\alpha-\nu+1}{2}\right)\Gamma\left(\frac{\alpha+\nu+1}{2}\right)\cos\frac{\pi(\alpha-\nu)}{2}}$$
$$+\frac{x^{\nu}}{2^{\alpha}}\sum_{k=0}^{\infty}\frac{(-1)^{k}x^{2k}\zeta(\alpha-\nu-2k)}{k!\,\Gamma(\nu+k+1)}. \tag{4.13}$$

Corollary 4.1 *Relying on (4.13), we can find alternating series over Bessel functions as a series over Dirichlet's eta functions*

$$\sum_{n=1}^{\infty}\frac{(-1)^{n-1}J_\nu(nx)}{n^\alpha}=\sum_{n=1}^{\infty}\frac{J_\nu(nx)}{n^\alpha}-\frac{1}{2^{\alpha-1}}\sum_{n=1}^{\infty}\frac{J_\nu(2nx)}{n^\alpha}$$
$$=\sum_{k=0}^{\infty}\frac{(-1)^k\left(\frac{x}{2}\right)^{\nu+2k}\eta(\alpha-\nu-2k)}{k!\,\Gamma(\nu+k+1)},\qquad(4.14)$$

where we used the relation $\eta(\alpha-\nu-2k)=\left(1-2^{1-(\alpha-\nu-2k)}\right)\zeta(\alpha-\nu-2k)$.

We take $\alpha=\nu+2m$ in (4.14). Since the eta function vanishes at even negative integers, the above right-hand series is brought into closed form

$$\sum_{n=1}^{\infty}\frac{(-1)^{n-1}J_\nu(nx)}{n^{\nu+2m}}=\sum_{k=0}^{m}\frac{(-1)^k\eta(2m-2k)\left(\frac{x}{2}\right)^{\nu+2k}}{k!\,\Gamma(\nu+k+1)},\ -\pi<x<\pi,\qquad(4.15)$$

where $m=1,2,3,\ldots;\ \mathrm{Re}\,\nu>-2m-\frac{1}{2}$. The entry 14 in [23, p. 681] has a different form in comparison with (4.15), but there holds.

Proposition 2 *The formula (4.15) is equivalent to entry 14 in [23, p. 681]*

$$\sum_{n=1}^{\infty}\frac{(-1)^nJ_\nu(nx)}{n^{2k+\nu}}=\frac{(-1)^{k+1}x^{2k+\nu}}{(2k)!2^{\nu+1}\sqrt{\pi}}\sum_{n=0}^{2k}\binom{2k}{n}B_n$$
$$\times\sum_{i=0}^{2k-n}\binom{2k-n}{i}\frac{2^n\pi^{n+i}\Gamma\left(\frac{k-(n+i-1)}{2}\right)}{x^{n+i}\Gamma\left(\frac{k+\nu+1-(n+i)}{2}\right)},\qquad-\pi<x<\pi,\qquad(4.16)$$

where $k=1,2,3,\ldots;\mathrm{Re}\,\nu>-2k-\frac{1}{2}$ *and* B_n *denote Bernoulli numbers.*

Proof. As in the case of (4.10) and entry 13, we can prove in a similar way the equality between (4.15) and (4.16). □

Note that Lorch and Szego [14] used mathematical induction to prove these two particular cases.

Corollary 4.2 *The summation formula for the series over Bessel functions containing odd arguments:*

$$\sum_{n=1}^{\infty} \frac{J_\nu((2n-1)x)}{(2n-1)^\alpha} = \frac{\pi\left(\frac{x}{2}\right)^{\alpha-1}}{4\Gamma\left(\frac{\alpha-\nu+1}{2}\right)\Gamma\left(\frac{\alpha+\nu+1}{2}\right)\cos\frac{\pi(\alpha-\nu)}{2}} + \sum_{k=0}^{\infty} \frac{(-1)^k\left(\frac{x}{2}\right)^{\nu+2k}\lambda(\alpha-\nu-2k)}{k!\,\Gamma(\nu+k+1)}. \tag{4.17}$$

Proof. Using the relation

$$\sum_{n=1}^{\infty} \frac{J_\nu((2n-1)x)}{(2n-1)^\alpha} = \sum_{n=1}^{\infty} \frac{J_\nu(nx)}{n^\alpha} - \frac{1}{2^\alpha}\sum_{n=1}^{\infty} \frac{J_\nu(2nx)}{n^\alpha}$$

and the equality $\left(1-2^{-(\alpha-\nu-2k)}\right)\zeta(\alpha-\nu-2k) = \lambda(\alpha-\nu-2k)$, we come to (4.17). □

The right-hand side series (4.17) truncates because the lambda function also vanishes at even negative integers. In addition, unlike the other two Dirichlet functions, there holds $\lambda(0) = 0$. So for $\alpha-\nu = 2m$, $m \in \mathbb{N}$, we have

$$\sum_{n=1}^{\infty} \frac{J_\nu((2n-1)x)}{(2n-1)^{\nu+2m}} = \frac{(-1)^m\pi\left(\frac{x}{2}\right)^{\nu+2m-1}}{4\Gamma\left(m+\frac{1}{2}\right)\Gamma\left(\nu+m+\frac{1}{2}\right)} + \sum_{k=0}^{m-1} \frac{(-1)^k\left(\frac{x}{2}\right)^{\nu+2k}\lambda(2m-2k)}{k!\,\Gamma(\nu+k+1)}.$$

Corollary 4.3 *The summation formula of an alternating series over Bessel functions with odd arguments is*

$$\sum_{n=1}^{\infty} \frac{(-1)^{n-1}J_\nu((2n-1)x)}{(2n-1)^\alpha} = \sum_{k=0}^{\infty} \frac{(-1)^k\left(\frac{x}{2}\right)^{\nu+2k}\beta(\alpha-\nu-2k)}{k!\Gamma(k+\nu+1))}. \tag{4.18}$$

Proof. Referring to the second formula in [37, p. 446], we obtain

$$\sum_{n=1}^{\infty}\frac{(-1)^{n-1}J_\nu((2n-1)x)}{(2n-1)^\alpha}$$

$$=\frac{2\left(\frac{x}{2}\right)^\nu}{\sqrt{\pi}\,\Gamma\left(\nu+\frac{1}{2}\right)}\int_0^{\frac{\pi}{2}}\sin^{2\nu}\theta\sum_{n=1}^{\infty}\frac{(-1)^{n-1}\cos((2n-1)x\cos\theta)}{(2n-1)^{\alpha-\nu}}\,d\theta$$

$$=\frac{2\left(\frac{x}{2}\right)^\nu}{\sqrt{\pi}\,\Gamma\left(\nu+\frac{1}{2}\right)}\sum_{k=0}^{\infty}\frac{(-1)^k\beta(\alpha-\nu-2k)\,x^{2k}}{\Gamma\left(2\left(k+\frac{1}{2}\right)\right)}\int_0^{\pi/2}\sin^{2\nu}\theta\cos^{2k}\theta\,d\theta.$$

After calculating the last integral, we come to (4.18). □

Based on the relation

$$\beta(-(2n-1))=\frac{4^{2n-1}}{2n}\left(-B_{2n}\left(\frac{1}{4}\right)+B_{2n}\left(\frac{3}{4}\right)\right)=0,$$

where B_{2n} are Bernoulli numbers, for $\alpha=\nu+2m-1$, $m\in\mathbb{N}$, the β function vanishes at negative odd integers and the closed-form formula of (4.18) is

$$\sum_{n=1}^{\infty}\frac{(-1)^{n-1}J_\nu((2n-1)x)}{(2n-1)^{\nu+2m-1}}=\sum_{k=0}^{m-1}\frac{(-1)^k\left(\frac{x}{2}\right)^{\nu+2k}\beta(2m-2k-1)}{k!\Gamma(k+\nu+1))}.$$

Formulas in Theorem 4.1 and Corollaries 4.1, 4.2, and 4.3 can be rewritten as one, encompassing all of them

$$\sum_{n=1}^{\infty}\frac{(s)^{n-1}J_\nu((an-b)x)}{(an-b)^\alpha}=\frac{c\pi\left(\frac{x}{2}\right)^{\alpha-1}}{2\Gamma\left(\frac{\alpha-\nu+1}{2}\right)\Gamma\left(\frac{\alpha+\nu+1}{2}\right)\cos\left(\frac{\alpha-\nu}{2}\pi\right)}+\sum_{k=0}^{\infty}\frac{(-1)^kF(\alpha-\nu-2k)\left(\frac{x}{2}\right)^{\nu+2k}}{k!\,\Gamma(\nu+k+1)},\tag{4.19}$$

where we choose the parameters involved by the following scheme

$$a=\left\{\begin{matrix}1\\2\end{matrix}\right\}b=\left\{\begin{matrix}0\\1\end{matrix}\right\}\left\{\frac{s=\left\{\begin{matrix}1\\-1\end{matrix}\right\}c=\left\{\begin{matrix}1\\0\end{matrix}\right\}F=\left\{\begin{matrix}\zeta\\\eta\end{matrix}\right\}}{s=\left\{\begin{matrix}1\\-1\end{matrix}\right\}c=\left\{\begin{matrix}1/2\\0\end{matrix}\right\}F=\left\{\begin{matrix}\lambda\\\beta\end{matrix}\right\}}\right\}.\tag{4.20}$$

In the case of $\alpha - \nu = 2m - 1, m \in \mathbb{N}$, we cannot immediately place this on the right-hand side of relation (4.3) because one encounters singularities, i.e., in $\cos\frac{\pi(\alpha-\nu)}{2}$ within the denominator of the first term and in the member of the right-hand series for the index $k = m - 1$.

Theorem 4.2 *For* $\alpha = \nu + 2m-1, m \in \mathbb{N}$, *there holds*

$$\sum_{n=1}^{\infty} \frac{J_\nu(nx)}{n^{\nu+2m-1}} = \frac{(-1)^{m-1}\left(\frac{x}{2}\right)^{\nu+2m-2}}{2\Gamma(m)\Gamma(\nu+m)}\left(H_{m-1} + H_{\nu+m-1} - 2\ln\frac{x}{2}\right)$$
$$+ \sum_{k=0}^{m-2} \frac{(-1)^k \zeta(2m-2k-1)\left(\frac{x}{2}\right)^{\nu+2k}}{k!\Gamma(\nu+k+1)}$$
$$+ \sum_{k=1}^{\infty} \frac{\Gamma(2k)\zeta(2k)\left(\frac{x}{4\pi}\right)^{2k}}{\Gamma(m+k)\Gamma(\nu+m+k)}. \tag{4.21}$$

Proof. Because of that, it is necessary to take the limiting value in (4.5) when $\alpha \to \nu + 2m - 1$ of the partial sum of the first m terms of the series from (4.7)

$$\lim_{\alpha\to\nu+2m-1}\left(\frac{\pi\left(\frac{x}{2}\right)^{\alpha-1}}{2\Gamma\left(\frac{\alpha-\nu+1}{2}\right)\Gamma\left(\frac{\alpha+\nu+1}{2}\right)\cos\frac{\pi(\alpha-\nu)}{2}} + \sum_{k=0}^{m-1}\frac{(-1)^k\zeta(\alpha-\nu-2k)\left(\frac{x}{2}\right)^{\nu+2k}}{k!\Gamma(\nu+k+1)}\right).$$

For $k = 0, 1, ..., m-2$, all the terms have no singularities if $\alpha = \nu + 2m - 1$, so it suffices to deal only with the term for $k = m - 1$, whereupon we find

$$\lim_{\alpha\to\nu+2m-1}\left(\frac{\pi\left(\frac{x}{2}\right)^{\alpha-1}}{2\Gamma\left(\frac{\alpha-\nu+1}{2}\right)\Gamma\left(\frac{\alpha+\nu+1}{2}\right)\cos\frac{\pi(\alpha-\nu)}{2}} + \frac{(-1)^{m-1}\left(\frac{x}{2}\right)^{\nu+2m-2}\zeta(\alpha-\nu-2m+2)}{(m-1)!\,\Gamma(\nu+m)}\right)$$
$$= \frac{(-1)^{m-1}\left(\frac{x}{2}\right)^{\nu+2m-2}}{2\Gamma(m)\Gamma(\nu+m)}\left(H_{m-1} + H_{\nu+m-1} - 2\log\frac{x}{2}\right),$$

where γ is Euler-Mascheroni's constant. Thus, we have

$$\sum_{n=1}^{\infty} \frac{J_\nu(nx)}{n^{\nu+2m-1}} = \frac{(-1)^{m-1}\left(\frac{x}{2}\right)^{\nu+2m-2}}{2\Gamma(m)\Gamma(\nu+m)} \left(H_{m-1} + H_{\nu+m-1} - 2\log\frac{x}{2}\right)$$
$$+ \sum_{k=0}^{m-2} \frac{(-1)^k \zeta(2m-2k-1)\left(\frac{x}{2}\right)^{\nu+2k}}{k!\Gamma(\nu+k+1)}$$
$$+ \sum_{k=m}^{\infty} \frac{(-1)^k \zeta(2m-2k-1)\left(\frac{x}{2}\right)^{\nu+2k}}{k!\,\Gamma(\nu+k+1)}.$$

As for the remainder, we transform it by shifting the index, so we introduce the substitution $k = m + j - 1$, then revert to k instead of using j, i.e.

$$\sum_{k=m}^{\infty} \frac{(-1)^k \zeta(2m-2k-1)\left(\frac{x}{2}\right)^{\nu+2k}}{k!\,\Gamma(\nu+k+1)}$$
$$= (-1)^{m-1}\left(\frac{x}{2}\right)^{\nu+2m-2} \sum_{k=1}^{\infty} \frac{(-1)^k \zeta(1-2k)\left(\frac{x}{2}\right)^{2k}}{\Gamma(m+k)\Gamma(\nu+m+k)}. \tag{4.22}$$

Recalling the relation

$$\zeta(1-2n) = (-1)^n \frac{2(2n-1)!\zeta(2n)}{(2\pi)^{2n}},$$

and applying (4.9) for $z = k$, equality (4.22) becomes

$$\sum_{k=m}^{\infty} \frac{(-1)^k \zeta(2m-2k-1)\left(\frac{x}{2}\right)^{\nu+2k}}{\Gamma(k+1)\,\Gamma(\nu+k+1)}$$
$$= 2(-1)^{m-1}\left(\frac{x}{2}\right)^{\nu+2m-2} \sum_{k=1}^{\infty} \frac{\Gamma(2k)\zeta(2k)\left(\frac{x}{4\pi}\right)^{2k}}{\Gamma(m+k)\Gamma(\nu+m+k)}, \tag{4.23}$$

which completes the proof. □

12.4.1 Series over Spherical Bessel Functions

Setting $\nu = \frac{1}{2}$ in (4.21), we have

$$\sum_{n=1}^{\infty} \frac{J_{1/2}(nx)}{n^{2m-1/2}} = \frac{(-1)^{m-1}\left(\frac{x}{2}\right)^{2m-3/2}}{2\Gamma(m)\Gamma(1/2+m)}\left(H_{m-1} + H_{m-1/2} - 2\log\frac{x}{2}\right)$$
$$+\sqrt{\frac{x}{2}}\sum_{k=0}^{m-2} \frac{(-1)^k \zeta(2m-2k-1)\left(\frac{x}{2}\right)^{2k}}{k!\Gamma(1/2+k+1)}$$
$$+\sum_{k=1}^{\infty} \frac{\Gamma(2k)\zeta(2k)\left(\frac{x}{4\pi}\right)^{2k}}{\Gamma(m+k)\Gamma(1/2+m+k)}. \tag{4.24}$$

For $\nu = \frac{1}{2}$, referring to (2.3) and applying the Choi-Srivastava theorem [5, 40], we bring the right-hand series in (4.23) to the closed form

$$\frac{(-1)^{m-1}}{\sqrt{2\pi}} 8x^{2m-3/2} \sum_{k=1}^{\infty} \frac{\zeta(2k)}{(2k)_{2m}}\left(\frac{x}{2\pi}\right)^{2k}$$
$$= \frac{(-1)^m 4x^{2m-3/2}}{(2m-1)!\sqrt{2\pi}}\left(H_{2m-1} - \log 2\pi\right)$$
$$- \frac{4}{\sqrt{2x\pi}} \sum_{k=0}^{m-2} \frac{(-1)^k \zeta(2m-2k-1)\, x^{2k+1}}{(2k+1)!}$$
$$+ \frac{(-1)^m 4(2\pi)^{2m-3/2}}{(2m-1)!\sqrt{x}}\left(\zeta'\left(1-2m, 1-\frac{x}{2\pi}\right) - \zeta'\left(1-2m, 1+\frac{x}{2\pi}\right)\right). \tag{4.25}$$

After rewriting the last term by differentiating both sides of Hurwitz's function's basic property

$$\zeta(s,a) = a^{-s} + \zeta(s, a+1)$$

with respect to s and putting then $a = \frac{x}{2\pi}$ and $s = 1-2m$, for the last term of (4.25), we find

$$-\zeta'\left(1-2m, 1+\frac{x}{2\pi}\right) = -\left(\frac{x}{2\pi}\right)^{2m-1} \log\frac{x}{2\pi} - \zeta'\left(1-2m, \frac{x}{2\pi}\right).$$

Using the result of (4.25), the right-hand side of (4.24) yields the closed form of the series (4.21)

$$\sum_{n=1}^{\infty} \frac{J_{1/2}(nx)}{n^{2m-1/2}} = \frac{(-1)^m 4(2\pi)^{2m-1}}{(2m-1)!\sqrt{2\pi x}} \left(\frac{1}{2}(H_{2m-1} - \log x)\left(\frac{x}{2\pi}\right)^{2m-1} + \zeta'\left(1-2m, 1-\frac{x}{2\pi}\right) - \zeta'\left(1-2m, \frac{x}{2\pi}\right)\right) + \sqrt{\frac{2x}{\pi}} \sum_{k=0}^{m-2} \frac{(-x^2)^k \zeta(2m-2k-1)}{(2k+1)!}$$

and we obtain the modified sum of the series (4.22).

We can generalize (4.24) by considering spherical Bessel functions of the first kind $j_k(z)$ linked to Bessel functions of the first kind as follows

$$j_p(x) = \sqrt{\frac{\pi}{2x}} J_{p+\frac{1}{2}}(x), p \in \mathbb{N}_0.$$

So, multiplying both sides of (4.19) by $\sqrt{\frac{\pi}{2x}}$ and setting $\nu = p + \frac{1}{2}, p \in \mathbb{N}_0$, we have

$$\sum_{n=1}^{\infty} \frac{(s)^{n-1} j_p((an-b)x)}{(an-b)^{\alpha}} = \frac{c\pi\sqrt{\pi}\left(\frac{x}{2}\right)^{\alpha-\frac{3}{2}}}{2\Gamma\left(\frac{\alpha-p}{2}+\frac{1}{4}\right)\Gamma\left(\frac{\alpha+p}{2}+\frac{3}{4}\right)\cos\left(\frac{\pi(\alpha-p)}{2}-\frac{\pi}{4}\right)} + \frac{\sqrt{\pi}}{2}\sum_{k=0}^{\infty} \frac{(-1)^k F\left(\alpha-p-\frac{1}{2}-2k\right)\left(\frac{x}{2}\right)^{p+2k}}{k!\,\Gamma\left(p+\frac{1}{2}+k+1\right)}. \tag{4.26}$$

Here, we are referring to Euler's reflection formula

$$\Gamma(1-s)\Gamma(s) = \frac{\pi}{\sin \pi s}, s \notin \mathbb{Z}, \tag{4.27}$$

with $s = \frac{\alpha-p}{2} + \frac{1}{4}$, and obtain

$$\Gamma\left(\frac{3}{4} - \frac{\alpha-p}{2}\right)\Gamma\left(\frac{\alpha-p}{2}+\frac{1}{4}\right) = \frac{\pi}{\cos\left(\frac{\pi}{4} - \frac{\pi(\alpha-p)}{2}\right)} = \frac{\pi}{\cos\left(\frac{\pi(\alpha-p)}{2} - \frac{\pi}{4}\right)}.$$

Replacing this in the first term of (4.26), and using (2.3), we obtain the summation formula for the series over spherical Bessel functions of the first kind

$$\sum_{n=1}^{\infty} \frac{(s)^{n-1} j_p\left((an-b)x\right)}{(an-b)^{\alpha}}$$
$$= \frac{c\sqrt{\pi}\, x^{\alpha-\frac{3}{2}}\Gamma\left(\frac{3}{4}-\frac{\alpha-p}{2}\right)}{2^{\alpha+\frac{1}{2}}\Gamma\left(\frac{3}{4}+\frac{\alpha+p}{2}\right)}$$
$$+ (2x)^p \sum_{k=0}^{\infty} \frac{(-1)^k \Gamma(p+k+1) F\left(\alpha-p-\frac{1}{2}-2k\right) x^{2k}}{k!\, \Gamma(2p+2k+2)}.$$

Now, for $a = 1, b = 0, s = 1, c = 1, F = \zeta$, we first evaluate

$$\lim_{\alpha\to p+2m-1/2} \Bigg(\frac{\sqrt{\pi}\, x^{\alpha-\frac{3}{2}}\Gamma\left(\frac{3}{4}-\frac{\alpha-p}{2}\right)}{2^{\alpha+\frac{1}{2}}\Gamma\left(\frac{3}{4}+\frac{\alpha+p}{2}\right)}$$
$$+ 2^p \frac{(-1)^{m-1}\Gamma(p+m)\zeta\left(\alpha-p-2m+\frac{3}{2}\right)x^{2m+p-2}}{(m-1)!\Gamma(2p+2m)}.$$

By bringing to the same denominator and applying L'Hôpital's rule, after the limiting process, we obtain

$$\sum_{n=1}^{\infty} \frac{j_p(nx)}{n^{p+2m-1/2}} = \frac{(-1)^{m-1}\left(\frac{x}{2}\right)^{p+2m-\frac{3}{2}}\sqrt{\frac{\pi}{8x}}}{\Gamma(m)\Gamma\left(p+\frac{1}{2}+m\right)} \left(H_{m-1} + H_{p+m-\frac{1}{2}} - 2\log\frac{x}{2}\right)$$
$$+ (2x)^p \sum_{k=0}^{m-2} \frac{(-1)^k \Gamma(p+k+1)\zeta(2m-1-2k)x^{2k}}{k!\,\Gamma(2p+2k+2)}$$
$$+ (-1)^{m-1} 2^{p+1} x^{p+2m-2} \sum_{n=1}^{\infty} \frac{(m+n)_p}{(2n)_{2p+2m}} \zeta(2n)\left(\frac{x}{2\pi}\right)^{2n}. \tag{4.28}$$

We consider a fractional decomposition of the ratio consisting of two Pochhammer symbols

$$\frac{(m+n)_p}{(2n)_{2p+2m}} = \frac{(m+n)\cdots(m+n+p-1)}{(2n)_{2m}(2n+2m)\cdots(2n+2m+2p-1)}$$
$$= \frac{1}{2^p(2n)_{2m}(2n+2m+1)(2n+2m+3)\cdots(2n+2m+2p-1)}.$$

By applying Heaviside's method, in the next step, we make a decomposition of the rational function

$$\frac{P(n)}{Q(n)} = \frac{1}{(2n+2m+1)(2n+2m+3)\cdots(2n+2m+2p-1)},$$

with $P(n)=1, Q(n)=(2n+2m+1)(2n+2m+3)\cdots(2n+2m+2p-1)$. Thus, we have

$$\frac{P(n)}{Q(n)} = \frac{C_1}{n-a_1} + \frac{C_2}{n-a_2} + \cdots + \frac{C_p}{n-a_p},$$

where the constants $C_k, k=1,\ldots,p$, are to be determined, and $Q(a_k)=0$. So, there holds

$$C_k = \frac{P(a_k)}{Q'(a_k)}, \; a_k = -(2m+2k-1)/2, \quad k=1,\ldots,p,$$

which means that the right-hand side series in (4.28)(4.28) can be expressed as follows

$$\sum_{n=1}^{\infty} \frac{(m+n)_p \zeta(2n)\left(\frac{x}{2\pi}\right)^{2n}}{(2n)_{2p+2m}} = \sum_{k=1}^{p} \frac{C_k}{2^{p-1}} \sum_{n=1}^{\infty} \frac{\zeta(2n)\left(\frac{x}{2\pi}\right)^{2n}}{(2n)_{2m}(2n+2m+2k-1)}. \tag{4.29}$$

Now, we are given constants $c_1,\ldots,c_{2k}$ in the representation

$$\frac{1}{2n+2m+2k-1} = \frac{c_1}{2n+2m} + \frac{c_2}{(2n+2m)_2} + \cdots + \frac{c_{2k}}{(2n+2m)_{2k}},$$

where $k=1,\ldots,p$. We multiply the numerator and denominator of the left-hand side fraction by the missing factors between $(2n+2m+2k-1)$ and $(2n+2m)$, including the latter. Also, we bring the right-hand side fractions to the same denominator. Thus, we obtain the equality of the numerators

$$(2n+2m)\cdots(2n+2m+2k-2) = c_1(2n+2m+1)\cdots(2n+2m+2k-1)$$
$$+c_2(2n+2m+2)\cdots(2n+2m+2k-1) + \cdots + c_{2k}, \quad k=1,\ldots,p.$$

The highest power on the left-hand side and the product at c_1 is $(2n)^{2k-1}$. There follows $c_1 = 1$. After rearrangements, we have

$$\begin{aligned}&-(2k-1)(2n+2m+1)\cdots(2n+2m+2k-2)\\&\quad= c_2(2n+2m+2)\cdots(2n+2m+2k-1)+c_3(2n+2m+3)\cdots\\&\qquad\cdots(2n+2m+2k-1)+\cdots+c_{2k-1}(2n+2m+2k-1)+c_{2k}.\end{aligned}$$

Regarding both sides as polynomial functions in n of the degree $2k-2$, we take the $(2k-2)$th derivative. As a result, we have

$$-(2k-1)\,2^{2k-2}(2k-2)! = c_2 2^{2k-2}(2k-2)!.$$

Hence, we find $c_2 = -(2k-1)$. Replacing this value in (4.30), we can determine c_3 similarly

$$\begin{aligned}&(2k-1)(2k-2)(2n+2m+2)\cdots(2n+2m+2k-2)\\&\quad= c_3(2n+2m+3)\cdots(2n+2m+2k-1)\\&\qquad+c_4(2n+2m+4)\cdots(2n+2m+2k-1)+\cdots+c_{2k-1}\\&\qquad(2n+2m+2k-1)+c_{2k}.\end{aligned}$$

Now, the polynomials on both sides are of the degree $2k-3$, and we take the $(2k-3)$th derivative, and obtain

$$(2k-1)(2k-2)\,2^{2k-3}(2k-3)! = c_3 2^{2k-3}(2k-3)!,$$

which implies $c_3 = (2k-1)(2k-2)$. By repeating this procedure, we arrive at the relation

$$\frac{1}{(2n)_{2m}(2n+2m+2k-1)} = \sum_{j=0}^{2k-1}\frac{(-1)^j}{(2n)_{2m+j+1}}\prod_{i=1}^{j}(2k-i).$$

Together with (4.29), this gives rise to the following form of the right-hand side series in (4.28)

$$\begin{aligned}&\sum_{n=1}^{\infty}\frac{(m+n)_p}{(2n)_{2p+2m}}\zeta(2n)\left(\frac{x}{2\pi}\right)^{2n}\\&= (-1)^{m-1}4x^{p+2m-2}\sum_{k=1}^{p}C_k\sum_{j=0}^{2k-1}(-1)^j\prod_{i=1}^{j}(2k-i)\sum_{n=1}^{\infty}\frac{\zeta(2n)\left(\frac{x}{2\pi}\right)^{2n}}{(2n)_{2m+j+1}}.\end{aligned}$$

The right-hand side of (4.28) becomes

$$\sum_{n=1}^{\infty} \frac{j_p(nx)}{n^{p+2m-1/2}}$$

$$= \frac{(-1)^{m-1}\left(\frac{x}{2}\right)^{p+2m-\frac{3}{2}}\sqrt{\frac{\pi}{8x}}}{\Gamma(m)\Gamma\left(p+\frac{1}{2}+m\right)}\left(H_{m-1}+H_{p+m-\frac{1}{2}}-2\log\frac{x}{2}\right)$$

$$+(2x)^p\sum_{k=0}^{m-2}\frac{(-1)^k\Gamma(p+k+1)\zeta(2m-1-2k)x^{2k}}{k!\,\Gamma(2p+2k+2)}$$

$$+(-1)^{m-1}4x^{p+2m-2}\sum_{k=1}^{p}C_k\sum_{j=0}^{2k-1}(-1)^j\prod_{i=1}^{j}(2k-i)\sum_{n=1}^{\infty}\frac{\zeta(2n)\left(\frac{x}{2\pi}\right)^{2n}}{(2n)_{2m+j+1}}.$$

We can apply the Choi-Srivastava theorem [5] to evaluate the sum of the series on the right side.

Example 4.1 By choosing $s=1, a=1, b=0$, according to (4.20), we have to set $c=1, F=\zeta$, and for $m=3$ and $p=2$, we apply (4.31) and obtain

$$\sum_{n=1}^{\infty}\frac{j_2(nx)}{n^{15/2}}$$

$$=\frac{x^6\left(\frac{3197}{1260}-\log x\right)}{15120}-\frac{x^4\zeta(3)}{420}+\frac{x^2\zeta(5)}{30}+\frac{4\pi^7\zeta'\left(-7,\frac{x}{2\pi}\right)}{315x}$$

$$-\frac{4\pi^7\zeta'\left(-7,1-\frac{x}{2\pi}\right)}{315x}-\frac{\pi^8\zeta'\left(-8,\frac{x}{2\pi}\right)}{105x^2}-\frac{\pi^8\zeta'\left(-8,1-\frac{x}{2\pi}\right)}{105x^2}$$

$$+\frac{2\pi^9\zeta'\left(-9,\frac{x}{2\pi}\right)}{945x^3}-\frac{2\pi^9\zeta'\left(-9,1-\frac{x}{2\pi}\right)}{945x^3}.$$

12.4.2 Series over Struve Functions

Repeating the same procedure as that for obtaining (4.3), but relying on series (4.2), we can obtain a similar result concerning the series over Struve functions.

Theorem 4.3 *The summation formula for the series over Struve functions is*

$$\sum_{n=1}^{\infty} \frac{\mathbf{H}_{\nu}(nx)}{n^{\alpha}} = \frac{\pi\left(\frac{x}{2}\right)^{\alpha-1}}{2\Gamma\left(\frac{\nu}{2}+\frac{1}{2}+\frac{\alpha}{2}\right)\Gamma\left(\frac{\alpha-\nu+1}{2}\right)\sin\frac{\pi(\alpha-\nu}{2})} + \sum_{k=0}^{\infty} \frac{(-1)^{k}\left(\frac{x}{2}\right)^{\nu+2k+1}\zeta(\alpha-\nu-2k-1)}{\Gamma\left(k+\frac{3}{2}\right)\Gamma\left(\nu+k+\frac{3}{2}\right)}. \tag{4.32}$$

Proof. We are going to apply the alternative integral representation (2.12) where we set nx instead of x, and after dividing by n^{α}, we consider the series

$$\sum_{n=1}^{\infty} \frac{\mathbf{H}_{\nu}(nx)}{n^{\alpha}} = \frac{2\left(\frac{x}{2}\right)^{\nu}}{\sqrt{\pi}\,\Gamma\left(\nu+\frac{1}{2}\right)} \sum_{n=1}^{\infty} \int_{0}^{1} (1-t^2)^{\nu-\frac{1}{2}} \frac{\sin nxt}{n^{\alpha-\nu}}\, dt.$$

We interchange summation and integration and apply (4.6), where we substitute xt for x and $\alpha-\nu$ for α

$$\sum_{n=1}^{\infty} \frac{\mathbf{H}_{\nu}(nx)}{n^{\alpha}} = \frac{2\left(\frac{x}{2}\right)^{\nu}}{\sqrt{\pi}\,\Gamma\left(\nu+\frac{1}{2}\right)} \int_{0}^{1} (1-t^2)^{\nu-\frac{1}{2}} \Bigg(\frac{\pi(xt)^{\alpha-\nu-1}}{2\Gamma(\alpha-\nu) f\left(\frac{\pi(\alpha-\nu)}{2}\right)} + \sum_{k=0}^{\infty} \frac{(-1)^{k}\zeta(\alpha-\nu-2k-1)}{(2k+1)!}(xt)^{2k+1}\Bigg)dt,$$

which gives rise to

$$\sum_{n=1}^{\infty} \frac{\mathbf{H}_{\nu}(nx)}{n^{\alpha}} = \frac{\left(\frac{x}{2}\right)^{\nu} x^{\alpha-\nu-1}\sqrt{\pi}}{2\Gamma\left(\nu+\frac{1}{2}\right)\Gamma(\alpha-\nu) f\left(\frac{\pi(\alpha-\nu)}{2}\right)} \int_{0}^{1} (1-t^2)^{\left(\nu+\frac{1}{2}\right)-1} (t^2)^{\frac{\alpha-\nu}{2}-1} dt^2$$
$$+ \sum_{k=0}^{\infty} \frac{(-1)^{k}\left(\frac{x}{2}\right)^{\nu}\zeta(\alpha-\nu-2k-1)\, x^{2k+1}}{(2k+1)!\sqrt{\pi}\,\Gamma\left(\nu+\frac{1}{2}\right)} \int_{0}^{1} (1-t^2)^{\left(\nu+\frac{1}{2}\right)-1} (t^2)^{(k+1)-1} dt^2.$$

Both right-hand integrals we calculate through the beta function and find

$$\sum_{n=1}^{\infty}\frac{\mathbf{H}_\nu(nx)}{n^\alpha}=\frac{\left(\frac{x}{2}\right)^\nu x^{\alpha-\nu-1}\sqrt{\pi}}{2\Gamma\left(\nu+\frac{1}{2}\right)\Gamma(\alpha-\nu)f\left(\frac{\pi(\alpha-\nu)}{2}\right)}\frac{\Gamma\left(\nu+\frac{1}{2}\right)\Gamma\left(\frac{\alpha-\nu}{2}\right)}{\Gamma\left(\nu+\frac{1}{2}+\frac{\alpha-\nu}{2}\right)}$$
$$+\frac{\left(\frac{x}{2}\right)^\nu}{\sqrt{\pi}}\sum_{k=0}^{\infty}\frac{(-1)^k\zeta(\alpha-\nu-2k-1)\,x^{2k+1}}{\Gamma(2k+2)\Gamma\left(\nu+\frac{1}{2}\right)}\frac{\Gamma\left(\nu+\frac{1}{2}\right)\Gamma(k+1)}{\Gamma\left(\nu+\frac{1}{2}+k+1\right)}.$$

Applying Legendre's duplication formula (4.9) twice, first with $z=\frac{\alpha-\nu}{2}$, then with $z=k+1$, we arrive at (4.32). □

Applying a similar procedure for the rest of the parameters, we obtain a general formula for the summation of the series (4.1) derived in [31]

$$\sum_{n=1}^{\infty}\frac{(s)^{n-1}\phi_\nu((an-b)x)}{(an-b)^\alpha}=\frac{c\pi\left(\frac{x}{2}\right)^{\alpha-1}}{2\Gamma\left(\frac{\alpha-\nu+1}{2}\right)\Gamma\left(\frac{\alpha+\nu+1}{2}\right)f\left(\frac{\pi(\alpha-\nu)}{2}\right)}$$
$$+\sum_{k=0}^{\infty}\frac{(-1)^k\left(\frac{x}{2}\right)^{\nu+2k+\delta}F(\alpha-\nu-2k-\delta)}{\Gamma\left(k+1+\frac{\delta}{2}\right)\Gamma\left(\nu+k+1+\frac{\delta}{2}\right)},\tag{4.33}$$

where $\phi_\nu=\left\{\begin{array}{c}J_\nu\\ \mathbf{H}_\nu\end{array}\right\}f=\left\{\begin{array}{c}\cos\\ \sin\end{array}\right\}\delta=\left\{\begin{array}{c}0\\ 1\end{array}\right\}$, and the parameters s, a, b, c are in Table 12.1, p. 272. So formula (4.3) follows from (4.33) by taking $s=1$, $a=1, b=0, c=1, F=\zeta$.

12.4.3 Summation Based on Poisson's Formula

There exist several definitions of Poisson's formula in the literature (see [6, 19, 27]). Here we assume that $f(x)$ is a continuous function, smooth and absolutely integrable on $(0,+\infty)$. We shall make use of the Fourier cosine transform of the function $f(x)$

$$\mathcal{F}_c(f(x),\omega)=F_c(\omega)=\sqrt{\frac{2}{\pi}}\int_0^\infty f(x)\cos\omega x\,dx.\tag{4.34}$$

Then, for $\alpha\beta=2\pi, \alpha>0$, there holds

$$\sqrt{\alpha}\left(\frac{1}{2}f(0)+\sum_{n=1}^{\infty}f(n\alpha)\right)=\sqrt{\beta}\left(\frac{1}{2}F_c(0)+\sum_{n=1}^{\infty}F_c(n\beta)\right),\tag{4.35}$$

where we assume that the left and right sides of (4.35) converge absolutely. This is Poisson's formula.

We shall now find the sums of series (4.1) by using quite a different procedure.

Theorem 4.4 *The summation formula for the series over the Bessel function based on Poisson's formula (4.35) has the following form*

$$\sum_{n=1}^{\infty}\frac{J_{\mu}(n\alpha)}{n^{\nu}}=\frac{\alpha^{\nu-1}\Gamma\left(\frac{1+\mu-\nu}{2}\right)}{2^{\nu}\Gamma\left(\frac{1+\mu+\nu}{2}\right)}+\frac{\left(\frac{\alpha}{2}\right)^{\mu}\cos\frac{\pi(\mu-\nu+1)}{2}\Gamma(\mu-\nu+1)}{(2\pi)^{\mu-\nu}\pi\Gamma(\mu+1)}$$
$$\times\sum_{k=0}^{\infty}\frac{\left(\frac{\mu-\nu+1}{2}\right)_k\left(\frac{\mu-\nu+2}{2}\right)_k}{(\mu+1)_k}\frac{\left(\frac{\alpha}{2\pi}\right)^{2k}}{k!}\zeta(2k+\mu-\nu+1)$$
$$-\frac{\alpha^{\nu}\delta_{\mu,\nu}}{2^{\mu+1}\Gamma(\mu+1)},\quad \mu>\nu. \tag{4.36}$$

Proof. First make use of the Bessel functions representation (2.2), where, after a certain rearrangement, we have

$$J_{\mu}(x)=\frac{x^{\mu}}{2^{\mu}}\sum_{m=0}^{\infty}\frac{(-1)^m x^{2m}}{2^{2m}\,m!\,\Gamma(\mu+m+1)},\qquad |\arg x|<\pi.$$

On this basis, we consider a function

$$g(x)=\frac{J_{\mu}(x)}{x^{\nu}}=\frac{x^{\mu-\nu}}{2^{\mu}}\sum_{k=0}^{\infty}\frac{(-1)^k x^{2k}}{2^{2k}\,k!\,\Gamma(\mu+k+1)},\qquad \mu>\nu, x\neq 0. \tag{4.37}$$

If we take the limit in (4.37), we find

$$\lim_{x\to 0}g(x)=\frac{x^{\mu-\nu}}{2^{\mu}}\left(\frac{1}{\Gamma(\mu+1)}-\frac{x^2}{2^2\Gamma(\mu+2)}+\frac{x^4}{2^4 2!\Gamma(\mu+3)}-\cdots\right)$$
$$=\frac{\delta_{\mu,\nu}}{2^{\mu}\Gamma(\mu+1)},$$

where

$$\delta_{\mu,\nu}=\begin{cases}1, & \mu=\nu\\ 0, & \mu>\nu.\end{cases}$$

On this basis, we define a function $f(x)$ as follows

$$f(x)=\begin{cases} g(x), & x\neq 0 \\ \frac{\delta_{\mu,\nu}}{2^{\mu}\Gamma(\mu+1)}, & x=0. \end{cases} \tag{4.38}$$

It is continuous, smooth, and absolutely integrable on $(0,+\infty)$, so we can apply the Fourier transform

$$\mathcal{F}_c(f(x),\omega)=F_c(\omega)=\sqrt{\frac{2}{\pi}}\int_0^{\infty}\frac{J_{\mu}(x)}{x^{\nu}}\cos\omega x\,dx,$$

where we set first $\omega=0$, then $n\beta$ instead of ω, then put the values $F_c(0)$ and $F_c(n\beta)$ in (4.35). Also, $f(n\alpha)=\frac{J_{\mu}(n\alpha)}{(n\alpha)^{\nu}}$, so that Poisson's formula becomes

$$\sqrt{\alpha}\left(\frac{1}{2}f(0)+\sum_{n=1}^{\infty}\frac{J_{\mu}(n\alpha)}{(n\alpha)^{\nu}}\right)=\sqrt{\frac{2\beta}{\pi}}\,(\frac{1}{2}\int_0^{\infty}\frac{J_{\mu}(x)}{x^{\nu}}dx$$
$$+\sum_{n=1}^{\infty}\int_0^{\infty}\frac{J_{\mu}(x)}{x^{\nu}}\cos(n\beta x)\,dx). \tag{4.39}$$

Since $\alpha\beta=2\pi$ in (4.35), it follows that $\beta=\frac{2\pi}{\alpha}$, so (4.39) becomes

$$\sum_{n=1}^{\infty}\frac{J_{\mu}(n\alpha)}{(n\alpha)^{\nu}}=\frac{2}{\alpha}\left(\frac{1}{2}\int_0^{\infty}\frac{J_{\mu}(x)}{x^{\nu}}dx\right.$$
$$\left.+\sum_{n=1}^{\infty}\int_0^{\infty}\frac{J_{\mu}(x)}{x^{\nu}}\cos\frac{2n\pi x}{\alpha}dx\right)-\frac{1}{2}f(0). \tag{4.40}$$

In [10, p. 676, 6.561.14] we find the integral

$$\int_0^{\infty}x^{\mu}J_{\nu}(ax)\,dx=2^{\mu}a^{-\mu-1}\frac{\Gamma\left(\frac{1+\nu+\mu}{2}\right)}{\Gamma\left(\frac{1+\nu-\mu}{2}\right)}, \tag{4.41}$$

with $-\mathrm{Re}\,\nu-1<\mathrm{Re}\,\mu<\frac{1}{2}, a>0$.

Also, we use the integral [23, p. 192]

$$\int_0^{\infty}\frac{J_{\mu}(x)}{x^{\nu}}\cos\frac{2n\pi x}{\alpha}dx=\frac{1}{2^{2\mu-\nu+1}}\left(\frac{\alpha}{n\pi}\right)^{\mu-\nu+1}\frac{\Gamma(\mu-\nu+1)}{\Gamma(\mu+1)}$$
$$\times\cos\frac{\pi}{2}(\mu-\nu+1)\ F12$$
$$\left(\frac{\mu-\nu+1}{2},\frac{\mu-\nu+2}{2};\mu+1;\left(\frac{\alpha}{2n\pi}\right)^2\right).$$

Here $\left(\frac{\alpha}{2n\pi}\right)^2 < 1$, because $\alpha < 2n\pi$. Making use of the Gauss hypergeometric function F12 (2.4), there follows

$$\int_0^\infty \frac{J_\mu(x)}{x^\nu} \cos\frac{2n\pi x}{\alpha} dx$$
$$= \frac{1}{2^{2\mu-\nu+1}}\left(\frac{\alpha}{n\pi}\right)^{\mu-\nu+1} \frac{\Gamma(\mu-\nu+1)}{\Gamma(\mu+1)}$$
$$\times \cos\frac{\pi}{2}(\mu-\nu+1) \sum_{k=0}^{\infty} \frac{\left(\frac{\mu-\nu+1}{2}\right)_k \left(\frac{\mu-\nu+2}{2}\right)_k}{(\mu+1)_k} \cdot \frac{\left(\frac{\alpha}{2n\pi}\right)^{2k}}{k!}. \tag{4.42}$$

with $\mu + 1 > \nu > -\frac{1}{2}$.

Now we put the right-hand sides of (4.41), where we set $a = \alpha$, swap μ and ν, then replace ν with $-\nu$, (4.42), and (4.38) instead of the corresponding expressions in (4.40), which gives rise to a double sum in (4.40). Interchanging the order of these two summations yields

$$\sum_{n=1}^{\infty} \frac{J_\mu(n\alpha)}{n^\nu} = \frac{\alpha^{\nu-1}\Gamma\left(\frac{1+\mu-\nu}{2}\right)}{2^\nu\,\Gamma\left(\frac{1+\mu+\nu}{2}\right)} + \frac{\left(\frac{\alpha}{2\pi}\right)^\mu \cos\frac{\pi(\mu-\nu+1)}{2}\Gamma(\mu-\nu+1)}{2^\mu\pi^{1-\nu}\Gamma(\mu+1)}$$
$$\times \sum_{k=0}^{\infty} \frac{\left(\frac{\mu-\nu+1}{2}\right)_k \left(\frac{\mu-\nu+2}{2}\right)_k}{(\mu+1)_k} \cdot \frac{\left(\frac{\alpha}{2\pi}\right)^{2k}}{k!} \sum_{n=1}^{\infty} \frac{1}{n^{2k+\mu-\nu+1}}$$
$$- \frac{\alpha^\nu \delta_{\mu,\nu}}{2^{\mu+1}\Gamma(\mu+1)}. \tag{4.43}$$

Since

$$\sum_{n=1}^{\infty} \frac{1}{n^{2k+\mu-\nu+1}} = \zeta(2k+\mu-\nu+1),$$

we have (4.36). If $\mu > \nu$, $2k + \mu - \nu + 1 > 1$ is valid for $k \in \mathbb{N}_0$. □

Unlike (4.3), closed-form cases of (4.36) do not ensue because Riemann's zeta function vanishes. Its argument in this case does not take negative even integers since $\mu > \nu$. But, for $\nu - \mu = -2p$, $p \in \mathbb{N}_0$, the expression

$$\frac{1}{\Gamma\left(\frac{\nu-\mu}{2}\right)}$$

becomes zero since Γ has poles at non-positive integers. Because of $\delta_{\mu,\nu} = 0$ for $\mu > \nu$, (4.36) takes a closed form

$$\sum_{n=1}^{\infty} \frac{J_\mu(n\alpha)}{n^{\mu-2p}} = \frac{\alpha^{\mu-2p-1}\Gamma\left(p+\frac{1}{2}\right)}{2^{\mu-2p}\,\Gamma\left(\mu-p+\frac{1}{2}\right)}. \tag{4.44}$$

But, for $\mu = \nu$, we have seen that $\delta_{\mu,\nu} = 1$, so in that case, (4.36) takes closed form

$$\sum_{n=1}^{\infty} \frac{J_\mu(n\alpha)}{n^{\mu}} = \frac{\alpha^{\mu-1}\sqrt{\pi}}{2^{\mu}\,\Gamma\left(\mu+\frac{1}{2}\right)} - \frac{\alpha^{\mu}}{2^{\mu+1}\Gamma(\mu+1)}. \tag{4.45}$$

Have a look at formula 9 in [23, p. 678]

$$\sum_{k=1}^{\infty} \frac{J_\nu(kx)}{k^{\nu-2n}} = \frac{\left(\frac{x}{2}\right)^{\nu-2n-1}\Gamma\left(n+\frac{3}{2}\right)}{(2n+1)\,\Gamma\left(\nu-n+\frac{1}{2}\right)},$$

where $n \in \mathbb{N}, \nu > 2n - \frac{1}{2}, 0 \le x < 2\pi$. Since

$$\Gamma\left(n+\frac{3}{2}\right) = \left(n+\frac{1}{2}\right)\Gamma\left(n+\frac{1}{2}\right),$$

after cancellation, this formula becomes (4.44), which in turn holds if $\mu - 2p = \nu > -\frac{1}{2}$, whence we have $\mu > 2p - \frac{1}{2}$, and we conclude that these formulas are identical.

The formula 8 in [23, p. 678]

$$\sum_{k=1}^{\infty} \frac{J_{2n+m}(kx)}{k^{m}} = \frac{x^{m-1}}{(2n+1)(2n+3)\cdots(2n+2m-1)}, \quad n \in \mathbb{N}, 0 < x < 2\pi,$$

is another particular case obtainable from the formula (4.36). If we set $\nu = m \in \mathbb{N}$ in (4.36) with $\mu - \nu = 2p \in \mathbb{N}$, after a rearrangement, we come to the last formula, which is a special case of formula (4.44).

So, as for the summation of series (4.1), if μ and ν are real numbers such that $\nu > \mu > -\frac{1}{2}$, we apply the summation formula (4.3) obtained by using Poisson's integral. If, for these numbers, it holds that $\nu - \mu = 2k$, $k \in \mathbb{N}_0$, we apply (4.10). If $\nu - \mu = 2k + 1$, $k \in \mathbb{N}_0$, we use (4.22).

12.5 SERIES OVER ANGER AND WEBER FUNCTIONS

There is, however, another summation formula for series (4.1) valid for both $\alpha \le \nu$ and $\alpha > \nu$ for $\nu = 2p - r$, where $p \in \mathbb{N}_0$, $r = \left\{ \begin{matrix} 1 \\ 0 \end{matrix} \right\}$ $g = \left\{ \begin{matrix} \sin \\ \cos \end{matrix} \right\}$, and $\alpha > 0$, obtained by relying on Anger and Weber functions (3.1). Placing now (3.3) in (4.1), and referring again to Dirichlet's test, (4.1), as in the case of the series (4.2), we similarly ascertain that the series

$$\sum_{n=1}^{\infty} \frac{g((an-b)x \sin\theta)}{(an-b)^{\alpha}}$$

converges uniformly with respect to θ for $\alpha > 0$ and x belonging convergence regions from Table 12.1, Appendix, p. 272.

Consequently, summation and integration are interchangeable

$$\begin{aligned} S_{\alpha}^{\phi} &= \frac{1}{\pi} \sum_{n=1}^{\infty} \int_0^{\pi} \frac{f((2p-r)\theta) g((an-b)x \cos\theta)}{(an-b)^{\alpha}} d\theta \\ &= \frac{1}{\pi} \int_0^{\pi} f((2p-r)\theta) \sum_{n=1}^{\infty} \frac{g((an-b)x \cos\theta)}{(an-b)^{\alpha}} d\theta. \end{aligned}$$

Afterwards, we follow the same procedure as when deriving (4.3). However, we first set $x \sin\theta$ instead of x in (4.6), and have

$$\begin{aligned} \sum_{n=1}^{\infty} \frac{g((an-b)x \sin\theta)}{(an-b)^{\alpha}} &= \frac{c\pi x^{\alpha-1} \sin^{\alpha-1}\theta}{2\Gamma(\alpha) g\left(\frac{\pi\alpha}{2}\right)} \\ &+ \sum_{j=0}^{\infty} \frac{(-1)^j F(\alpha-2j-\delta)}{(2j+\delta)!} x^{2j+\delta} \sin^{2j+\delta}\theta, \end{aligned}$$

where $g = \left\{ \begin{matrix} \cos \\ \sin \end{matrix} \right\}$ $\delta = \left\{ \begin{matrix} 0 \\ 1 \end{matrix} \right\}$. Now, there holds

$$\begin{aligned} S_{\alpha}^{\phi} = \frac{1}{\pi} \int_0^{\pi} f((2p-r)\theta) \Bigg(& \frac{c\pi x^{\alpha-1} \cos^{\alpha-1}\theta}{2\Gamma(\alpha) g\left(\frac{\pi\alpha}{2}\right)} \\ &+ \sum_{j=0}^{\infty} \frac{(-1)^j F(\alpha-2j-\delta)}{(2j+\delta)!} x^{2j+\delta} \sin^{2j+\delta}\theta \Bigg) d\theta. \end{aligned}$$

After integrating and then interchanging summation and integration, we have

$$S_\alpha^\phi = \frac{c\,x^{\alpha-1}}{2\Gamma(\alpha)g\left(\frac{\pi\alpha}{2}\right)} \int_0^\pi f((2p-r)\,\theta)\,\sin^{\alpha-1}\theta\,d\theta$$
$$+\frac{1}{\pi}\sum_{j=0}^{\infty}\frac{(-1)^j F(\alpha-2j-\delta)}{(2j+\delta)!}x^{2j+\delta}\int_0^\pi f((2p-r)\,\theta)\,\sin^{2j+\delta}\theta\,d\theta.$$

To evaluate both integrals, we apply the entries 3.631.1. and 3.631.8. in [10, p. 397]

$$\int_0^\pi f(ax)\,\sin^{\nu-1}x\,\mathrm{d}x = \frac{\pi f\left(\frac{a\pi}{2}\right)}{2^{\nu-1}\nu\,\mathrm{B}\left(\frac{\nu+a+1}{2},\frac{\nu-a+1}{2}\right)},\quad \mathrm{Re}\,\nu>0,$$

where $f=\sin$ or $f=\cos$, so that we find

$$\int_0^\pi f((2p-r)\,\theta)\,\sin^{\alpha-1}\theta\,d\theta = \frac{2^{-\alpha+1}\pi f\left(p\pi-\frac{r\pi}{2}\right)\Gamma(\alpha)}{\Gamma\left(\frac{\alpha-r+1}{2}+p\right)\Gamma\left(\frac{\alpha+r+1}{2}-p\right)}$$
$$\int_0^\pi f((2p-r)\,\theta)\,\sin^{(2j+1+\delta)-1}\theta\,d\theta = \frac{2^{-2j-\delta}\pi f\left(p\pi-\frac{r\pi}{2}\right))\Gamma(2j+1+\delta)}{\Gamma\left(j+p+1+\frac{\delta-r}{2}\right)\Gamma\left(j-p+1+\frac{\delta+r}{2}\right)},$$

and we arrive at the summation formula for series (4.1) in terms of Anger or Weber functions

$$S_\alpha^\phi = \frac{f\left(p\pi-\frac{r\pi}{2}\right)\,c\pi\left(\frac{x}{2}\right)^{\alpha-1}}{2g\left(\frac{\pi\alpha}{2}\right)\Gamma\left(\frac{\alpha-r+1}{2}+p\right)\Gamma\left(\frac{\alpha+r+1}{2}-p\right)}$$
$$+\sum_{j=0}^{\infty}\frac{(-1)^j\left(\frac{x}{2}\right)^{2j+\delta}f\left(p\pi-\frac{r\pi}{2}\right)F(\alpha-2j-\delta)}{\Gamma\left(j+p+1+\frac{\delta-r}{2}\right)\Gamma\left(j-p+1+\frac{\delta+r}{2}\right)}, \qquad (5.1)$$

with $\alpha\in\mathbb{R}^+$, $\phi_{2p-r} = \left\{\begin{array}{c}\mathbf{J}_{2p-r}\\ \mathbf{E}_{2p-r}\end{array}\right\}g = \left\{\begin{array}{c}f\\ \tilde{f}\end{array}\right\}$, and $r = \left\{\begin{array}{c}0\\ 1\end{array}\right\}f = \left\{\begin{array}{c}\cos\\ \sin\end{array}\right\}\tilde{f} = \left\{\begin{array}{c}-\sin\\ \cos\end{array}\right\}, g = \left\{\begin{array}{c}\cos\\ \sin\end{array}\right\}\delta = \left\{\begin{array}{c}0\\ 1\end{array}\right\}$. We read the parameters s, a, b, c, F from Table 12.1, Appendix.

Unlike formula (4.33), where the condition $\alpha > 2p - r$ is required, formula (5.1) holds true even though $\alpha < 2p - r$. Since Anger and Bessel functions are equal if their index is a positive integer, formulas (4.33) and (5.1) are equal then.

If $\alpha - 2p + r \in \mathbb{N}$ or $2p - r - \alpha \in \mathbb{N}$, formula (5.1) takes the closed form, and then $\alpha - 2p + r$ or $2p - r - \alpha \in \mathbb{N}$ replaces α in Table 12.1.

If $\phi_{2p-r} = \mathbf{J}_{2p-r}$, then $g = f$, implying $r = \left\{ \begin{matrix} 0 \\ 1 \end{matrix} \right\} g = \left\{ \begin{matrix} \cos \\ \sin \end{matrix} \right\} \delta = \left\{ \begin{matrix} 0 \\ 1 \end{matrix} \right\} g = \left\{ \begin{matrix} \cos \\ \sin \end{matrix} \right\}$. That all means $\delta = r$. According to (5.1), we have

$$\sum_{n=1}^{\infty} \frac{\mathbf{J}_{2p-r}(nx)}{n^{\alpha}} = \frac{f\left(p\pi - \frac{r\pi}{2}\right)\pi\left(\frac{x}{2}\right)^{\alpha-1}}{2f\left(\frac{\pi\alpha}{2}\right)\Gamma\left(\frac{\alpha-r+1}{2} + p\right)\Gamma\left(\frac{\alpha+r+1}{2} - p\right)} + \sum_{j=p-r}^{\infty} \frac{(-1)^j\left(\frac{x}{2}\right)^{2j+\delta} f\left(p\pi - \frac{r\pi}{2}\right)\zeta(\alpha - 2j - \delta)}{\Gamma(j+p+1)\Gamma(j-p+1+r)}. \tag{5.2}$$

The summation index j begins with $p-r$, because $\Gamma(j-p+1+r)$ has poles at $j = 0, ..., p-r$, so its reciprocal values at these points equal zero, meaning that the first $p - r$ terms vanish.

The right-hand side series presents a finite sum if $g = \left\{ \begin{matrix} \sin \\ \cos \end{matrix} \right\} \alpha = \left\{ \begin{matrix} 2m-1 \\ 2m \end{matrix} \right\}$, $m \in \mathbb{N}$. If we take $g = \sin$, then $\alpha = 2m - 1$, $\delta = 1$, $r = 1$. So, for $2m - 1 \geqslant 2p - 1$, i.e., $m \geqslant p$, formula (5.2) becomes

$$\sum_{n=1}^{\infty} \frac{\mathbf{J}_{2p-1}(nx)}{n^{2m-1}} = \frac{(-1)^{p+m}\pi\left(\frac{x}{2}\right)^{2m-2}}{2\Gamma\left(m+p-\frac{1}{2}\right)\Gamma\left(m-p+\frac{1}{2}\right)} - \sum_{j=p-1}^{m-1} \frac{(-1)^{p+j}\zeta(2m-2j-2)\left(\frac{x}{2}\right)^{2j+1}}{\Gamma(j-p+2)\,(j+p)!}. \tag{5.3}$$

But for $m < p$, since $j \geqslant p - 1$, there follows $2m - 2j - 2 < 2m - 2p < 0$, meaning that $\zeta(2m - 2j - 2) = 0$, so the whole infinite sum in (5.2) vanishes

$$\sum_{n=1}^{\infty} \frac{\mathbf{J}_{2p-1}(nx)}{n^{2m-1}} = \frac{(-1)^{p+m}\pi\left(\frac{x}{2}\right)^{2m-2}}{2\Gamma\left(m+p-\frac{1}{2}\right)\Gamma\left(m-p+\frac{1}{2}\right)}. \tag{5.4}$$

In the case $g = \cos$, then $\alpha = 2m$, $\delta = 0$, $r = 0$, for $m \geqslant p$ we obtain a formula similar to (5.3), and for $m < p$ another one similar to (5.4).

Formulas (5.3) and (5.4) present the sums of the series in terms of Bessel functions, since Anger functions are Bessel ones for non-negative integers.

For a particular choice of parameters, the closed-form formula (4.44) can be reduced to (5.4) or a similar closed-form case resulting from (5.2). If we take $\nu = 2l-1$, $\mu = 2k+1$, and require $l < k+1$, which are conditions for (5.4), we further have $\mu - \nu = 2p$, where $p = k-l+1 \in \mathbb{N}$. Replacing these values in (4.44), setting afterwards $z = k-l+\frac{3}{2}$ and applying Euler's reflection formula (4.27), we easily arrive at (5.4). So, by using different methods, we obtain the same formula. Yet, note that (4.44) holds whenever the real numbers μ and ν ($\mu > \nu > -\frac{1}{2}$) satisfy $\mu - \nu = 2p$ ($p \in \mathbb{N}$), whereas (5.4) holds only for positive integers μ and ν.

Now, if $\phi_{2p-r} = \mathbf{E}_{2p-r}$, then $g = \tilde{f}$, which implies $r = \left\{\begin{matrix} 0 \\ 1 \end{matrix}\right\} \tilde{f} = \left\{\begin{matrix} -\sin \\ \cos \end{matrix}\right\} f = \left\{\begin{matrix} \cos \\ \sin \end{matrix}\right\}$, $\delta = \left\{\begin{matrix} 1 \\ 0 \end{matrix}\right\} g = \left\{\begin{matrix} \sin \\ \cos \end{matrix}\right\}$, and that means $r = \left\{\begin{matrix} 1 \\ 0 \end{matrix}\right\} \delta = \left\{\begin{matrix} 0 \\ 1 \end{matrix}\right\}$, so (5.1) yields

$$\sum_{n=1}^{\infty} \frac{\mathbf{E}_{2p-r}(nx)}{n^{\alpha}} = \frac{f\left(p\pi - \frac{r\pi}{2}\right)\pi\left(\frac{x}{2}\right)^{\alpha-1}}{2\tilde{f}\left(\frac{\pi\alpha}{2}\right)\Gamma\left(\frac{\alpha-r+1}{2}+p\right)\Gamma\left(\frac{\alpha+r+1}{2}-p\right)}$$
$$+\sum_{j=0}^{\infty} \frac{(-1)^j\left(\frac{x}{2}\right)^{2j+\delta} f\left(p\pi - \frac{r\pi}{2}\right)\zeta(\alpha-2j-\delta)}{\Gamma\left(j+p+1+\frac{\delta-r}{2}\right)\Gamma\left(j-p+\frac{3}{2}\right)}. \tag{5.5}$$

The right-hand side series is in closed form, if $\tilde{f} = \left\{\begin{matrix} -\sin \\ \cos \end{matrix}\right\} \alpha = \left\{\begin{matrix} 2m-1 \\ 2m \end{matrix}\right\}$, $m \in \mathbb{N}$. So, if we take $\tilde{f} = \cos$, then $\alpha = 2m$, $\delta = 0$, $r = 1$. So, for $2m > 2p - 1$, i.e., $m \geqslant p$, we find

$$\sum_{n=1}^{\infty} \frac{\mathbf{E}_{2p-1}(nx)}{n^{2m}} = \frac{(-1)^{m+p-1}\left(\frac{x}{2}\right)^{2m-1}\pi}{2\Gamma(m+p)\Gamma(m-p+1)} + \sum_{j=0}^{m} \frac{(-1)^{p+j-1}\left(\frac{x}{2}\right)^{2j}\zeta(2m-2j)}{\Gamma\left(j+p+\frac{1}{2}\right)\Gamma\left(j-p+\frac{3}{2}\right)}. \tag{5.6}$$

Nothing will change if set $(-1)^{p-j}$ instead of $(-1)^{p+j}$ in (5.6).

From (3.4), we can express the series over Struve functions as that over Weber functions

$$\sum_{n=1}^{\infty} \frac{(s)^{n-1}\mathbf{H}_{\nu}((an-b)x)}{(an-b)^{\alpha}} = \sum_{n=1}^{\infty} \frac{1}{\pi} \sum_{k=0}^{\left[\frac{\nu-1}{2}\right]} \frac{(s)^{n-1}\Gamma\left(k+\frac{1}{2}\right)\left(\frac{x}{2}\right)^{\nu-2k-1}}{(an-b)^{\alpha-\nu+2k+1}\Gamma\left(\nu-k+\frac{1}{2}\right)} - \sum_{n=1}^{\infty} \frac{(s)^{n-1}\mathbf{E}_{\nu}((an-b)x)}{(an-b)^{\alpha}},$$

whence we have

$$\sum_{n=1}^{\infty} \frac{(s)^{n-1}\mathbf{H}_{\nu}((an-b)x)}{(an-b)^{\alpha}} = \frac{1}{\pi} \sum_{k=0}^{\left[\frac{\nu-1}{2}\right]} \frac{\Gamma\left(k+\frac{1}{2}\right)\left(\frac{x}{2}\right)^{\nu-2k-1} F(\alpha-\nu+2k+1)}{\Gamma\left(\nu-k+\frac{1}{2}\right)} - \sum_{n=1}^{\infty} \frac{(s)^{n-1}\mathbf{E}_{\nu}((an-b)x)}{(an-b)^{\alpha}}, \tag{5.7}$$

where F is one of the functions $\zeta, \eta, \lambda, \beta$.

Applying now (5.7) for $a = 1$, $b = 0$, $s = 1$, $F = \zeta$, we have

$$\sum_{n=1}^{\infty} \frac{\mathbf{H}_{2p-1}(nx)}{n^{2m}} = \frac{1}{\pi} \sum_{j=0}^{p-1} \frac{\Gamma\left(j+\frac{1}{2}\right)\left(\frac{x}{2}\right)^{2p-2j-2} \zeta(2m-2p+2j+2)}{\Gamma\left(2p-j-\frac{1}{2}\right)} + \frac{(-1)^{m+p}\left(\frac{x}{2}\right)^{2m-1}\pi}{2\Gamma(m+p)\Gamma(m-p+1)} + \sum_{j=0}^{m} \frac{(-1)^{p-j}\left(\frac{x}{2}\right)^{2j}\zeta(2m-2j)}{\Gamma\left(j+p+\frac{1}{2}\right)\Gamma\left(j-p+\frac{3}{2}\right)}.$$

Taking $z = j + \frac{1}{2}$ in Euler's reflection formula (4.27), we find

$$\Gamma\left(j + \frac{1}{2}\right)\Gamma\left(\frac{1}{2} - j\right) = (-1)^j \pi,$$

whence we express $\Gamma\left(j + \frac{1}{2}\right)$ and replace it in the first sum on the right-hand side

$$\sum_{j=0}^{p-1} \frac{(-1)^j \left(\frac{x}{2}\right)^{2p-2j-2} \zeta(2m - 2p + 2j + 2)}{\Gamma\left(\frac{1}{2} - j\right)\Gamma\left(2p - j - \frac{1}{2}\right)}.$$

By changing the summation index, introducing the substitution $i = p - j - 1$, this sum becomes

$$-\sum_{i=0}^{p-1} \frac{(-1)^{p-i} \left(\frac{x}{2}\right)^{2i} \zeta(2m - 2i)}{\Gamma\left(i + p + \frac{1}{2}\right)\Gamma\left(i - p + \frac{3}{2}\right)},$$

and it cancels out the first p terms of the second sum, where we previously replaced j with i

$$\sum_{i=p}^{m} \frac{(-1)^{p-i} \left(\frac{x}{2}\right)^{2i} \zeta(2m - 2i)}{\Gamma\left(i + p + \frac{1}{2}\right)\Gamma\left(i - p + \frac{3}{2}\right)}.$$

After shifting the index $i = p + j$, and including the middle term, we finally obtain

$$\sum_{n=1}^{\infty} \frac{\mathbf{H}_{2p-1}(nx)}{n^{2m}} = \frac{(-1)^{m+p-1} \left(\frac{x}{2}\right)^{2m-1} \pi}{2\Gamma(m + p)\Gamma(m - p + 1)}$$
$$+ \sum_{j=0}^{m-p} \frac{(-1)^j \left(\frac{x}{2}\right)^{2j+2p} \zeta(2m - 2p - 2j)}{\Gamma\left(j + \frac{3}{2}\right)\Gamma\left(j + 2p + \frac{1}{2}\right)},$$

which we can find directly from (4.33) by a suitable choice of parameters.

If μ is a nonnegative integer and ν is a positive real number, then, irrespective of being $\nu > \mu$ or $\nu \leq \mu$, we apply the summation formula (5.2) obtained through Bessel's integral. In addition, if $\nu \in \mathbb{N}$, then closed-form cases might ensue. Some examples are given by (5.3) and (5.4).

Conversely, if μ and ν are real numbers such that $\mu > \nu > -\frac{1}{2}$, we use (4.36) obtained by means of a power series representation of Bessel functions and Poisson's formula. In particular, if there holds $\mu - \nu = 2p, p \in \mathbb{N}_0$, we apply (4.44) or (4.45).

In special cases, when μ and ν are odd positive integers, we may use (5.4) or a similar formula obtainable from (5.2), when they are even positive integers.

12.6 SERIES OVER BOURGET FUNCTIONS

Bourget functions are a generalization of Bessel functions. They are defined by Bourget [4, 38], investigated by Gorowara [9], and represented by the integral by (see [41, p. 326])

$$J_{p,q}(z) = \frac{1}{\pi}\int_0^{\pi} (2\cos\theta)^q \cos(p\,\theta - z\sin\theta)\, d\theta, \quad p \in \mathbb{N}_0, \quad q \in \mathbb{N}. \tag{6.1}$$

For $q = 0$ they become Anger functions defined by (3.1), whence, after a rearrangement, one obtains (2.11) for $\nu = p$. So Bourget functions can be considered a generalization of Anger functions and Bessel functions for integer order.

One can write Bourget's functions in a more suitable form.

Lemma 6.1 *The formula combining both cases of (6.1) for $p + q$ odd and even is given as follows*

$$J_{p,q}(z) = \frac{1}{\pi}\int_0^{\pi} (2\cos\theta)^q f(p\theta) f(z\sin\theta)\, d\theta, \tag{6.2}$$

where $f = \left\{ \begin{array}{c} \cos \\ \sin \end{array} \right\} p + q = \left\{ \begin{array}{c} 2m \\ 2m{-}1 \end{array} \right\}, \quad m \in \mathbb{N}.$

Proof. Using the cosine formula for a difference of angles, (6.1) yields

$$\begin{aligned}
J_{p,q}(z) &= \frac{1}{\pi}\int_0^{\pi} (2\cos\theta)^q (\cos p\theta \cos(z\sin\theta) + \sin p\theta \sin(z\sin\theta))\, d\theta \\
&= \frac{1}{\pi}\int_0^{\pi} (2\cos\theta)^q \cos p\theta \cos(z\sin\theta)\, d\theta \\
&\quad + \frac{1}{\pi}\int_0^{\pi} (2\cos\theta)^q \sin p\theta \sin(z\sin\theta)\, d\theta.
\end{aligned}$$

Let I_c and I_s denote the first and second integral. Taking $p = 2n,\ n \in \mathbb{N}_0$, after the substitution $\theta = \pi - t$, we find

$$I_c = (-1)^q \int_0^{\pi} (2\cos t)^q \cos(2n\pi - 2nt) \cos(z \sin t)\, dt,$$

and seeing that the cosine function is even, we obtain

$$I_c = (-1)^q \int_0^{\pi} (2\cos t)^q \cos 2nt \cos(z \sin t)\, dt = (-1)^q I_c.$$

So $I_c = -I_c$, if q is a positive odd integer, implying $I_c = 0$. Hence, for p even and q odd, meaning that $p + q$ is odd, (6.1) becomes

$$J_{p,q}(z) = \frac{1}{\pi} \int_0^{\pi} (2\cos\theta)^q \sin p\theta \sin(z \sin\theta)\, d\theta.$$

Let now $p = 2n + 1, p \in \mathbb{N}_0$. We are acting in the same manner, but using I_s instead. If we substitute $\pi - t$ for θ, we shall come, for q odd, to a relation

$$I_s = (-1)^q \int_0^{\pi} (2\cos t)^q \sin(2n+1)t \sin(z \sin t)\, dt = (-1)^q I^s = -I^s,$$

So $I_s = 0$ for q odd, that is if $p + q$ is even, (6.1) becomes

$$J_{p,q}(z) = \frac{1}{\pi} \int_0^{\pi} (2\cos\theta)^q \cos p\theta \cos(z \sin\theta)\, d\theta.$$

Thereby, we arrive at (6.2). □

In [28], H. M. Srivastava defined another type of Bourget function analogous to (6.1)

$$I_{p,q}(z) = \frac{1}{\pi} \int_0^{\pi} (2\cos\theta)^q \sin(p\theta - z\sin\theta)\, d\theta, \quad p \in \mathbb{N}_0, \quad q \in \mathbb{N}. \tag{6.3}$$

Note that by taking $q = 0$ in (6.3), one obtains the Weber function $\mathbf{E}_p(z)$ of integer order $p \in \mathbb{N}_0$ (see [41]).

Lemma 6.2 *The formula combining both cases of (6.3) for $p+q$ odd and even is given as follows*

$$I_{p,q}(z) = \frac{1}{\pi} \int_0^{\pi} (2\cos\theta)^q f(p\theta)\, \tilde{f}\,(z \sin\theta)\, d\theta, \tag{6.4}$$

$$\textit{where } f = \left\{ \begin{array}{c} \cos \\ \sin \end{array} \right\} \tilde{f} = \left\{ \begin{array}{c} -\sin \\ \cos \end{array} \right\} p + q = \left\{ \begin{array}{c} 2m \\ 2m{-}1 \end{array} \right\}, m \in \mathbb{N}.$$

Proof. Following a similar procedure as in the preceding proof, but using the sine formula for a difference of angles instead, we rewrite (6.3) in the form of two integrals

$$I_{p,q}(z) = \frac{1}{\pi} \int_0^{\pi} (2 \cos \theta)^q \sin p\theta \cos (z \sin \theta)\, d\theta$$
$$- \frac{1}{\pi} \int_0^{\pi} (2 \cos \theta)^q \cos p\theta \sin (z \sin \theta)\, d\theta = \frac{1}{\pi} I_s - \frac{1}{\pi} I_c.$$

Assuming that for an even p, after the change $\theta = \pi - t$, q is even too, we easily conclude $I_s = 0$, meaning that if $p+q$ is even, only the second integral remains, i.e.

$$I_{p,q}(z) = \frac{1}{\pi} \int_0^{\pi} (2 \cos \theta)^q \cos p\theta\, (-\sin (z \sin \theta))\, d\theta.$$

Conversely, if p is odd, after the change $\theta = \pi - t$ in the second integral, and if q is even, we find $I_c = 0$, so that if $p + q$ is odd, we have

$$I_{p,q}(z) = \frac{1}{\pi} \int_0^{\pi} (2 \cos \theta)^q \sin p\theta \cos (z \sin \theta)\, d\theta\, d\theta.$$

Combining both expressions, we obtain (6.4). □

Integral representations (6.2) and (6.4) have similar forms, which enables us to represent them by means of a single formula

$$\phi_{p,q}(z) = \frac{1}{\pi} \int_0^{\pi} (2 \cos \theta)^q f(p\theta)\, h(z \sin \theta)\, d\theta, \tag{6.5}$$

where $\phi_{p,q} = \left\{ \begin{array}{c} J_{p,q} \\ I_{p,q} \end{array} \right\} h = \left\{ \begin{array}{c} f \\ \tilde{f} \end{array} \right\}, f = \left\{ \begin{array}{c} \cos \\ \sin \end{array} \right\} \tilde{f} = \left\{ \begin{array}{c} -\sin \\ \cos \end{array} \right\} p + q =$ $\left\{ \begin{array}{c} 2m \\ 2m - 1 \end{array} \right\}, m \in \mathbb{N}$. For $q = 0$ in (6.5), the Bourget functions $J_{p,q}$ and $I_{p,q}$ become Anger $\mathbf{J}_p$ and Weber $\mathbf{E}_p$ functions, respectively (see (3.1)).

If we substitute $\phi_{p,q}$ for Bessel functions J_ν in (4.3), we are dealing with a new type of series

$$S_\alpha^{\phi_{p,q}} = \sum_{n=1}^{\infty} \frac{(s)^{n-1} \phi_{p,q}((an - b)\, x)}{(an - b)^\alpha}, \tag{6.6}$$

where $\alpha > 0, a = \left\{ \begin{array}{c} 1 \\ 2 \end{array} \right\} b = \left\{ \begin{array}{c} 0 \\ 1 \end{array} \right\}, s = 1$ or $s = -1$.

Theorem 6.1 *The sum of the series (6.6) is given by the formula*

$$S_\alpha^{\phi_{p,q}} = \frac{2^{q-1} c\, x^{\alpha-1}}{\Gamma(\alpha)\, h\left(\frac{\pi\alpha}{2}\right)} I_{\alpha-1}^{p,q} + \frac{2^q}{\pi} \sum_{k=0}^{\infty} \frac{(-1)^k x^{2k+\delta} F(\alpha - 2k - \delta)}{(2k+\delta)!} I_{2k+\delta}^{p,q}, \quad (6.7)$$

where $\phi_{p,q} = \left\{ \begin{array}{c} J_{p,q} \\ I_{p,q} \end{array} \right\} h = \left\{ \begin{array}{c} f \\ \tilde{f} \end{array} \right\}, f = \left\{ \begin{array}{c} \cos \\ \sin \end{array} \right\} \tilde{f} = \left\{ \begin{array}{c} -\sin \\ \cos \end{array} \right\} p + q = \left\{ \begin{array}{c} 2m \\ 2m-1 \end{array} \right\}$, $m \in \mathbb{N}$, $h = \left\{ \begin{array}{c} \cos \\ \sin \end{array} \right\} \delta = \left\{ \begin{array}{c} 0 \\ 1 \end{array} \right\}$, $I_{\alpha-1}^{p,q}$, $I_{2k+\delta}^{p,q}$ *are integrals to be calculated. The rest of the parameters are read from* Table 12.1.

Proof. First, we replace $\phi_{p,q}$ here with the right-hand side of (6.5)

$$\frac{1}{\pi} \sum_{n=1}^{\infty} \frac{(s)^{n-1}}{(an-b)^\alpha} \int_0^\pi (2\cos\theta)^q f(p\,\theta)\, h((an-b)\,x\sin\theta)\, d\theta,$$

then interchange the summation and integration

$$S_\alpha^{\phi_{p,q}} = \frac{2^q}{\pi} \int_0^\pi \cos^q\theta f(p\,\theta) \left(\sum_{n=1}^{\infty} \frac{(s)^{n-1} h((an-b)\,x\sin\theta)}{(an-b)^\alpha} \right) d\theta, \quad (6.8)$$

which is allowed because of the uniform convergence of the right-hand side series in (6.8) with respect to $\theta \in [0, \pi]$ (see [35]). The convergence regions with respect to x are the same as those in Table 12.1.

Further, we have

$$S_\alpha^{\phi_{p,q}} = \frac{2^q}{\pi} \int_0^\pi \cos^q\theta f(p\theta) \Big(\frac{c\pi (x\sin\theta)^{\alpha-1}}{2\Gamma(\alpha)\, h\left(\frac{\pi\alpha}{2}\right)} + \sum_{k=0}^{\infty} \frac{(-1)^k F(\alpha - 2k - \delta)}{(2k+\delta)!} (x\sin\theta)^{2k+\delta} \Big) d\theta,$$

with $h = \left\{ \begin{array}{c} \sin \\ \cos \end{array} \right\} \delta = \left\{ \begin{array}{c} 1 \\ 0 \end{array} \right\}$, where have we applied the formula (4.6) for the summation of trigonometric series, setting there $x \sin\theta$ instead of x.

Thus the previous formula becomes

$$S_{\alpha}^{\phi_{p,q}} = \frac{2^{q-1} c\, x^{\alpha-1}}{\Gamma(\alpha)\, h\left(\frac{\pi\alpha}{2}\right)} \int_0^{\pi} \cos^q\theta f(p\,\theta)\, \sin^{\alpha-1}\theta d\theta$$
$$+ \frac{2^q}{\pi} \sum_{k=0}^{\infty} \frac{(-1)^k F(\alpha - 2k - \delta)}{(2k+\delta)!} x^{2k+\delta} \int_0^{\pi} \cos^q\theta f(p\,\theta)\, \sin^{2k+\delta}\theta\, d\theta, \tag{6.9}$$

where c, F, and convergence regions are read from Table 12.1.

We continue by calculating the right-hand side integrals, and in this regard, consider

$$I_{\alpha-1}^{p,q} = \int_0^{\pi} \cos^q\theta\, f(p\,\theta)\, \sin^{\alpha-1}\theta\, d\theta, \tag{6.10}$$

where $p, q \in \mathbb{N}, \alpha > -1, f = \sin$ or $f = \cos$ and use of the formulas for $\cos p\,\theta$ and $\sin p\,\theta$ (see [10]) which can be written as a single formula

$$f(p\,\theta) = \sum_{j=0}^{[p/2]} (-1)^j \binom{p}{2j+d} \cos^{p-2j-d}\theta\, \sin^{2j+d}\theta\, d\theta, \tag{6.11}$$

where $f = \left\{ \begin{matrix} \cos \\ \sin \end{matrix} \right\}$ $d = \left\{ \begin{matrix} 0 \\ 1 \end{matrix} \right\}$. Replacing $f(p\,\theta)$ in (6.10) with the right-hand side sum from the preceding formula, we have

$$I_{\alpha-1}^{p,q} = \sum_{j=0}^{[p/2]} (-1)^j \binom{p}{2j+d} \int_0^{\pi} \cos^{p+q-2j-d}\theta\, \sin^{\alpha+2j+d-1}\theta\, d\theta. \tag{6.12}$$

Let J denote the integral in (6.12). Rewriting it as a sum of two integrals

$$J = \int_0^{\frac{\pi}{2}} \cos^{p+q-2j-d}\theta\, \sin^{\alpha+2j+d-1}\theta\, d\theta + \int_{\frac{\pi}{2}}^{\pi} \cos^{p+q-2j-d}\theta\, \sin^{\alpha+2j+d-1}\theta\, d\theta,$$

and after introducing the substitution $\theta = \pi - t$ in the second integral, we have

$$J = \left(1 + (-1)^{p+q-2j-d}\right) \int_0^{\pi/2} \cos^{p+q-2j-d}\theta\, \sin^{\alpha+2j+d-1}\theta\, d\theta.$$

However, because of the entries (5.12.1) and (5.12.2) in [1, p. 142]

$$2\int_0^{\pi/2} \cos^{2a-1}\theta \sin^{2b-1}\theta \, d\theta = \mathrm{B}(a,b) = \frac{\Gamma(a)\Gamma(a)}{\Gamma(a+b)},$$

there follows

$$\int_0^{\pi/2} \cos^{p+q-2j-d}\theta \sin^{\alpha+2j+d-1}\theta \, d\theta = \frac{\Gamma\left(\frac{p+q-d+1}{2}-j\right)\Gamma\left(\frac{\alpha+d}{2}+j\right)}{2\Gamma\left(\frac{p+q+\alpha+1}{2}\right)}$$

By virtue of this, (6.12) now becomes

$$I_{\alpha-1}^{p,q} = \sum_{j=0}^{[p/2]} (-1)^j \binom{p}{2j+d} \left(1+(-1)^{p+q-2j-d}\right) \frac{\Gamma\left(\frac{p+q-d+1}{2}-j\right)\Gamma\left(\frac{\alpha+d}{2}+j\right)}{2\Gamma\left(\frac{p+q+\alpha+1}{2}\right)},$$

where $d = 0$ if we set $f = \cos$ in (6.10), and $d = 1$ if $f = \sin$. If we replace $\alpha - 1$ with $2k + \delta$ in (6.10), we shall obtain quite a similar formula for $I_{2k+\delta}$, so that we finally obtain (6.7). □

Placing $q = 0$ in (6.7), we are dealing with its particular cases, depending on p being odd or even, knowing that $f = \left\{ \begin{array}{c} \cos \\ \sin \end{array} \right\} p = \left\{ \begin{array}{c} 2m \\ 2m+1 \end{array} \right\}$, $m \in \mathbb{N}_0$.

Corollary 6.1 *For $q = 0, p = 2m, m \in \mathbb{N}_0$, the particular case of the formula (6.7) is the series over Anger functions*

$$S_\alpha^{\mathbf{J}_{2m}} = \sum_{n=1}^{\infty} \frac{(s)^{n-1}\mathbf{J}_{2m}((an-b)x)}{(an-b)^\alpha} = \frac{c\pi(-1)^m\left(\frac{x}{2}\right)^{\alpha-1}}{2\Gamma\left(\frac{\alpha+1}{2}-m\right)\Gamma\left(\frac{\alpha+1}{2}+m\right)\cos\frac{\pi\alpha}{2}} + \sum_{k=m}^{\infty} \frac{(-1)^{m+k}\left(\frac{x}{2}\right)^{2k} F(\alpha-2k)}{\Gamma(k-m+1)\Gamma(k+m+1)}. \tag{6.13}$$

with $h = f = \cos, p = 2m, m \in \mathbb{N}$, implying $\delta = 0$. The rest of the parameters are read from Table 12.1.

Proof. Relying on the entry 5.12.6 in [1, p. 142], we calculate the integral (6.10)

$$I_{\alpha-1}^{2m,0} = \int_0^{\pi} \cos(2m\,\theta)\ \sin^{\alpha-1}\theta \, d\theta = \frac{\pi\Gamma(\alpha)\cos(m\pi)}{2^{\alpha-1}\Gamma\left(\frac{\alpha-2m+1}{2}\right)\Gamma\left(\frac{\alpha+2m+1}{2}\right)}$$

In addition, replacing $\alpha - 1$ with $2k$, and setting both values in (6.7), we obtain (6.13). □

If, for $m \in \mathbb{N}, \alpha - \delta = 2m$ and F is Riemann's ζ function or one of the Dirichlet functions η and λ or $\alpha - \delta = 2m - 1$ and F is the Dirichlet function β, the right-hand side of (6.7) presents a finite sum because the functions ζ, η and λ vanish at negative even numbers, and the function β vanishes at negative odd numbers. So we write $\alpha = 2m - r + \delta$, where $r = 0$ for $F = \zeta, \eta, \lambda$ and $r = 1$ for $F = \beta$. Thereby we have proved.

Corollary 6.2 *Formula (6.7) takes closed form and is given by*

$$S_{2m+\delta-r}^{\phi_{p,q}} = \sum_{n=1}^{\infty} \frac{(s)^{n-1} \phi_{p,q}((an-b)x)}{(an-b)^{2m-r+\delta}} = \frac{2^{q-1} c\, x^{2m-r+\delta-1} I_{2m-r+\delta-1}^{p,q}}{(2m-r+\delta)!\, h\left(m\pi + \frac{\delta-r}{2}\right)}$$
$$+ \frac{2^q}{\pi} \sum_{k=0}^{m} \frac{(-1)^k x^{2k+\delta} F(2m-2k-r)}{(2k+\delta)!} I_{2k+\delta}^{p,q},$$

where $\phi_{p,q} = \left\{ \begin{matrix} J_{p,q} \\ I_{p,q} \end{matrix} \right\} h = \left\{ \begin{matrix} f \\ \tilde{f} \end{matrix} \right\}; f = \left\{ \begin{matrix} \cos \\ \sin \end{matrix} \right\} \tilde{f} = \left\{ \begin{matrix} -\sin \\ \cos \end{matrix} \right\} p + q = \left\{ \begin{matrix} 2m \\ 2m-1 \end{matrix} \right\}, m \in \mathbb{N}, h = \left\{ \begin{matrix} \cos \\ \sin \end{matrix} \right\} \delta = \left\{ \begin{matrix} 0 \\ 1 \end{matrix} \right\}$. *The rest of the parameters we read from* Table 12.1.

Example 6.1 If $a = 1, b = 0, s = -1$, then $c = 0$ and $F = \eta$ (refer to Table 12.1), meaning that $r = 0$. If $\phi_{p,q} = I_{p,q}$, there follows that $h = \tilde{f}$, and if we take $f = \sin$, then $\delta = 1$ and $h = \cos$, correspondingly there must be $\delta = 0$. For $p = 2, q = 3, m = 2$, the above formula becomes

$$\sum_{n=1}^{\infty} \frac{(-1)^n I_{2,3}(nx)}{n^4} = \frac{8}{\pi} \sum_{i=0}^{2} \frac{(-1)^i \eta(4-2i)}{(2i)!} x^{2i} I_{2i} = \frac{14\pi^3}{225} - \frac{8\pi x^2}{105} + \frac{16x^4}{945\pi},$$

where the convergence region is $-\pi < x < \pi$.

12.7 SERIES OVER A PRODUCT OF BESSEL FUNCTIONS

In the sequel, we are going to derive a summation formula for the series over a product of Bessel functions [36, 33, 7]

$$S_{\alpha}^{J,J} = \sum_{n=1}^{\infty} \frac{(s)^{n-1} J_{\mu}((an-b)x) J_{\nu}((an-b)x)}{(an-b)^{\alpha}}. \tag{7.1}$$

and we consider all the aspects of relations among the parameters μ, ν, and α arising from different representations. Closed-form cases, as well as those of limiting values, are given in detail.

12.7.1 Summation Based on the Gegenbauer Integral

The Austrian mathematician Leopold Gegenbauer derived a formula [9] for the product of two Bessel functions $J_{\mu}(z)$ and $J_{\nu}(z)$, where $\mu, \nu \in \mathbb{R}$ with $\mu + \nu > -1$, i.e.

$$J_{\mu}(z) J_{\nu}(z) = \frac{2}{\pi} \int_{0}^{\pi/2} J_{\mu+\nu}(2z \cos\theta) \cos(\mu - \nu)\theta \, d\theta, \tag{7.2}$$

and we call (7.2) Gegenbauer's integral.

Theorem 7.1 *The summation formula for the series (7.1) based on Gegenbauer's integral (7.2) is*

$$S_{\alpha}^{J,J} = \frac{c\,\Gamma(\alpha)\,\Gamma\left(\frac{\mu+\nu-\alpha+1}{2}\right)\left(\frac{x}{2}\right)^{\alpha-1}}{2\,\Gamma\left(\frac{\alpha+\mu+\nu+1}{2}\right)\Gamma\left(\frac{\alpha+\mu-\nu+1}{2}\right)\Gamma\left(\frac{\alpha-\mu+\nu+1}{2}\right)} + \sum_{k=0}^{\infty} \frac{(-1)^{k}\Gamma(2k+\mu+\nu+1)\,F(\alpha-\mu-\nu-2k)\left(\frac{x}{2}\right)^{2k+\mu+\nu}}{k!\,\Gamma(k+\mu+1)\,\Gamma(k+\nu+1)\,\Gamma(k+\mu+\nu+1)}, \tag{7.3}$$

where $\alpha, \mu, \nu \in \mathbb{R}$, $\alpha > 0$, $\alpha > \mu + \nu > -\frac{1}{2}$, *and convergence region is* $0 < x < \pi$.

Proof. In order to find a summation formula for the series (7.1), we apply an integral representation of the product of Bessel functions (7.2). As in earlier procedures, we put (7.2), with $(an-b)x$ instead of z, in (7.1)

$$S_{\alpha}^{J,J}=\frac{2}{\pi}\sum_{n=1}^{\infty}\frac{(s)^{n-1}}{(an-b)^{\alpha}}\int_{0}^{\pi/2}\cos(\mu-\nu)\theta J_{\mu+\nu}(2(an-b)x\cos\theta)\,d\theta.$$

In the next step, we need to change the order of summation and integration, but we are allowed to do that because the right-hand series in (4.3), where $\alpha>0,\nu>-\frac{1}{2},\alpha>\nu$, is, by Abel's theorem, absolutely convergent in the region is $0<x<2\pi$. Since $0\leq\cos\theta\leq 1$, we have $0\leq 2x\cos\theta\leq 2x<2\pi$, for $x\in(0,\pi)$, whence there follows that the series

$$\sum_{n=1}^{\infty}\frac{J_{\mu+\nu}(2(an-b)x\cos\theta)}{(an-b)^{\alpha}},\quad \mu+\nu>-\frac{1}{2},\quad \alpha>0,$$

uniformly converges with respect to $\theta\in\left[0,\frac{\pi}{2}\right]$ for $x\in(0,\pi)$.

So, we have

$$S_{\alpha}^{J,J}=\frac{2}{\pi}\int_{0}^{\pi/2}\cos(\mu-\nu)\theta\sum_{n=1}^{\infty}\frac{(s)^{n-1}J_{\mu+\nu}(2(an-b)x\cos\theta)}{(an-b)^{\alpha}}\,d\theta. \quad (7.4)$$

Now we employ formula (4.33), where we set $\phi=J$, so $\alpha>\mu+\nu>-\frac{1}{2}$, and replacing x with $2x\cos\theta$ and ν with $\mu+\nu$, we obtain

$$S_{\alpha}^{J,J}=\frac{2}{\pi}\int_{0}^{\pi/2}\cos(\mu-\nu)\theta\left(\frac{c\pi(x\cos\theta)^{\alpha-1}}{2\Gamma\left(\frac{\alpha-\mu-\nu+1}{2}\right)\Gamma\left(\frac{\alpha+\mu+\nu+1}{2}\right)\cos\left(\frac{\pi(\alpha-\mu-\nu)}{2}\right)}\right.$$
$$\left.+\sum_{k=0}^{\infty}\frac{(-1)^{k}(x\cos\theta)^{\mu+\nu+2k}F(\alpha-\mu-\nu-2k)}{\Gamma(k+1)\Gamma(\mu+\nu+k+1)}\right)d\theta.$$

Interchanging integration and summation, and by using entry 9 in [10, p. 397], we solve the integrals

$$\int_{0}^{\pi/2}\cos^{\alpha-1}\theta\cos(\mu-\nu)\theta\,d\theta=\frac{\pi\Gamma(\alpha)}{2^{\alpha}\Gamma\left(\frac{\alpha+\mu-\nu+1}{2}\right)\Gamma\left(\frac{\alpha-\mu+\nu+1}{2}\right)},$$

$$\int_{0}^{\pi/2}\cos^{\mu+\nu+2k}\theta\cos(\mu-\nu)\theta\,d\theta=\frac{\pi\Gamma(\mu+\nu+2k+1)}{2^{\mu+\nu+2k+1}\Gamma(\mu+k+1)\Gamma(\nu+k+1)}.$$

After applying Euler's reflection formula (4.27) for the first term, we obtain (7.3). □

If $\alpha - \mu - \nu = 2k$, $k \in \mathbb{N}_0$, the right-hand side series in (7.3) is brought into closed form, since by virtue of

$$\zeta(-2n) = \frac{2(2n)!}{(2\pi)^z}\zeta(2n+1)\cos\frac{\pi}{2}(2n+1) = 0,$$

the Riemann zeta function equals zero at negative even integers, so $\zeta(2k-2j) = 0$ for $j > k$. That means the series (7.3) reduces to a finite number of terms

$$\sum_{n=1}^{\infty}\frac{J_\mu(nx)J_\nu(nx)}{n^{\mu+\nu+2k}} = \frac{\Gamma(\mu+\nu+2k)\Gamma\left(\frac{1}{2}-k\right)\left(\frac{x}{2}\right)^{\mu+\nu+2k-1}}{2\Gamma\left(\mu+\nu+k+\frac{1}{2}\right)\Gamma\left(\mu+k+\frac{1}{2}\right)\Gamma\left(\nu+k+\frac{1}{2}\right)} + \sum_{j=0}^{k}\frac{(-1)^j\Gamma(2j+\mu+\nu+1)\,\zeta(2k-2j)\left(\frac{x}{2}\right)^{2j+\mu+\nu}}{i!\,\Gamma(j+\mu+\nu+1)\,\Gamma(j+\mu+1)\Gamma(j+\nu+1)}. \tag{7.5}$$

Example 7.1 If we set $\mu = \frac{1}{3}, \nu = \frac{2}{3}, \alpha = 1$ in (7.5), we have

$$\sum_{n=1}^{\infty}\frac{J_{\frac{1}{3}}(nx)J_{\frac{2}{3}}(nx)}{n} = \frac{1}{\Gamma\left(\frac{5}{6}\right)\Gamma\left(\frac{7}{6}\right)} - \frac{x}{4\Gamma\left(\frac{4}{3}\right)\Gamma\left(\frac{5}{3}\right)} = \frac{3}{\pi}\left(1-\frac{x\sqrt{3}}{2}\right).$$

12.7.2 Summation Based on the Anger-Weber Integral

Except for (7.3), one can derive another summation formula for series (7.1) valid for both $\mu+\nu \le \alpha$ and $\mu+\nu > \alpha$ $(\alpha > 0)$, providing that $\mu+\nu = 2p-r$, where $p \in \mathbb{N}_0$, $r = 0$ or $r = 1$. We use summation formula (5.2), noting that (4.3) involves Anger functions, obtained by means of integral (3.3) [31], but in the case of dealing with an integer order, we take Bessel functions instead, so that we have

$$\sum_{n=1}^{\infty}\frac{J_{\mu+\nu}(nx)}{n^{\alpha}}=\sum_{n=1}^{\infty}\frac{J_{2p-r}(nx)}{n^{\alpha}}$$

$$=\frac{(-1)^{p-r}\pi\left(\frac{x}{2}\right)^{\alpha-1}}{2f\left(\frac{\pi\alpha}{2}\right)\Gamma\left(\frac{\alpha-r+1}{2}+p\right)\Gamma\left(\frac{\alpha+r+1}{2}-p\right)}$$

$$+\sum_{j=p-r}^{\infty}\frac{(-1)^{j+p-r}\left(\frac{x}{2}\right)^{2j+\delta}\zeta(\alpha-2j-\delta)}{\Gamma(j+p+1)\Gamma(j-p+1+r)}, \tag{7.6}$$

where $f=\left\{\begin{array}{c}\sin\\ \cos\end{array}\right\}\delta=\left\{\begin{array}{c}1\\ 0\end{array}\right\}r=\left\{\begin{array}{c}1\\ 0\end{array}\right\}$. So to find the sum of products $J_{\mu}(nx)$ and $J_{\nu}(nx)$, $n\in\mathbb{N}$, we use (7.4), where we apply (7.6), and get

$$S_{\alpha}^{J,J}=\frac{2}{\pi}\int_{0}^{\pi/2}\cos(\mu-\nu)\theta\left(\frac{(-1)^{p-r}\pi x^{\alpha-1}\cos^{\alpha-1}\theta}{2f\left(\frac{\pi\alpha}{2}\right)\Gamma\left(\frac{\alpha-r+1}{2}+p\right)\Gamma\left(\frac{\alpha+r+1}{2}-p\right)}\right.$$

$$\left.+\sum_{j=p-r}^{\infty}\frac{(-1)^{j+p-r}x^{2j+\delta}\cos^{2j+\delta}\theta\zeta(\alpha-2j-\delta)}{\Gamma(j+p+1)\Gamma(j-p+1+r)}\right)d\theta.$$

After integrating, we have to solve the integrals

$$S_{\alpha}^{J,J}=\frac{(-1)^{p-r}\left(\frac{x}{2}\right)^{\alpha-1}}{f\left(\frac{\pi\alpha}{2}\right)\Gamma\left(\frac{\alpha-r+1}{2}+p\right)\Gamma\left(\frac{\alpha+r+1}{2}-p\right)}\int_{0}^{\pi/2}\cos(\mu-\nu)\theta\cos^{\alpha-1}\theta\,d\theta$$

$$+\sum_{j=p-r}^{\infty}\frac{(-1)^{j+p-r}\left(\frac{x}{2}\right)^{2j+\delta}\zeta(\alpha-2j-\delta)}{\Gamma(j+p+1)\Gamma(j-p+1+r)}\frac{2}{\pi}$$

$$\int_{0}^{\pi/2}\cos(\mu-\nu)\theta\cos^{2j+\delta}\theta\,d\theta.$$

Relying on the integral (entry 3.631.9 in [10, p. 397])

$$\int_{0}^{\pi/2}\cos b\theta\cos^{\alpha-1}\theta\,d\theta=\frac{\pi\Gamma(a)}{2^{a}\Gamma\left(\frac{a+b+1}{2}\right)\Gamma\left(\frac{a-b+1}{2}\right)},$$

and considering that $\delta = r$, we obtain

$$S_{\alpha}^{J,J} = \frac{(-1)^{p-r}\left(\frac{x}{2}\right)^{\alpha-1} \pi\Gamma(\alpha)}{f\left(\frac{\pi\alpha}{2}\right) G_{\alpha,p}} + \sum_{j=p-r}^{\infty} \frac{(-1)^{j+p-r} \binom{2j+\delta}{j+p} \left(\frac{x}{2}\right)^{2j+\delta} \zeta(\alpha - 2j - \delta)}{\Gamma(j+1+p-\nu)\Gamma(j+1+\nu-p+r)}, \tag{7.7}$$

where $\alpha > 0, \mu, \nu \in \mathbb{R}, \mu + \nu = 2p - r, p \in \mathbb{N}_0, f = \left\{ \begin{array}{c} \sin \\ \cos \end{array} \right\}$ $\delta = \left\{ \begin{array}{c} 1 \\ 0 \end{array} \right\} r = \left\{ \begin{array}{c} 1 \\ 0 \end{array} \right\}$, and $G_{\alpha,p} = 2\Gamma\left(\frac{\alpha-r+1}{2} + p\right)\Gamma\left(\frac{\alpha+r+1}{2} - p\right)$ $\Gamma\left(\frac{\alpha+\mu-\nu+1}{2}\right)\Gamma\left(\frac{\alpha-\mu+\nu+1}{2}\right)$.

When $f = \left\{ \begin{array}{c} \sin \\ \cos \end{array} \right\} \alpha = \left\{ \begin{array}{c} 2m-1 \\ 2m \end{array} \right\}$, $m \in \mathbb{N}$, (7.7) takes closed form. We consider first $\alpha = 2m-1$. Then $f = \sin$, $\delta = 1$ and $r = 1$. If $2m-1 \geqslant 2p-1$, we obtain

$$\sum_{n=1}^{\infty} \frac{J_{\mu}(nx) J_{\nu}(nx)}{n^{2m-1}} = \frac{\pi(-1)^{m+p-r}\left(\frac{x}{2}\right)^{2m-2} (2m-2)!}{G_{m,p}} + \sum_{j=p-1}^{m-1} \frac{(-1)^{j+p-1} \binom{2j+\delta}{j+p} \left(\frac{x}{2}\right)^{2j+1} \zeta(2m-2j-2)}{\Gamma(j+1+p-\nu)\Gamma(j+1+\nu-p+1)}, \tag{7.8}$$

where

$$G_{m,p} = 2\Gamma\left(m+p-\frac{1}{2}\right)\Gamma\left(m-p+\frac{1}{2}\right)\Gamma\left(m+p-\nu-\frac{1}{2}\right)\Gamma\left(m-p+\nu+\frac{1}{2}\right).$$

Example 7.2 For $m = 3, p = 2, \mu = \frac{5}{2}, \nu = \frac{1}{2}$, we find

$$\sum_{n=1}^{\infty} \frac{J_{\frac{5}{2}}(nx) J_{\frac{1}{2}}(nx)}{n^5} = \frac{\pi x^3}{45} - \frac{4x^4}{105} + \frac{x^5}{63\pi}.$$

On the other hand, the condition $m < p$ implies $2m - 2j - 2 < 2p - 2j - 2 < 0$ for $j > p - 1$, meaning that $\zeta(2m - 2j - 2) = 0$, so the whole right-hand series (7.7) vanishes, yielding

$$\sum_{n=1}^{\infty} \frac{J_\mu(nx) J_\nu(nx)}{n^{2m-1}} = \frac{(-1)^{p-m+1}\left(\frac{x}{2}\right)^{2m-2}(2m-2)!\,\pi}{G_{m,p}}. \tag{7.9}$$

Example 7.3 Taking $m = 2, p = 3, \mu = \frac{13}{3}, \nu = \frac{2}{3}$ results in the sum

$$\sum_{n=1}^{\infty} \frac{J_{\frac{13}{3}}(nx) J_{\frac{2}{3}}(nx)}{n^3} = \frac{2x^2}{105\Gamma\left(\frac{1}{6}\right)\Gamma\left(\frac{23}{6}\right)} = \frac{2x^2}{105 \cdot \frac{17}{6} \cdot \frac{11}{6} \cdot \frac{5}{6} \cdot \Gamma\left(\frac{1}{6}\right)\Gamma\left(\frac{5}{6}\right)}$$
$$= \frac{144x^2}{32725 \cdot \frac{\pi}{\sin\frac{\pi}{6}}} = \frac{72x^2}{32725\pi},$$

where, in the penultimate expression, we have applied Euler's reflection formula (4.27).

Now, we consider $\alpha = 2m$. Then $f = \cos$, $\delta = 0$, $r = 0$. If $m \geqslant p$, we obtain a formula similar to (7.8)

$$\sum_{n=1}^{\infty} \frac{J_\mu(nx) J_\nu(nx)}{n^{2m}} = \frac{(-1)^{p-l+1}\left(\frac{x}{2}\right)^{2m-1}(2m-1)!}{G_{m,p}}$$
$$+ \sum_{j=p}^{m} \frac{(-1)^{j+p}\binom{2i}{j+p}\zeta(2m-2j)\left(\frac{x}{2}\right)^{2j}}{\Gamma(j+1+p-\nu)\Gamma(j+1-p+\nu)}, \tag{7.10}$$

and for $m < p$ another one similar to (7.9)

$$\sum_{n=1}^{\infty} \frac{J_\mu(nx) J_\nu(nx)}{n^{2m}} = \frac{(-1)^{p-m+1}\left(\frac{x}{2}\right)^{2m-1}(2m-1)!\pi}{G_{m,p}}. \tag{7.11}$$

12.7.3 Application of Poisson's Formula

Here we are applying Poisson's formula (4.35) once more, relying on a power series representation of the product of Bessel functions [41, p. 147]

$$J_\mu(x) J_\nu(x) = \frac{x^{\mu+\nu}}{2^{\mu+\nu}} \sum_{n=0}^{\infty} \frac{(-1)^n\left(\frac{x}{2}\right)^{2n}\Gamma(\mu+\nu+2n+1)}{n!\,\Gamma(\mu+n+1)\Gamma(\nu+n+1)\Gamma(\mu+\nu+n+1)}.$$

On this basis, we construct a function involving the product of Bessel functions

$$g(x) = \frac{J_\mu(x) J_\nu(x)}{x^\sigma} \sum_{n=0}^{\infty} \frac{(-1)^n x^{\mu+\nu-\sigma+2n} \Gamma(\mu+\nu+2n+1) 2^{-\mu-\nu-2n}}{n! \Gamma(\mu+n+1) \Gamma(\nu+n+1) \Gamma(\mu+\nu+n+1)},$$

where $\mu + \nu > \sigma, x \neq 0$. If we take the limit, we find $\lim_{x\to 0} g(x) = 0$ for $\mu + \nu > \sigma$, and

$$\lim_{x\to 0} g(x) = \frac{1}{2^{\mu+\nu}\Gamma(\mu+1)\Gamma(\nu+1)}, \quad \mu+\nu=\sigma,$$

which we write in the form of

$$\lim_{x\to 0} g(x) = \frac{\delta_{\mu+\nu,\sigma}}{2^{\mu+\nu}\Gamma(\mu+1)\Gamma(\nu+1)}, \delta_{\mu+\nu,\sigma} = \begin{cases} 1, \mu+\nu=\sigma \\ 0, \mu+\nu>\sigma. \end{cases}$$

Because of this, we may introduce a continuous, smooth, and absolutely integrable on $(0, +\infty)$ function $f(x)$ as follows

$$f(x) = \begin{cases} g(x), x \neq 0 \\ \frac{\delta_{\mu+\nu,\sigma}}{2^{\mu+\nu}\Gamma(\mu+1)\Gamma(\nu+1)}, x = 0. \end{cases} \tag{7.12}$$

So (4.34), p. 4.34, now becomes

$$F_c(\omega) = \sqrt{\frac{2}{\pi}} \int_0^\infty f(x) \cos \omega x \, \mathrm{d}x = \sqrt{\frac{2}{\pi}} \int_0^\infty \frac{J_\mu(x) J_\nu(x)}{x^\sigma} \cos \omega x \, \mathrm{d}x,$$

and, considering that $\beta = \frac{2\pi}{\alpha}$, from Poisson's formula (4.35), multiplying the latter previously by α^σ, there follows

$$\sum_{n=1}^{\infty} \frac{J_\mu(n\alpha) J_\nu(n\alpha)}{n^\sigma} = \alpha^{\sigma-1} \left(\int_0^\infty \frac{J_\mu(x) J_\nu(x)}{x^\sigma} \mathrm{d}x \right. \\ \left. + 2 \sum_{n=1}^{\infty} \int_0^\infty \frac{J_\mu(x) J_\nu(x)}{x^\sigma} \cos \frac{2n\pi x}{\alpha} \mathrm{d}x - \frac{\sigma}{2} f(0) \right). \tag{7.13}$$

We use the integral (see [23, p. 211])

$$\int_0^\infty \frac{J_\mu(x) J_\nu(x)}{x^\sigma} dx = \frac{2^{-\sigma}\Gamma(\sigma)\Gamma\left(\frac{1+\mu+\nu-\sigma}{2}\right)}{\Gamma\left(\frac{1+\mu-\nu+\sigma}{2}\right)\Gamma\left(\frac{1+\mu+\nu+\sigma}{2}\right)\Gamma\left(\frac{1-\mu+\nu+\sigma}{2}\right)}, \tag{7.14}$$

where $1+\mu+\nu>\sigma>0$, as well as the integral (see [23, p. 226])

$$\int_0^\infty \frac{J_\mu(x)J_\nu(x)}{x^\sigma}\cos\frac{2n\pi x}{\alpha}dx = {}_4F_3\left(a_1,a_2,a_3,a_4;b_1,b_2,b_3;\left(\frac{\alpha}{n\pi}\right)^2\right)$$
$$\times\frac{\Gamma(1-\sigma+\mu+\nu)}{2^{\nu+\mu}\Gamma(\mu+1)\Gamma(\nu+1)}\left(\frac{\alpha}{2n\pi}\right)^{\mu+\nu+1-\sigma}$$
$$\cos\frac{\pi}{2}(1-\sigma+\mu+\nu), \tag{7.15}$$

with $0<\alpha<n\pi, \mu+\nu+1>\sigma>0$, where ${}_4F_3$ is the hypergeometric function with the Pochhammer symbols, respectively

$$a_1=\left(\frac{\mu+\nu-\sigma+2}{2}\right)_k,\ a_2=\left(\frac{\mu+\nu-\sigma+1}{2}\right)_k,\ a_3=\left(\frac{\mu+\nu+2}{2}\right)_k,$$
$$a_4=\left(\frac{\mu+\nu+1}{2}\right)_k,$$
$$b_1=(\mu+\nu+1)_k,\ b_2=(\nu+1)_k,\ b_3=(\mu+1)_k.$$

We shall write ${}_4F_3$ as a power series, then place the right-hand sides of (7.14), (7.15), and (7.12) in (7.13), and use afterward Euler's reflection formula in the form of

$$\Gamma\left(\frac{1}{2}-s\right)\Gamma\left(\frac{1}{2}+s\right)=\frac{\pi}{\cos\pi s}, s\notin\mathbb{Z},$$

so that (7.13) becomes

$$\sum_{n=1}^\infty\frac{J_\mu(n\alpha)J_\nu(n\alpha)}{n^\sigma}=\frac{\alpha^{\sigma-1}\Gamma(\sigma)\Gamma\left(\frac{1-\sigma+\mu+\nu}{2}\right)}{2^\sigma\Gamma\left(\frac{1+\mu-\nu+\sigma}{2}\right)\Gamma\left(\frac{1+\mu+\nu+\sigma}{2}\right)\Gamma\left(\frac{1-\mu+\nu+\sigma}{2}\right)}$$
$$+\frac{\left(\frac{\alpha}{4\pi}\right)^{\mu+\nu}(2\pi)^\sigma\Gamma(1-\sigma+\mu+\nu)}{\Gamma(\mu+1)\Gamma(\nu+1)\Gamma\left(\frac{2+\mu+\nu-\sigma}{2}\right)\Gamma\left(\frac{\sigma-\mu-\nu}{2}\right)}$$
$$\times\sum_{n=1}^\infty\frac{1}{n^{1-\sigma+\mu+\nu}}\sum_{k=0}^\infty\frac{a_1\,a_2a_3a_4}{b_1\,b_2b_3}\frac{\alpha^{2k}}{n^{2k}\pi^{2k}k!}.$$

Also, we can swap the right-hand sums, bearing in mind that $\mu+\nu>\sigma$ implies $2k+\mu+\nu-\sigma+1>1, k\in\mathbb{N}_0$, and we make use of

$$\sum_{n=1}^\infty\frac{1}{n^{2k+\mu+\nu-\sigma+1}}=\zeta(2k+\mu+\nu-\sigma+1).$$

By virtue of Legendre's duplication formula (4.9) and the Gamma function's basic property $\Gamma(s+1) = s\Gamma(s)$, we finally obtain

$$\sum_{n=1}^{\infty} \frac{J_\mu(n\alpha) J_\nu(n\alpha)}{n^\sigma}$$

$$= \frac{\alpha^{\sigma-1}\Gamma(\sigma)\Gamma\left(\frac{1+\mu+\nu-\sigma}{2}\right)}{2^\sigma\Gamma\left(\frac{1+\mu-\nu+\sigma}{2}\right)\Gamma\left(\frac{1+\mu+\nu+\sigma}{2}\right)\Gamma\left(\frac{1-\mu+\nu+\sigma}{2}\right)}$$

$$- \frac{\alpha^\sigma \delta_{\mu+\nu,\sigma}}{2^{\mu+\nu+1}\Gamma(\mu+1)\Gamma(\nu+1)} + \frac{\left(\frac{\alpha}{2\pi}\right)^{\mu+\nu} \pi^{\sigma-\frac{1}{2}}\Gamma\left(\frac{1+\mu+\nu-\sigma}{2}\right)}{\Gamma(1+\mu)\Gamma(1+\nu)\Gamma\left(\frac{\sigma-\mu-\nu}{2}\right)}$$

$$\times \sum_{k=0}^{\infty} \frac{\zeta(2k+1-\sigma+\mu+\nu)\left(\frac{\alpha}{\pi}\right)^{2k} G_k}{k!}, \tag{7.16}$$

where $0 < \alpha < \pi, \mu+\nu+1 > \sigma > 0$, and for the sake of brevity and simplicity, we have introduced a denotation

$$G_k = \prod_{j=1}^{k} \frac{\left(j+\frac{\mu+\nu-\sigma}{2}\right)\left(j+\frac{\mu+\nu}{2}\right)\left(j+\frac{\mu+\nu-\sigma-1}{2}\right)\left(j+\frac{\mu+\nu-1}{2}\right)}{(j+\mu+\nu)(i+\mu)(j+\nu)}.$$

Unlike (7.3), closed-form cases of (7.16) do not ensue because the ζ function vanishes, since it does not equal zero for $\sigma-\mu-\nu = -2p, p \in \mathbb{N}_0$, but because Γ has poles at non-positive integers $\frac{\sigma-\mu-\nu}{2} = -p, p \in \mathbb{N}_0$, so the reciprocal value of $\Gamma\left(\frac{\sigma-\mu-\nu}{2}\right)$ is zero, and consequently, the last term in (7.16) disappears.

Here, we distinguish between $p > 0$ and $p = 0$. If $p > 0$, then $\mu+\nu > \sigma$, implying $\delta_{\mu+\nu,\sigma} = 0$, and (7.16) takes closed form

$$\sum_{n=1}^{\infty} \frac{J_\mu(n\alpha) J_\nu(n\alpha)}{n^{\mu+\nu-2p}} = \frac{\left(\frac{\alpha}{2}\right)^{\mu+\nu-2p-1}\Gamma(\mu+\nu-2p)\Gamma\left(p+\frac{1}{2}\right)}{2\Gamma\left(\mu-p+\frac{1}{2}\right)\Gamma\left(\mu+\nu-p+\frac{1}{2}\right)\Gamma\left(\nu-p+\frac{1}{2}\right)}. \tag{7.17}$$

However, if $p = 0$, then $\mu + \nu = \sigma$, meaning that $\delta_{\mu+\nu,\sigma} = 1$, so in that case (7.16) takes closed form

$$\sum_{n=1}^{\infty} \frac{J_\mu(n\alpha) J_\nu(n\alpha)}{n^{\mu+\nu}} = \frac{\sqrt{\pi}\left(\frac{\alpha}{2}\right)^{\mu+\nu-1} \Gamma(\mu+\nu)}{2\Gamma\left(\mu+\frac{1}{2}\right)\Gamma\left(\mu+\nu+\frac{1}{2}\right)\Gamma\left(\nu+\frac{1}{2}\right)} - \frac{\left(\frac{\alpha}{2}\right)^{\mu+\nu}}{2\Gamma(\mu+1)\Gamma(\nu+1)},$$

which is identical to formula 9 in [23], p. 683.

In order to compare the formula 1 in [23, p. 683], which is

$$\sum_{k=1}^{\infty} \frac{J_{2m+1}(kx) J_{2n(2m+1)}(kx)}{k} = \frac{(-1)^{m+n+1} 2}{(2m+1)^2 (4n^2-1)\pi}, \tag{7.18}$$

where $0 < x < \pi$, $n \in \mathbb{N}$, $m \in \mathbb{N}_0$, with (7.17), we take there $\mu = 2m+1$, $\nu = 2n(2m+1)$, and $\mu + \nu - 2p = 1$, and there follows $2p = \mu + \nu - \sigma = 2(2mn + m + n)$, i.e., $p = 2mn + m + n$. Placing these values in (7.17), and using the relations

$$\Gamma\left(n+\frac{3}{2}\right) = \left(n+\frac{1}{2}\right)\Gamma\left(n+\frac{1}{2}\right)$$

and

$$\Gamma(z)\Gamma(1-z)\sin \pi z = \pi, \quad z = 2mn - m + n + \frac{1}{2},$$

after cancellation, we obtain (7.18).

So for $\mu + \nu - \sigma = 2n, n \in \mathbb{N}_0$, (7.16) coincides with some of the closed form cases of (7.7). Namely, for a particular choice of parameters, the closed-form formula (7.17) can be reduced to (7.9) or (7.11). If we take $\sigma = 2l - 1$, $\mu + \nu = 2k + 1$, and require $l < k + 1$ (these are the conditions for holding (7.9)), we further have $\mu + \nu - \sigma = 2p$, where $p = k - l + 1 \in \mathbb{N}$. Replacing these values in (7.17), setting afterwards $z = k - l + \frac{3}{2}$ and applying the property

$$\Gamma(z)\Gamma(1-z) = \frac{\pi}{\sin \pi z},$$

we easily come to (7.9). So, by using different methods, we obtain the same formula. Yet, note that (7.17) holds whenever real numbers μ, ν, and σ satisfy $\mu + \nu + 1 > \sigma > 0$ and $\mu + \nu = \sigma - 2p$ $(p \in \mathbb{N})$, whereas (7.9) holds only for positive integers $\mu + \nu$ and σ.

12.8 PRODUCT OF A TRIGONOMETRIC AND A SPECIAL FUNCTION

Relying on trigonometric series, we shall determine a general formula for the series over the product of a trigonometric function and a special function

$$S_\alpha^{\phi,g} = \sum_{n=1}^{\infty} \frac{(s)^{n-1}\phi_\nu((an-b)x)g((an-b)y)}{(an-b)^\alpha}, \quad \alpha \in \mathbb{R}^+, \tag{8.1}$$

where $a = \left\{ \begin{array}{c} 1 \\ 2 \end{array} \right\} b = \left\{ \begin{array}{c} 0 \\ 1 \end{array} \right\}$, $s = 1$ or $s = -1$, $g = \sin$ or $g = \cos$, and ϕ_ν is the Bessel, Struve, Anger, or Weber function.

Also, we consider series over the product of one trigonometric function and two Bessel functions

$$S_\alpha^{J,J,g} = \sum_{n=1}^{\infty} \frac{(s)^{n-1}J_\mu((an-b)x)J_\nu((an-b)x)}{(an-b)^\alpha} g((an-b)y). \tag{8.2}$$

12.8.1 Product of Bessel or Struve Functions and a Trigonometric Function

We first regard ϕ_ν in the series (8.1) as Bessel or Struve functions, using their representation through Poisson's integral (2.12)

$$\phi_\nu(x) = \frac{2\left(\frac{x}{2}\right)^\nu}{\sqrt{\pi}\,\Gamma\left(\nu+\frac{1}{2}\right)} \int_0^1 \left(1-t^2\right)^{\nu-\frac{1}{2}} f(xt)\,dt, \quad \operatorname{Re}\nu > -\frac{1}{2} \tag{8.3}$$

$$\phi_\nu = \left\{ \begin{array}{c} J_\nu \\ \mathbf{H}_\nu \end{array} \right\} f = \left\{ \begin{array}{c} \cos \\ \sin \end{array} \right\}.$$

Theorem 8.1 *The summation formula for the series (8.1) is*

$$\begin{aligned} S_\alpha^{\phi,g} =& \frac{\left(\frac{x}{2}\right)^\nu \tau}{\sqrt{\pi}} \left(\sum_{k=0}^{\infty} \binom{\alpha-\nu-1}{2k+d} \frac{c\pi y^{\alpha-\nu-1-2k-d} x^{2k+d}}{2\,\Gamma(\alpha-\nu)\, h\left(\frac{\pi}{2}(\alpha-\nu)\right)} \frac{\Gamma\left(k+\frac{d+1}{2}\right)}{\Gamma\left(\nu+k+1+\frac{d}{2}\right)} \right. \\ & \left. + \sum_{k=0}^{\infty} \frac{(-1)^k F(\alpha-\nu-2k-\delta)}{(2k+\delta)!} \sum_{j=0}^{k-\nu} \binom{2k+\delta}{2j+d} \frac{y^{2k+\delta-2j-d} x^{2j+d} \Gamma\left(j+\frac{d+1}{2}\right)}{\Gamma\left(\nu+j+1+\frac{d}{2}\right)} \right), \end{aligned} \tag{8.4}$$

where $\phi_\nu = \left\{ \begin{array}{c} J_\nu \\ \mathbf{H}_\nu \end{array} \right\} f = \left\{ \begin{array}{c} \cos \\ \sin \end{array} \right\} d = \left\{ \begin{array}{c} 0 \\ 1 \end{array} \right\}, \left\{ \begin{array}{c} f = g \\ f \neq g \end{array} \right\} \delta = \left\{ \begin{array}{c} 0 \\ 1 \end{array} \right\} h = \left\{ \begin{array}{c} \cos \\ \sin \end{array} \right\}$, $\tau = -1, \nu = 1$ *only iff* $f = g = \sin$, *otherwise* $\tau = 1, \nu = 0$. *For the parameters* a, b, s, c, F, *we refer to* Table 12.2, p. 273, *and convergence regions are* K_1, K_2, K_3, K_4.

Proof. After putting (8.3) into (8.1), one may interchange summation and integration because the series over the producs unifort of trigonometric functions convergemly. Thus, we have

$$S_\alpha^{\phi,g} = \frac{2\left(\frac{x}{2}\right)^\nu}{\sqrt{\pi}\,\Gamma\left(\nu + \frac{1}{2}\right)} \int_0^1 \left(1 - t^2\right)^{\nu - \frac{1}{2}} \sum_{n=1}^{\infty} \frac{(s)^{n-1} f((an-b)xt)\, g((an-b)y)}{(an-b)^{\alpha-\nu}} dt,$$

where $\alpha - \nu > 0$. For this series, we make use of the summation formula for the series under the integration sign (see [34])

$$T_\alpha^{f,g} = \frac{\tau c \pi}{2\Gamma(\alpha)\, h\left(\frac{\pi\alpha}{2}\right)} \sum_{k=0}^{\infty} \binom{\alpha - 1}{2k + d} y^{\alpha-1-2k-d} x^{2k+d}$$

$$+\tau \sum_{k=0}^{\infty} \frac{(-1)^k F(\alpha - 2k - \delta)}{(2k+\delta)!} \sum_{j=0}^{k-\nu} \binom{2k+\delta}{2j+d} y^{2k+\delta-2j-d} x^{2j+d}, \qquad (8.5)$$

where $f = \left\{ \begin{array}{c} \sin \\ \cos \end{array} \right\} d = \left\{ \begin{array}{c} 1 \\ 0 \end{array} \right\}, \left\{ \begin{array}{c} f = g \\ f \neq g \end{array} \right\} \delta = \left\{ \begin{array}{c} 0 \\ 1 \end{array} \right\} h = \left\{ \begin{array}{c} \cos \\ \sin \end{array} \right\}$, and $\tau = -1, \nu = 1$ only if $f = g = \sin$, otherwise $\tau = 1, \nu = 0$. Depending on the choice of parameters, the convergence region is K_1, K_2, K_3 or K_4. With xt in place of x and $\alpha - \nu$ instead of α, for $S_\alpha^{\phi,g}$ we obtain

$$S_\alpha^{\phi,g} = \frac{2\left(\frac{x}{2}\right)^\nu \tau}{\sqrt{\pi}} \int_0^1 \frac{\left(1-t^2\right)^{\nu-\frac{1}{2}}}{\Gamma\left(\nu+\frac{1}{2}\right)} \left(\sum_{k=0}^{\infty} \binom{\alpha-\nu-1}{2k+d} \frac{c\pi y^{\alpha-\nu-1-2k-d} (xt)^{2k+d}}{2\Gamma(\alpha-\nu)\, h\left(\frac{\pi}{2}(\alpha-\nu)\right)} \right.$$

$$\left. + \sum_{k=0}^{\infty} \frac{(-1)^k F(\alpha-\nu-2k-\delta)}{(2k+\delta)!} \sum_{j=0}^{k-\nu} \binom{2k+\delta}{2j+d} y^{2k+\delta-2j-d} (xt)^{2j+d} \right) dt.$$

where $h = \sin$ or $h = \cos$, $g = \left\{ \begin{array}{c} \sin \\ \cos \end{array} \right\} d = \left\{ \begin{array}{c} 1 \\ 0 \end{array} \right\}$, and $\tau = -1$ only if $f = g = \sin$, otherwise $\tau = 1$.

By swapping the order of integration and summation, we solve the Euler beta integrals [8] and obtain

$$\int_0^1 (1-t^2)^{\nu-\frac{1}{2}} t^{2m+d} dt = \frac{1}{2}\int_0^1 (1-t^2)^{\left(\nu+\frac{1}{2}\right)-1} (t^2)^{\left(m+\frac{d+1}{2}\right)-1} dt^2$$
$$= \frac{1}{2}\mathrm{B}\left(\nu+\frac{1}{2}, m+\frac{d+1}{2}\right) = \frac{\Gamma\left(\nu+\frac{1}{2}\right)\Gamma\left(m+\frac{d+1}{2}\right)}{2\,\Gamma\left(\nu+m+1+\frac{d}{2}\right)},$$

with $m = k$ or $m = j$, and obtain (8.4). □

Formula (8.4) comprises and includes some results presented in papers [8], [14], [29], [36], and [30]. Some particular cases of these results can be found in well-known books [17, 23].

12.8.2 Closed-Form Cases

Closed-form formulas concerned with the summation formula (8.4) can be obtained for $\alpha - \nu = 2m - r$, where $m \in \mathbb{N}$, r is 0 or 1. In this way, we can arrive at closed-form cases even for positive non-integers.

As we know, the functions Riemann's zeta, Dirichlet's eta, and lambda truncate at even negative integers, whereas Dirichlet's beta truncates at odd negative integers. So the closed-form formula corresponding to (8.4) is as follows

$$\begin{aligned} S_{\nu+2m-r}^{\phi,g} = &\frac{c\,\tau\left(\frac{x}{2}\right)^{\nu}\sqrt{\pi}}{2\,\Gamma(2m-r)\,h\left(\frac{\pi}{2}(2m-r)\right)} \sum_{k=0}^{m-r} \binom{2m-r-1}{2k+d} \\ &\times \frac{y^{2m-r-1-2k-d} x^{2k+d}\,\Gamma\left(k+\frac{d+1}{2}\right)}{\Gamma\left(\nu+k+1+\frac{d}{2}\right)} \\ &+ \frac{\tau\left(\frac{x}{2}\right)^{\nu}}{\sqrt{\pi}} \sum_{k=0}^{m-r} \frac{(-1)^k F(2m-r-2k-\delta)}{(2k+\delta)!} \\ &\times \sum_{j=0}^{k-\nu} \binom{2k+\delta}{2j+d} \frac{y^{2k+\delta-2j-d} x^{2j+d}\,\Gamma\left(j+\frac{d+1}{2}\right)}{\Gamma\left(\nu+j+1+\frac{d}{2}\right)}, \end{aligned} \tag{8.6}$$

where $\phi_\nu = \left\{ \begin{array}{c} J_\nu \\ \mathbf{H}_\nu \end{array} \right\} f = \left\{ \begin{array}{c} \cos \\ \sin \end{array} \right\} d = \left\{ \begin{array}{c} 0 \\ 1 \end{array} \right\}, \left\{ \begin{array}{c} f=g \\ f\neq g \end{array} \right\} \delta = \left\{ \begin{array}{c} 0 \\ 1 \end{array} \right\} h = \left\{ \begin{array}{c} \cos \\ \sin \end{array} \right\} r = \left\{ \begin{array}{c} 0 \\ 1 \end{array} \right\}$, $\tau = -1, v = 1$ only if $f = g = \sin$, otherwise $\tau = 1, v = 0$.

Example 8.1 For $\phi_\nu = J_\nu, g = \cos, a = 1, b = 0, s = 1, \alpha - \nu = 2$ from (8.6) there follows

$$\sum_{n=1}^{\infty} \frac{J_\nu(nx)}{n^{\nu+2}} \cos ny = \frac{\left(\frac{x}{2}\right)^\nu}{24\Gamma(\nu+2)} \left(3x^2 + (\nu+1)\left(6y^2 - 12\pi y + 4\pi^2\right)\right),$$

where $K_1 = \{(x, y) \mid -\pi < x < \pi, |x| < y < 2\pi - |x|\}$ is the convergence region.

Consider the formula (74.1.19) in [11], for $\alpha - \nu = 2$

$$\sum_{n=1}^{\infty} \frac{J_\nu(nx)}{n^{\nu+2}} \cos nx = \frac{\left(\frac{x}{2}\right)^\nu}{24\Gamma(\nu+2)} \left(3x^2 + (\nu+1)\left(6x^2 - 12\pi x + 4\pi^2\right)\right),$$

where $0 < x < \pi$, $\operatorname{Re} \nu > -\frac{3}{2}$. The same result is obtained by using the above formula for $y = x$ and $\operatorname{Re} \nu > -\frac{1}{2}$.

Example 8.2 With $\phi_\nu = \mathbf{H}_\nu, g = \sin, \alpha = 5, \nu = 2, a = 2, b = 1, s = -1$, the formula (8.6) takes the form of

$$\sum_{n=1}^{\infty} \frac{(-1)^{n-1} \mathbf{H}_2((2n-1)x)}{(2n-1)^5} \sin(2n-1)y = \frac{x^3 y}{30},$$

and the convergence region is $K_4 = \left\{(x, y), |x| < \frac{\pi}{2} \ \& \ |x| - \frac{\pi}{2} < y < \frac{\pi}{2} - |x|\right\}$.

12.8.3 Product of Anger or Weber Functions and a Trigonometric Function

Now we determine the sum of the series (8.1) involving Anger or Weber functions, and we recall their representation (3.3), i.e.

$$\phi_{2p-r}(x) = \frac{1}{\pi} \int_0^\pi f((2p-r)\,\theta)\, g(x \sin\theta)\, d\theta,$$

where $\phi_{2p-r} = \left\{ \begin{array}{c} \mathbf{J}_{2p-r} \\ \mathbf{E}_{2p-r} \end{array} \right\} g = \left\{ \begin{array}{c} f \\ \tilde{f} \end{array} \right\}$, where $r = \left\{ \begin{array}{c} 0 \\ 1 \end{array} \right\} f = \left\{ \begin{array}{c} \cos \\ \sin \end{array} \right\} \tilde{f} = \left\{ \begin{array}{c} -\sin \\ \cos \end{array} \right\}$.

Theorem 8.2 *The summation formula for the series (8.1) is as follows*

$$S_{\alpha}^{\phi,h} = \frac{c\pi\tau f\left(\pi p - \frac{r\pi}{2}\right)}{2\,\Gamma(\alpha)\, q\left(\frac{\pi\alpha}{2}\right)} \sum_{k=0}^{\infty} \frac{(\alpha-1)^{(2k+d)} y^{\alpha-1-2k-d}\left(\frac{x}{2}\right)^{2k+d}}{\Gamma\left(k+p+1+\frac{d-r}{2}\right)\Gamma\left(k-p+1+\frac{d+r}{2}\right)}$$
$$+ \sum_{k=0}^{\infty}\sum_{j=0}^{k-\nu} \frac{(-1)^k \tau f\left(\pi p - \frac{r\pi}{2}\right) F(\alpha-2k-\delta)\, y^{2k+\delta-2j-d}\left(\frac{x}{2}\right)^{2j+d}}{(2k-2j+\delta-d)!\,\Gamma\left(j+p+1+\frac{d-r}{2}\right)\Gamma\left(j-p+1+\frac{d+r}{2}\right)}. \tag{8.7}$$

Proof. Placing $\phi_{2p-r}(x)$ in the formula (8.1), where $\nu = 2p - r$, we have

$$S_{\alpha}^{\phi,h} = \frac{1}{\pi}\sum_{n=1}^{\infty} \frac{(s)^{n-1} h((an-b)y)}{(an-b)^{\alpha}} \int_0^{\pi} f((2p-r)\,\theta)\, g((an-b)\,x\,\sin\theta)\, d\theta.$$

After interchanging integration and summation, we further reduce the derivation of the formula to dealing with the series over the product of two trigonometric functions

$$S_{\alpha}^{\phi,h} = \frac{1}{\pi}\int_0^{\pi} f((2p-r)\,\theta) \sum_{n=1}^{\infty} \frac{(s)^{n-1} g((an-b)\,x\,\sin\theta)\; h((an-b)\,y)}{(an-b)^{\alpha}}\, d\theta.$$

Here we are referring to the formula where we substitute $x \sin\theta$ for x, the function g for f, h for g, and q for h, and we have

$$T_{\alpha}^{g,h} = \tau c\pi \frac{(x\sin\theta + y)^{\alpha-1} + (-1)^d (y - x\sin\theta)^{\alpha-1}}{4\Gamma(\alpha)\, q\left(\frac{\pi\alpha}{2}\right)}$$
$$+\,\tau \sum_{k=0}^{\infty} \frac{(-1)^k F(\alpha-2k-\delta)}{(2k+\delta)!} \sum_{j=0}^{k-\nu} \binom{2k+\delta}{2j+d} y^{2k+\delta-2j-d} x^{2j+d} \sin^{2j+d}\theta,$$

where $g = \left\{ \begin{array}{c} \sin \\ \cos \end{array} \right\} d = \left\{ \begin{array}{c} 1 \\ 0 \end{array} \right\}, \left\{ \begin{array}{c} g = h \\ g \neq h \end{array} \right\} \delta = \left\{ \begin{array}{c} 0 \\ 1 \end{array} \right\} q = \left\{ \begin{array}{c} \cos \\ \sin \end{array} \right\}$, $\tau = -1, \nu = 1$ only if $g = h = \sin$, otherwise $\tau = 1, \nu = 0$, and the convergence regions are K_1', K_2', K_3', K_4', in Table 12.2, p. 273. So by applying the binomial series expansion

$$(1+x)^{\alpha} = \sum_{n=0}^{\infty} \binom{\alpha}{n} x^n = \sum_{n=0}^{\infty} \frac{\alpha^{(n)}}{n!} x^n = \sum_{n=0}^{\infty} \frac{(-1)^n (-\alpha)_n}{n!} x^n \tag{8.8}$$

and the falling factorial

$$\alpha^{(n)} = \alpha(\alpha-1)(\alpha-2)\cdots(\alpha-n+1) = \frac{\Gamma(\alpha+1)}{\Gamma(\alpha+1-n)}, \quad \alpha \in \mathbb{C}.$$

then replacing the above series with $T_{\alpha}^{g,h}$, and interchanging summation and integration, we find

$$\begin{aligned} S_{\alpha}^{\phi,h} &= \frac{c\tau}{2\Gamma(\alpha) q\left(\frac{\pi\alpha}{2}\right)} \sum_{k=0}^{\infty} \frac{(\alpha-1)^{(2k+d)}}{(2k+d)!} y^{\alpha-1-2k-d} x^{2k+d} \\ &\times \int_0^{\pi} \sin^{2k+d}\theta f((2p-r)\theta)\, d\theta + \frac{\tau}{\pi} \sum_{k=0}^{\infty} \frac{(-1)^k F(\alpha-2k-\delta)}{(2k+\delta)!} \\ &\times \sum_{j=0}^{k-\nu} \binom{2k+\delta}{2j+d} y^{2k+\delta-2j-d} x^{2j+d} \int_0^{\pi} \sin^{2j+d}\theta f((2p-r)\theta)\, d\theta. \end{aligned}$$

We solve these integrals by means of the reciprocal beta function (see entries 3.631.1. and 3.631.8. in [10, p. 397])

$$\int_0^{\pi} \sin^{\nu-1}\theta f((2p-r)\theta)\, d\theta = \frac{\pi}{2^{\nu-1}} \frac{f\left(\pi p - \frac{r\pi}{2}\right)\Gamma(\nu)}{\Gamma\left(\frac{\nu-r+1}{2}+p\right)\Gamma\left(\frac{\nu+r+1}{2}-p\right)},$$

where $\nu = 2k+1+d$ or $\nu = 2j+1+d$, and $f = \sin$ or $f = \cos$, so we obtain (8.7). □

If $f = \sin$ and $r = 0$ or $f = \cos$ and $r = 1$, the right-hand side of (8.7) equals zero. Also, all the terms equal zero, provided that $k \leqslant p-2$ or $k \leqslant p-1$ and $j \leqslant p-2$ or $j \leqslant p-1$, depending on d and r being both 1 or 0.

12.8.4 Product of a Trigonometric Function and Two Bessel Functions

Dealing with a series over a product of trigonometric and two Bessel functions $J_{\mu}(x)$ and $J_{\nu}(x)$, $\mu, \nu \in \mathbb{R}$, $\mu+\nu > -1$ requires Gegenbauer's integral representation (7.2) of the product of two Bessel functions, and we can make use of it to find the sum of the series (8.2), i.e.

$$S_{\alpha}^{J,J,g} = \frac{2}{\pi}\int_0^{\pi/2} \cos(\mu-\nu)\theta \sum_{n=1}^{\infty} \frac{(s)^{n-1} J_{\mu+\nu}(2(an-b)x\cos\theta)}{(an-b)^{\alpha}} g((an-b)y)\, d\theta.$$

Theorem 8.3 *The summation formula for series (8.2) is*

$$S_{\alpha}^{J,J,g} = \sum_{k=0}^{\infty} \frac{c\pi\binom{\alpha-\mu-\nu-1}{2k} y^{\alpha-\mu-\nu-2k-1}\left(\frac{x}{2}\right)^{\mu+\nu+2k}}{2\Gamma(\alpha-\mu-\nu)\, h\left(\frac{\pi}{2}(\alpha-\mu-\nu)\right)} \Gamma_k^{\mu,\nu}$$
$$+ \sum_{k=0}^{\infty} \frac{(-1)^k F(\alpha-\mu-\nu-2k-\delta)}{(2k+\delta)!}$$
$$\sum_{j=0}^{k} \binom{2k+\delta}{2j} y^{2k-2j+\delta}\left(\frac{x}{2}\right)^{\mu+\nu+2j} \Gamma_j^{\mu,\nu}, \tag{8.9}$$

where $\mu, \nu \in \mathbb{R}$, $\alpha > \mu+\nu > -\frac{1}{2}$, $\left\{\begin{array}{c} f=g \\ f\neq g \end{array}\right\}\delta = \left\{\begin{array}{c} 0 \\ 1 \end{array}\right\} h = \left\{\begin{array}{c} \cos \\ \sin \end{array}\right\}$, $\Gamma_k^{\mu,\nu}$ *and* $\Gamma_j^{\mu,\nu}$ *are given by the equality*

$$\Gamma_m^{\mu,\nu} = \frac{(2m)!\,\Gamma(\mu+\nu+2m+1)}{m!\,\Gamma(\mu+\nu+m+1)\,\Gamma(\mu+m+1)\,\Gamma(\nu+m+1)}, \qquad m = k, j.$$

Proof. To determine the sum of (8.2), we use formula (8.4), where we take $\phi_\nu = J_\nu$, replacing then throughout that formula ν with $\mu+\nu$, setting $2x\cos\theta$ in place of x as well. Seeing that $f = \cos$ in (2.11), we have to take $d = 0$, $\tau = 1$, $\nu = 0$ in (8.4).

Thus, for this choice of parameters, after a rearrangement, formula (8.4) becomes

$$S_{\alpha}^{J,J,g} = \sum_{k=0}^{\infty} \frac{c\sqrt{\pi}\binom{\frac{\alpha-\mu-\nu-1}{2k}}{} y^{\alpha-\mu-\nu-1-2k}}{2\,\Gamma(\alpha-\mu-\nu)\,h\left(\frac{\pi}{2}(\alpha-\mu-\nu)\right)} \frac{(2x\cos\theta)^{\mu+\nu+2k}\Gamma\left(k+\frac{1}{2}\right)}{2^{\mu+\nu}\Gamma(\mu+\nu+k+1)}$$
$$+\sum_{k=0}^{\infty} \frac{(-1)^k F(\alpha-\mu-\nu-2k-\delta)}{(2k+\delta)!}$$
$$\times \sum_{j=0}^{k} \binom{2k+\delta}{2j} \frac{y^{2k-2j+\delta}(2x\cos\theta)^{\mu+\nu+2j}\Gamma\left(j+\frac{1}{2}\right)}{2^{\mu+\nu}\sqrt{\pi}\,\Gamma(\mu+\nu+j+1)},$$

where $\alpha > \mu+\nu > -\frac{1}{2}, \left\{\begin{array}{c} f=g \\ f\neq g \end{array}\right\} \delta = \left\{\begin{array}{c} 0 \\ 1 \end{array}\right\} h = \left\{\begin{array}{c} \cos \\ \sin \end{array}\right\}$. For parameters a, b, s, c, and F, we refer to Table 12.2, p. 273, and convergence regions are K_1, K_2, K_3, K_4.

In addition, making use of the relation $\Gamma\left(m+\frac{1}{2}\right) = \frac{(2m)!\sqrt{\pi}}{2^{2m}m!}, m = k, j,$ now, we have

$$S_{\alpha}^{J,J,g} = \frac{2}{\pi}\int_0^{\pi/2} \left(\sum_{k=0}^{\infty} \frac{c\pi \binom{\frac{\alpha-\mu-\nu-1}{2k}}{} y^{\alpha-\mu-\nu-1-2k}}{2\,\Gamma(\alpha-\mu-\nu)\,h\left(\frac{\pi}{2}(\alpha-\mu-\nu)\right)}\right.$$
$$\times \frac{(2x\cos\theta)^{\mu+\nu+2k}(2k)!}{2^{\mu+\nu+2k}k!\,\Gamma(\mu+\nu+k+1)} + \sum_{k=0}^{\infty} \frac{(-1)^k F(\alpha-\mu-\nu-2k-\delta)}{(2k+\delta)!}$$
$$\left.\times \sum_{j=0}^{k} \binom{2k+\delta}{2j} \frac{y^{2k-2j+\delta}(2x\cos\theta)^{\mu+\nu+2j}(2j)!}{2^{\mu+\nu+2j}j!\,\Gamma(\mu+\nu+j+1)}\right) \cos(\mu-\nu)\,\theta d\theta.$$

After interchanging integration and summation again, we solve integrals by applying the reciprocal beta function [25, entry 5.12.5]

$$\frac{2}{\pi}\int_0^{\pi/2} (2x\cos\theta)^{\mu+\nu+2m} \cos(\mu-\nu)\,\theta\, d\theta = \frac{x^{\mu+\nu+2m}\Gamma(\mu+\nu+2m+1)}{\Gamma(\mu+m+1)\,\Gamma(\nu+m+1)}, \tag{8.10}$$

with Re $(\mu+\nu) > -1$ and $m = k, j$. For the sake of simplicity, we introduce abbreviations in (8.10) and obtain the summation formula for the series (8.2). □

We refer to Table 12.3 for the other relevant parameters. As for the convergence regions, we consider the ones in Table 12.2, p. 273, taking there $2x$ instead of x, so that we obtain K_5, K_6, K_7, K_8 in Table 12.3 in the Appendix.

If both $\mu+\nu$ and α are integers, i.e., $\mu+\nu = p \in \mathbb{N}_0, \alpha = m, m \in \mathbb{N}$, $m > p$, formula (8.9) is brought into closed form. If $m < p$, the right-hand side of (8.9) is zero.

Example 8.3 Taking $\alpha = 4, \mu = \frac{1}{2}, \nu = \frac{5}{2}, a = 2, b = 1, s = -1, g = \cos$, from (8.9) we find

$$\sum_{n=1}^{\infty} \frac{(-1)^{n-1} J_{\frac{1}{2}}((2n-1)x) J_{\frac{5}{2}}((2n-1)x)}{(2n-1)^4} \cos(2n-1)y = \frac{x^3}{30}.$$

Example 8.4 Taking $\alpha = 2, \mu = \frac{1}{2}, \nu = -\frac{1}{2}, a = 1, b = 0, s = 1, g = \cos$, from (8.9) we find

$$\sum_{n=1}^{\infty} \frac{J_{\frac{1}{2}}(nx) J_{-\frac{1}{2}}(nx)}{n^2} \cos ny = -\frac{\pi}{6} + \frac{2x^2}{3\pi}.$$

12.9 SERIES OVER NEUMANN OR MACDONALD FUNCTIONS

One can apply the summation formula (4.3) for obtaining sums of new series. For instance, we can consider [23, entry 35, p. 182]

$$\int_0^{\infty} \frac{x^{\nu+1}}{x^2-y^2} J_\nu(nx)\, dx = -\frac{y^\nu \pi}{2} Y_\nu(ny),\ n = 1, 2, ...; y > 0; -1 < \mathrm{Re}\,\nu < \frac{3}{2},$$

where Y_ν is Bessel function of the second kind, also called the Neumann function. By multiplying both sides by $1/n^\alpha$, then adding together the terms of both sequences, we form an infinite series, i.e.

$$\sum_{n=1}^{\infty}\int_0^{\infty}\frac{x^{\nu+1}}{x^2-y^2}\frac{J_\nu(nx)}{n^\alpha}\,dx=\int_0^{\infty}\frac{x^{\nu+1}}{x^2-y^2}\sum_{n=1}^{\infty}\frac{J_\nu(nx)}{n^\alpha}\,\mathrm{d}x$$
$$=-\frac{y^\nu\pi}{2}\sum_{n=1}^{\infty}\frac{Y_\nu(ny)}{n^\alpha},\tag{9.1}$$

whereby we are allowed to interchange summation and integration because of the uniform convergence of the series over the Bessel functions.

Now we apply (4.3) to the left-hand side sum and substitute there $2m$ for $\alpha-\nu$, $m\in\mathbb{N}_0$, so $\alpha=\nu+2m$. Then, since $\zeta(2m-2k)=0$, if $k>m$, the series becomes a finite sum.

$$\sum_{n=1}^{\infty}\frac{J_\nu(nx)}{n^{\nu+2m}}=\frac{(-1)^{m+1}\pi\left(\frac{x}{2}\right)^{\nu+2m-1}}{2\Gamma\left(m+\frac{1}{2}\right)\Gamma\left(m+\nu+\frac{1}{2}\right)}+\sum_{k=0}^{m}\frac{(-1)^{k+1}x^{\nu+2k}\zeta(2m-2k)}{2^{\nu+2k}k!\,\Gamma(\nu+k+1)}\tag{9.2}$$

As a result of substituting the right-hand side sum of (9.2) for the series in (9.1), there follows

$$\int_0^{\infty}\frac{x^{\nu+1}}{x^2-y^2}\left(\frac{(-1)^{m+1}\pi\left(\frac{x}{2}\right)^{\nu+2m-1}}{2\Gamma\left(m+\frac{1}{2}\right)\Gamma\left(m+\nu+\frac{1}{2}\right)}\right.$$
$$\left.+\sum_{k=0}^{m}\frac{(-1)^{k+1}x^{\nu+2k}\zeta(2m-2k)}{2^{\nu+2k}k!\,\Gamma(\nu+k+1)}\right)\mathrm{d}x=\frac{y^\nu\pi}{2}\sum_{n=1}^{\infty}\frac{Y_\nu(ny)}{n^{2m+\nu}}.$$

Further, we have

$$\frac{y^\nu\pi}{2}\sum_{n=1}^{\infty}\frac{Y_\nu(ny)}{n^{2m+\nu}}=\frac{(-1)^{m+1}\pi}{2^{\nu+2m}\Gamma\left(m+\frac{1}{2}\right)\Gamma\left(m+\nu+\frac{1}{2}\right)}\int_0^{\infty}\frac{x^{2\nu+2m}}{x^2-y^2}\,dx$$
$$+\sum_{k=0}^{m}\frac{(-1)^{k+1}\zeta(2m-2k)}{2^{\nu+2k}k!\,\Gamma(\nu+k+1)}\int_0^{\infty}\frac{x^{2\nu+2k+1}}{x^2-y^2}\,dx.\tag{9.3}$$

Here we need to evaluate the right-hand integrals. We find an entry in [23, p. 423], containing the hypergeometric series ${}_3F_2$ which we write here in an expanded form

$$\int_0^a \frac{x^{\alpha-1}}{x^2-y^2} P_{2n+\epsilon}\left(\frac{x}{a}\right) dx$$
$$= \frac{(-1)^{\epsilon+1}\pi}{2} y^{\alpha-2}\, tg\frac{\alpha\pi}{2} 2\epsilon - 1 P_{2n+\epsilon}\left(\frac{y}{a}\right)$$
$$+ \frac{(-1)^n a^{\alpha-2}\left(\frac{3-\alpha+\epsilon}{2}\right)_n}{2\left(\frac{\alpha+\epsilon}{2}-1\right)_{n+1}} \sum_{k=0}^{\infty} \frac{\left(1-\frac{\alpha+\epsilon}{2}-n\right)_k \left(\frac{3-\alpha+\epsilon}{2}+n\right)_k}{\left(\frac{3-\alpha}{2}\right)_k \left(2-\frac{\alpha}{2}\right)_k} \left(\frac{y}{a}\right)^{2k},$$

where $P_n(x)$ is the Legendre polynomial, $\epsilon = 0$ or $\epsilon = 1$; $a > y > 0$, $\operatorname{Re}\alpha > -\epsilon$ and $(p)_n$ is the Pochhammer symbol. Knowing that $P_0(x) = 1$, we take $\epsilon = 0, n = 0$, and after a simplification, the previous formula becomes

$$\int_0^a \frac{x^{\alpha-1}}{x^2-y^2} dx = \frac{\pi y^{\alpha-2}}{2} \operatorname{ctg}\frac{\alpha\pi}{2} - a^{\alpha-2} \sum_{k=0}^{\infty} \frac{1}{2k+2-\alpha} \left(\frac{y}{a}\right)^{2k}.$$

If $0 < \alpha < 2$, we let $a \to +\infty$, and find

$$\int_0^{+\infty} \frac{x^{\alpha-1}}{x^2-y^2} dx = \frac{y^{\alpha-2}\pi}{2} ctg\frac{\alpha\pi}{2}. \tag{9.4}$$

So, replacing α first with $2\nu+2m+1$, then with $2\nu+2k+2$, we can evaluate the integrals in (9.3) by means of (9.4), so that we have

$$\frac{y^\nu \pi}{2} \sum_{n=1}^{\infty} \frac{Y_\nu(ny)}{n^{2m+\nu}} = \frac{(-1)^{m+1}\pi^2 y^{2\nu+2m-1}}{2^{\nu+2m+1}\Gamma\left(m+\frac{1}{2}\right)\Gamma\left(m+\nu+\frac{1}{2}\right)} \operatorname{ctg}\frac{(2\nu+2m+1)\pi}{2}$$
$$+ \frac{\pi y^{2\nu}}{2} \sum_{k=0}^{m} \frac{(-1)^{k+1}\zeta(2m-2k)}{2^{\nu+2k}k!\,\Gamma(\nu+k+1)} y^{2k} \operatorname{ctg}(\nu+k+1)\pi. \tag{9.5}$$

and obtain the required sum of the series over the Bessel functions of the second kind

$$\sum_{n=1}^{\infty} \frac{Y_\nu(ny)}{n^{2m+\nu}} = \frac{(-1)^m \pi y^{\nu+2m-1} \operatorname{tg}\nu\pi}{2^{\nu+2m}\Gamma\left(m+\frac{1}{2}\right)\Gamma\left(m+\nu+\frac{1}{2}\right)}$$
$$+ \sum_{k=0}^{m} \frac{(-1)^{k+1}\zeta(2m-2k)}{2^{\nu+2k}k!\,\Gamma(\nu+k+1)} y^{\nu+2k} \operatorname{ctg}\nu\pi.$$

As another example, we take (see entry 28 in [23, p. 182])

$$\int_0^{+\infty} \frac{x^{\nu+1}}{(x^2+z^2)^{\varrho}} J_\nu(nx)\,dx = \frac{n^{\varrho-1} z^{\nu-\varrho+1}}{2^{\varrho-1}\Gamma(\varrho)} K_{\nu-\varrho+1}(nz), \, n \in \mathbb{N},$$

where $K_\nu(z)$ ($\operatorname{Re} z > 0, -1 < \operatorname{Re}\nu < 2\operatorname{Re}\varrho - \frac{1}{2}$) is the modified Bessel function of the third kind, also called the MacDonald function. Repeating the preceding procedure, taking into account again (9.2), we have

$$\frac{z^{\nu-\varrho+1}}{2^{\varrho-1}\Gamma(\varrho)} \sum_{n=1}^{\infty} \frac{K_{\nu-\varrho+1}(nz)}{n^{\nu+2m-\varrho+1}} = \frac{(-1)^m \pi}{2^{\nu+2m}\Gamma\left(m+\frac{1}{2}\right)\Gamma\left(m+\nu+\frac{1}{2}\right)} \times$$

$$\int_0^{+\infty} \frac{x^{2\nu+2m}}{(x^2+z^2)^{\varrho}} + \sum_{k=0}^{m} \frac{(-1)^k \zeta(2m-2k)}{2^{\nu+2k} k!\,\Gamma(\nu+k+1)} \int_0^{+\infty} \frac{x^{2\nu+2k+1}}{(x^2+z^2)^{\varrho}}\,dx. \quad (9.6)$$

In [10, p. 325], entry 11, we find the integral

$$\int_0^{+\infty} x^{\mu-1}(1+\beta x^p)^{-\nu} dx = \frac{1}{p}\beta^{-\frac{\mu}{p}} \mathrm{B}\left(\frac{\mu}{2}, \nu - \frac{\mu}{2}\right), |\arg\beta| < \pi, \, p > 0,$$

with $0 < \operatorname{Re}\mu < p \operatorname{Re}\nu$, where we set $\beta = 1, p = 0, \nu = \varrho$, and take μ instead of $\mu - 1$, which yields

$$\int_0^{+\infty} \frac{x^{\mu}}{(1+x^2)^{\varrho}}\,dx = \frac{\Gamma\left(\frac{1+\mu}{2}\right)\Gamma\left(\varrho-\frac{1+\mu}{2}\right)}{2\Gamma(\varrho)}.$$

After the substitution $x z = y$, we get

$$\int_0^{+\infty} \frac{y^{\mu}}{(y^2+z^2)^{\varrho}}\,dy = \frac{z^{\mu+1-2\varrho}\Gamma\left(\frac{1+\mu}{2}\right)\Gamma\left(\varrho-\frac{1+\mu}{2}\right)}{2\Gamma(\varrho)}. \quad (9.7)$$

Now, we replace the variable y with x in (9.7), then take $\mu = 2\nu + 2m$ for the first integral in (9.6), and $\mu = 2\nu + 2k + 1$ for the second one. Thus, one obtains

$$\sum_{n=1}^{\infty} \frac{K_{\nu-\varrho+1}(nz)}{n^{2m+\nu-\varrho+1}} = \frac{(-1)^m \pi z^{2m+\nu-\varrho}\Gamma\left(\varrho-m-\nu-\frac{1}{2}\right)}{2^{2m+\nu-\varrho+2}\Gamma\left(m+\frac{1}{2}\right)}$$

$$+ \sum_{k=0}^{m} \frac{(-1)^k \zeta(2m-2k)\, z^{\nu+2k-\varrho+1}\Gamma(\varrho-\nu-k-1)}{2^{\nu+2k-\varrho+2} k!}.$$

In the sequel, we deal with the series over the product of modified Bessel functions of the first and second kinds. Here, we start with the formula (see entry 30 in [23, p. 182])

$$\int_0^\infty \frac{x^{-\nu}}{\left(x^2+z^2\right)^{\nu+\frac{1}{2}}} J_\nu(nx)\, dx = \frac{(2n)^\nu \Gamma(\nu+1)}{z^{2\nu}\Gamma(2\nu+1)} I_\nu\left(\frac{nz}{2}\right) K_\nu\left(\frac{nz}{2}\right), \quad n \in \mathbb{N},$$

where $I_\nu(z), K_\nu(z)$ ($\operatorname{Re} z > 0, \operatorname{Re}\nu > -\frac{1}{2}$) are modified Bessel functions of the first and second kinds. Applying the method as for the series over Neumann functions, we obtain

$$\int_0^\infty \frac{x^{-\nu}}{\left(x^2+z^2\right)^{\nu+\frac{1}{2}}} \sum_{n=1}^\infty \frac{J_\nu(nx)}{n^\alpha}\, \mathrm{d}x = \frac{2^{\nu-1}\Gamma(\nu)}{z^{2\nu}\Gamma(2\nu)} \sum_{n=1}^\infty \frac{I_\nu\left(\frac{nz}{2}\right) K_\nu\left(\frac{nz}{2}\right)}{n^{\alpha-\nu}}.$$

Taking $\alpha = \nu + 2m$ ($m \in \mathbb{N}_0$) and relying on the formula (9.2), we have

$$\begin{aligned}\frac{2^{\nu-1}\Gamma(\nu)}{z^{2\nu}\Gamma(2\nu)} \sum_{n=1}^\infty \frac{I_\nu\left(\frac{nz}{2}\right) K_\nu\left(\frac{nz}{2}\right)}{n^{2m}} &= \frac{(-1)^m \pi}{2^{\nu+2m}\Gamma\left(m+\frac{1}{2}\right)\Gamma\left(m+\nu+\frac{1}{2}\right)} \\ &\quad\times \int_0^\infty \frac{x^{2m-1}\, \mathrm{d}x}{\left(x^2+z^2\right)^{\nu+\frac{1}{2}}} \\ &\quad+ \sum_{k=0}^m \frac{(-1)^k \zeta(2m-2k)}{2^{\nu+2k} k!\, \Gamma(\nu+k+1)} \int_0^\infty \frac{x^{2k}\, \mathrm{d}x}{\left(x^2+z^2\right)^{\nu+\frac{1}{2}}}.\end{aligned} \tag{9.8}$$

We make use of the integral (9.7) with $\varrho = \nu + \frac{1}{2}$ and x instead of y to find

$$\int_0^\infty \frac{x^\mu\, \mathrm{d}x}{\left(x^2+z^2\right)^{\nu+\frac{1}{2}}} = \frac{z^{\mu-2\nu}\Gamma\left(\frac{1+\mu}{2}\right)\Gamma\left(\nu-\frac{\mu}{2}\right)}{2\Gamma\left(\nu+\frac{1}{2}\right)}$$

By applying this and setting $\mu = 2m-1$ in the first integral of (9.8), and $\mu = 2k$ in the second one, there follows

$$\begin{aligned}\sum_{n=1}^\infty \frac{I_\nu\left(\frac{nz}{2}\right) K_\nu\left(\frac{nz}{2}\right)}{n^{2m}} = \frac{\Gamma(2\nu)}{\Gamma(\nu)\Gamma\left(\nu+\frac{1}{2}\right)} &\left(\frac{(-1)^m z^{2m-1}\pi\Gamma(m)\Gamma\left(\nu-m+\frac{1}{2}\right)}{2^{2\nu+2m}\Gamma\left(m+\frac{1}{2}\right)\Gamma\left(m+\nu+\frac{1}{2}\right)} \right. \\ &\left. + \sum_{k=0}^m \frac{(-1)^k z^{2k}\zeta(2m-2k)\Gamma\left(k+\frac{1}{2}\right)\Gamma(\nu-k)}{2^{2\nu+2k} k!\, \Gamma(\nu+k+1)} \right).\end{aligned}$$

By using Legendre's duplication formula (4.9), we have

$$\frac{\Gamma(2\nu)}{\Gamma(\nu)\Gamma\left(\nu+\frac{1}{2}\right)}=\frac{2^{2\nu-1}}{\sqrt{\pi}},$$

so that the sum of the series over the product of modified Bessel functions of the first and second kinds becomes

$$\sum_{n=1}^{\infty}\frac{I_\nu\left(\frac{nz}{2}\right)K_\nu\left(\frac{nz}{2}\right)}{n^{2m}}=\frac{(-1)^m z^{2m-1}(m-1)!\,\Gamma\left(\nu-m+\frac{1}{2}\right)}{2^{m+1}(2m-1)!!\,\Gamma\left(m+\nu+\frac{1}{2}\right)}+\sum_{k=0}^{m}\frac{(-1)^k z^{2k}(2k-1)!!\zeta(2m-2k)\Gamma(\nu-k)}{2^{3k+1}k!\,\Gamma(\nu+k+1)}. \tag{9.9}$$

12.10 APPENDIX – TABLES

TABLE 12.1 General Formula and Closed-Form Cases, $\alpha = 2m + p - 1$

a	b	s	c	F	f	δ	p	Convergence region
					sin	1	0	
		1	1	ζ	cos	0	1	$0 < x < 2\pi$
					sin	1	0	
1	0	-1	0	η	cos	0	1	$-\pi < x < \pi$
					sin	1	0	
		1	$\frac{1}{2}$	λ	cos	0	1	$0 < x < \pi$
					sin	1	1	
2	1	-1	0	β	cos	0	0	$-\frac{\pi}{2} < x < \frac{\pi}{2}$

TABLE 12.2 Product of Two Functions Sine and Cosine – Closed Forms

a	b	s	c	F	f	g	α	Dual convergence regions
					sin	cos	$2m-1$	
					cos	sin	$2m-1$	
		1	1	ζ	sin	sin	$2m$	$K_1 = \{(x,y) \mid -\pi < x < \pi, \lvert x\rvert < y < 2\pi - \lvert x\rvert\}$
					cos	cos	$2m$	
					sin	cos	$2m-1$	
1	0				cos	sin	$2m-1$	
		-1	0	η	sin	sin	$2m$	$K_2 = \{(x,y) \mid -\pi < x < \pi, \lvert x\rvert - \pi < y < \pi - \lvert x\rvert\}$
					cos	cos	$2m$	
					sin	cos	$2m-1$	
					cos	sin	$2m-1$	
		1	$\frac{1}{2}$	λ	sin	sin	$2m$	$K_3 = \left\{(x,y) \mid -\frac{\pi}{2} < x < \frac{\pi}{2}, \lvert x\rvert < y < \pi - \lvert x\rvert\right\}$
					cos	cos	$2m$	
					sin	cos	$2m$	
2	1				cos	sin	$2m$	
		-1	0	β	sin	sin	$2m-1$	$K_4 = \left\{(x,y) \mid -\frac{\pi}{2} < x < \frac{\pi}{2}, \lvert x\rvert - \frac{\pi}{2} < y < \frac{\pi}{2} - \lvert x\rvert\right\}$
					cos	cos	$2m-1$	

TABLE 12.3 Convergence regions

a	b	s	c	F	Convergence regions
1	0	1	1	ζ	$K_5 = \left\{(x,y) \mid -\frac{\pi}{2} < x < \frac{\pi}{2}, 2\lvert x\rvert < y < 2\pi - 2\lvert x\rvert\right\}$
1	0	-1	0	η	$K_6 = \left\{(x,y) \mid -\frac{\pi}{2} < x < \frac{\pi}{2}, 2\lvert x\rvert - \pi < y < \pi - 2\lvert x\rvert\right\}$
2	1	1	$\frac{1}{2}$	λ	$K_7 = \left\{(x,y) \mid -\frac{\pi}{4} < x < \frac{\pi}{4}, 2\lvert x\rvert < y < \pi - 2\lvert x\rvert\right\}$
2	1	-1	0	β	$K_8 = \left\{(x,y) \mid -\frac{\pi}{4} < x < \frac{\pi}{4}, 2\lvert x\rvert - \frac{\pi}{2} < y < \frac{\pi}{2} - 2\lvert x\rvert\right\}$

REFERENCES

1. M. Abramowitz, A. Stegun, *Handbook of Mathematical Functions*, Cambridge Universiy Press, New York, 2010.
2. T. Arakawa, T. Ibukiyama, M. Kaneko, *Bernoulli Numbers and Zeta Functions*, Springer, Heidelberg, 2014.
3. F. W. Bessel, Untersuchung des Teils der planetarischen Störungen welcher aus der Bewegung der Sonne entsteht, *Berliner Abh.* (1826), 1–52.
4. J. Bourget, Mémoire sur les nombres de Cauchy et leur application à divers problèmes de mécanique céleste, *Journal de Math.* **6** (1861), 33–54.
5. J. Choi, H. M. Srivastava, A certain family of series associated with the

Zeta and related functions, *Hiroshima Math. J.* **32** (2002), 417–429.

6. S. Colombo, *Les Transformations de Mellin et de Hankel: Applications à la physique mathematique,* Centre National Recherche Scientifique, Paris, 1959.
7. P. J. De Doelder, On a series of a product of Bessel functions of integer order, *Simon Stevin, Wissen Natur* **34** Jahrgang, Aflev. II (okt 1960), 54–57.
8. M. L. Glasser, A Classclass of Besselbessel Summationssummations, *Mathematics of Computation* **37**, No. 156 (1981), 499–501
9. K. K. Gorowara, On Bourget's function $J_{n,k}(z)$, *Ganita* **22**, No. 1 (1971), 21–26.
10. I. S. Gradshtein, I. M. Ryzhik, *Tables of Integrals, Series and Products,* A. Jeffrey, D. Zwillinger (Eds.), Elsevier, Academic Press, 2007.
11. E. R. Hansen, *A Table of Series and Products*, Prentice-Hall, Inc., Englewood Cliffs, NJ, 1975.
12. K. Knopp, *Theory and Application of Infinite Series*, Dover, New York, 1990.
13. E. C. J. von Lommel, *Studien über die Besselschen Funktionen*, Teubner, Leipzig, 1868.
14. L. Lorch, P. Szego, *Integrals of Bessel functions*, McGraw-Hill Book Company, Inc, New York, 1962.
15. Y. L. Luke, Closed expressions for some infinite series of Bessel and Struve functions, *J. Math. Anal. Appl.* **122** (1987), 47–57.
16. Y. L. Luke, *The Special Functions and Their Approximations, Volume I,* Academic Press, New York and London, 1969.
17. V. Mangulis, *Handbook of Series for Scientists and Engineers*, TRG, Control Data Corporation, Melville, New York, 1965.
18. D. S. Mitrinović, *Special Functions, Collection of Problems*, Naučna knjiga, Beograd, 1986. (in Serbian)
19. L. J. Mordell, Poisson's summation formula and the Riemann zeta function, *J. Lond. Math. Soc.* **4** (1929) 285–291.
20. R. B. Paris, *Digital Library of Mathematical Functions, §11.10 Anger-Weber Functions*, https://dlmf.nist.gov/11.10.
21. R. B. Paris, *Digital Library of Mathematical Functions, §3.6 Linear Difference Equations*, https://dlmf.nist.gov/3.6.
22. S. D. Poisson, Mémoire sur la distribution de la chaleur dans les corps solides, *Journal de l'École Polytechnique* **XII** (cahier 19) (1823), 249–403.

23. A. P. Prudnikov, Y. A. Brychkov, O. I. Marichev, *Integrals and Series, Vol. 2, Special Functions*, Taylor Francis, London, 1998.
24. J. W. S. Rayleigh, On a physical interpretation of Schlömilch's theorem in Bessel's functions, *The London, Edinburgh, and Dublin Phil. Mag. J. Sci.* **21** (1911), 567–571.
25. P. Rusev, Representation of holomorphic functions by Schlömilch's series, *Frac. Calc. Appl. Math.* **16**, No. 2 (2013), 431–435.
26. O. X. Schlömilch, Über die Besselschen Funktion, *Zeitschr. Math. Phys. Zweiter Jahrgang* (1857), 137–165.
27. I. N. Sneddon, *The Use of Integral Transforms*, Mc.Graw Hill, New York, 1972.
28. H. M. Srivastava, A note on functions analogous to Bourget's function, *Ganita* **19**, No. 2 (1968), 45–48.
29. M. S. Stanković, D. M. Petković, M. V. Djurić, Short table of summable series of Bessel functions Conf. *Appl. Math.* **5**, Ljubljana (1986), 147–152.
30. M. S. Stanković, M. V. Djurić, D. M. Petković, Closed form expressions for some series over product of two trigonometric functions, *Facta Universitatis, Ser. Math. Inform.* **9** (1994), 69–82.
31. M. S. Stanković, M. V. Vidanović, S. B. Tričković, On the sSummation of sSeries iInvolving bBessel or sStruve functions, *J. Math. Anal. Appl.* **247** (2000), 15–26.
32. N. M. Temme, *Special Functions: An Introduction to the Classical Functions of Mathematical Physics* (2nd print ed.), Wiley, New York, 1996, 228–231.
33. D. -D. Tošić, Some series of product of Bessel functions, *Univ. Beograd. Publ. Elektrotehn. Fak. Ser. Mat. Fiz.* **678–715** (1980), 105–110.
34. S. B. Tričković, M. V. Vidanović, M.S. Stanković, Some series over the product of two trigonometric functions and series involving Bessel functions, *Z. Anal. Anwen.* **20**, No. 1 (2001), 235–246.
35. S. B. Tričković, M. V. Vidanović, M.S. Stanković, On the summation of series in terms of Bessel functions, *Z. Anal. Anwen.* **25**, No. 3 (2006), 393–406.
36. S. B. Tričković, M. V. Vidanović, M.S. Stanković, On the summation of series over a product of Bessel functions, *Int. Trans. Spec. Func.* **17**, No. 10 (2006), 749–758.
37. S. B. Tričković, M. S. Stanković, M.V. Vidanović, On the summation of trigonometric series, *Int. Trans. Spec. Func.* **19**, No. 6 (2008), 441–452.

38. S. B. Tričković, M.V. Vidanović M.S. Stanković, Summation of series over Bourget functions, *Canad. Math. Bull.* **51**, No. 4 (2008), 627–636.
39. S. B. Tričković, M. S. Stanković, M. V. Vidanović, On the summation of Schlömilch series, *Int. Trans. Spec. Func.* **31**, No. 5 (2020), 339–367.
40. S. B. Tričković, M. S. Stanković, On the closed form of Clausen functions, *Int. Trans. Spec. Func.* **34**, No. 6 (2023), 469–477.
41. G. N. Watson, *A Treatise on the Theory of Bessel Functions*, Cambridge University Press, Cambridge, 1966.

CHAPTER 13

Cross-Diffusion-Driven Instability and Non-Linear Analysis in a Spatio-Temporal Oncolytic Therapeutic Model

Fatiha Najm, Radouane Yafia, and M. A. Aziz Alaoui

13.1 INTRODUCTION AND MATHEMATICAL MODEL

Oncolytic virotherapy uses viruses as weapons to kill cancer cells. These viruses are known for their ability to select and infect cancer cells, within the cancer cells, to cause viral burden to the cell which in turn leads to cell death [5, 6] (this process is called oncolysis). The subsequent release of these additional infectious viruses causes the infection of surrounding cancer cells. Oncolytic virotherapy started in the beginning of the twentieth century (1910 [17]), when the first patient presented a significant regression after receiving the live attenuated rabies vaccine. In the 1940s, a few studies were done on different viruses [4], but the period 1990–2015 was very successful for developing some selective replicating viruses, and in this period, no oncolytic virus was used on humans. In 2015, the first oncolytic virus was approved for human use by the FDA [14, 16]. Oncolytic virotherapy can take place in two ways. Here, we consider that the virus spreads between cancer cells by direct transmission (cell to cell), and the growth of tumor

DOI: 10.1201/9781003530602-13

cells follows the at a rate r (for uninfected cells u) and s (for infected cells v). The maximum sizes of the two tumor populations u and v are given by the same carrying capacity k. β is the spread rate of the virus into the tumor site. The infected tumour cell population is killed by the virus at rate a. Naturally, it's known that the populations of infected cancer cells and uninfected cancer cells exist in a so-called well-mixed system, that is, each population of cells can interact with one another regardless of their spatial location. For this reason, we assume that their dynamics follow an activator-inhibitor model. The model is given by (see, [11])

$$\begin{cases} \frac{\partial u}{\partial t}(t,X) = d_{11}\Delta u(t,X) + d_{12}\Delta v(t,X) + F(u(t,X),\, v(t,X)) \\ \frac{\partial v}{\partial t}(t,X) = d_{21}\Delta u(t,X) + d_{22}\Delta v(t,X) + G(u(t,X),\, v(t,X)) \\ \frac{\partial u}{\partial \nu}(t,X) = \frac{\partial v}{\partial \nu}(t,X) = 0,\ (t,X) \in (0,+\infty) \times \partial\Omega, \\ u(0,X) = u_0(X) > 0,\ v(0,X) = v_0(X) > 0,\ X \in \Omega, \end{cases} \tag{1.1}$$

where

$$\begin{cases} F(u,v) = ru\left(1 - \frac{N}{k}\right) - du - \beta uv, \\ G(u,v) = \beta uv + sv\left(1 - \frac{N}{k}\right) - av, \\ N = u + v, \end{cases}$$

u is Tumor cells that are not infected by the virus, v is Tumor cells that are infected by the virus, and $N = u + v$ is Total number of cells in the tumor.

Model (1.1) was introduced and studied biologically and numerically. Based on this model, other authors introduced indirect transmission with the simplified case of exponential growth of infected tumor cells (see [9]). In [10], the authors introduced a modification of model (1.1) (without self/cross-diffusion) by integrating the viral life-cycle "delay," and they proved the existence of solutions using their theories. The rest of the chapter is organized as follows: In Section 13.2, we investigate the existence of possible steady states and their stability. In Section 13.3, we prove the global stability of the non-trivial equilibrium and the non-occurrence of Turing instability in the presence of self-diffusion and absence of cross-diffusion, and its occurrence in the presence of both. In Section 13.4, we study nonlinear analysis of the full model. We derive the corresponding equations and we study the stability of these. In Section 13.5, we illustrate some of the theoretical results.

13.2 STABILITY WITHOUT DIFFUSION

Let

$$R_1 = \beta k(a-s) + ar - sd;\ R = \beta(r+\beta k) - s;\ R_2 = \beta k(r-d) - ar + sd \tag{2.1}$$

The possible expressions are given by: $E_0 = (0,0)$ (tumor-free equilibrium TFE), $E_1 = (u_1, 0) = \left(\frac{k}{r}(r-d), 0\right)$, $E_2 = (0, v_2) = \left(0, \frac{k(s-a)}{s}\right)$, $E^*(= (u^*, v^*) = \left(\frac{R_1}{R}, \frac{R_2}{R}\right)$. Without a diffusion system. (1.1) is written as

$$\begin{cases} \frac{du(t)}{dt} = ru(t)\left(1 - \frac{N(t)}{k}\right) - du(t) - \beta u(t)v(t) = F(u(t), v(t)) \\ \frac{dv(t)}{dt} = \beta u(t)v(t) + sv(t)\left(1 - \frac{N(t)}{k}\right) - av(t) = G(u(t), v(t)) \\ u(0) = u_0 \geq 0,\ v(0) = v_0 \geq 0, \\ N = u + v. \end{cases} \tag{2.2}$$

By linearizing System (2.2) around an arbitrary equilibrium point $E = (u, v)$, we get

$$\frac{dU(t)}{dt} = \mathcal{A}^E U(t),\ U = \begin{pmatrix} u \\ v \end{pmatrix} \tag{2.3}$$

where

$$\mathcal{A}^E = \begin{pmatrix} F_u(E) = r\left(1 - \frac{N}{k}\right) - \frac{r}{k}u - d - \beta v & F_v(E) = -\frac{r}{k}u - \beta u \\ G_u(E) = \beta v - \frac{s}{k}v & G_v(E) = \beta u + s\left(1 - \frac{N}{k}\right) - \frac{s}{k}v - a \end{pmatrix}.$$

Then we deduce the stability of E_0, E_1, and E_2.

Proposition 2.1 *i) If $r < d$ and $s < a$, TFE E_0 is stable. ii) If $s > a$ and $r\frac{a}{s} - d - \beta v_2 < 0$, UFE equilibrium E_2 is and otherwise. iii) If $r > d$ and $\beta u_1 + s\frac{d}{r} - a < 0$, IFE equilibrium E_1 is stable .*

Consider the following hypotheses

(H_1)

$$\frac{\beta(r+\beta k)}{s} > 1,\ \frac{a(\beta k + r)}{s(\beta k + d)} > 1,\ \frac{\beta kr + sd}{\beta kd + ar} > 1.$$

(H_2)

$$\frac{\beta(r+\beta k)}{s}<1,\quad \frac{a(\beta k+r)}{s(\beta k+d)}<1,\quad \frac{\beta kr+sd}{\beta kd+ar}<1.$$

Proposition 2.2 *If one of the hypotheses* (H_1) *or* (H_2) *holds, the UIE equilibrium* E^* *exists and is unique.*

Since $tr\left(\mathcal{A}^{E^*}\right) = -\left(\frac{r}{k}u^* + \frac{s}{k}v^*\right) < 0$, then the expression for E^* is deduced from

$$\det\left(\mathcal{A}^{E^*}\right) = \beta\left(\beta + \frac{s-r}{k}\right)u^*v^* = P(\beta)\,u^*v^*.$$

$\beta_1 = \frac{r-s}{k}$ is a root of the polynomial P, and we get the following result.

Proposition 2.3 *Suppose that one of the hypotheses* (H_1) *or* (H_2) *is satisfied. i) If* $\beta_1 > 0$ *and* $0 < \beta < \beta_1$, *we have* $\det(\mathcal{L}) < 0$, *and the* E^* *is true for System (2.2), ii) If* $\beta > \max(0, \beta_1)$, *we have* $\det(\mathcal{L}) > 0$, *and the* E^* *is true for System (2.2).*

Proposition 2.4 *Suppose one of the hypotheses* (H_1) *or* (H_2) *and* $\beta > \frac{s}{k}$ *are satisfied, then* E^* *is true for System (2.2).*

Proof: Let

$$\begin{aligned} V(u,v) =& \left(\frac{r}{k}+\beta\right)^{-1}\left(u-u^*-u^*\ln\left(\frac{u}{u^*}\right)\right) \\ &+\left(\beta-\frac{s}{k}\right)^{-1}\left(v-v^*-v^*\ln\left(\frac{v}{v^*}\right)\right), \end{aligned} \tag{2.4}$$

V defines a base for System (2.2). Differentiating V along the domain of (2.2) with respect to time t and as $\beta > \frac{s}{k}$, we have

$$\frac{dV}{dt}(u,v) = -\left(\frac{r}{k}+\beta\right)^{-1}(u-u^*)^2 - \left(\beta-\frac{s}{k}\right)^{-1}(v-v^*)^2 \leq 0.$$

□

13.3 TURING INSTABILITY

13.3.1 Non-Turing Bifurcation without Cross-Diffusion

Consider Model (1.1) by linearizing System (1.1) around E^*

$$\frac{\partial \mathcal{Z}}{\partial t}(t, X) = \mathcal{A}^{E^*}\mathcal{Z}(t, X) + \mathcal{D}\Delta\mathcal{Z}(t, X), \tag{3.1}$$

where $\mathcal{D} = \begin{pmatrix} d_{11} & d_{12} \\ d_{21} & d_{22} \end{pmatrix}$ and $\mathcal{Z} = \begin{pmatrix} u - u^* \\ v - v^* \end{pmatrix}$. To obtain the conditions for bifurcation, we introduce the following for E^*

$$u(t, X) = u^* + \epsilon_1 \exp(\lambda_\kappa t + i\kappa_X X), \tag{3.2}$$

$$v(t, X) = v^* + \epsilon_2 \exp(\lambda_\kappa t + i\kappa_X X), \tag{3.3}$$

where $\kappa_X X = \kappa_x x + \kappa_y y$ and $\epsilon_j, j = 1, 2$ are two positive small real numbers, λ_κ is the of perturbations in time t, and $\kappa_X = (\kappa_x, \kappa_y)$, where κ_x, κ_y are the wave numbers in the x, y direction, respectively. The equation is given as

$$\lambda^2 - l(\kappa^2)\lambda + h(\kappa^2) = 0, \tag{3.4}$$

where

$$l(\kappa^2) = \mathrm{Tr}(\mathcal{A}_\kappa) = \mathrm{Tr}(\mathcal{A}^{E^*}) - \kappa^2(d_{11} + d_{22}) = \mathrm{Tr}(\mathcal{A}) - \kappa^2 \mathrm{Tr}(\mathcal{D}) < 0 \tag{3.5}$$

and

$$h(\kappa^2) = \mathrm{Det}(\mathcal{A}_\kappa) = d_{11}d_{22}\kappa^4 - \kappa^2(\mathcal{A}_{11}d_{22} + \mathcal{A}_{22}d_{11}) + \mathrm{Det}(\mathcal{A}^{E^*}) > 0. \tag{3.6}$$

From inequalities (3.5) and (3.6), we deduce that the UIE E^* is also for all values, which means there is no Turing instability.

13.3.2 Turing Instability Induced by Cross-Diffusion

Consider now Model (1.1) with cross-diffusion, and by the same arguments as before, we get the corresponding

$$\lambda^2 + \varrho_1(\kappa^2)\lambda + \varrho_2(\kappa^2) = 0 \tag{3.7}$$

where

$$\varrho_1\left(\kappa^2\right) = -\kappa^2\left(d_{11} + d_{22}\right) + tr\left(\mathcal{A}^{E^*}\right)$$

and

$$\begin{aligned}\varrho_2\left(\kappa^2\right) &= \det\left(\mathcal{D}\right)\kappa^4 - \left(d_{22}\mathcal{A}_{11} + d_{11}\mathcal{A}_{22} - d_{21}\mathcal{A}_{12} - d_{12}\mathcal{A}_{21}\right)k^2 + \det\left(\mathcal{A}^{E^*}\right)\\ &= \Phi\kappa^4 - \Psi k^2 + \Theta\\ &= \left(\kappa^2\sqrt{\Phi} - \frac{\Psi}{2\sqrt{\Phi}}\right)^2 + \left(\Theta - \frac{\Phi^2}{4\Psi}\right)\end{aligned}$$

where $\Phi = \det\left(\mathcal{D}\right)$, $\Psi = d_{22}\mathcal{A}_{11} + d_{11}\mathcal{A}_{22} - d_{21}\mathcal{A}_{12} - d_{12}\mathcal{A}_{21}$ and $\Theta = \det\left(\mathcal{A}^{E^*}\right)$.

$$\lambda_\kappa = \frac{1}{2}\left[-\varrho_1\left(\kappa^2\right) \pm \sqrt{\varrho_1(\kappa^2)^2 - 4\varrho_2\left(\kappa^2\right)}\right]. \tag{3.8}$$

Note that, Hopf bifurcation occurs if

$$\mathrm{Im}\left(\lambda_\kappa\right) \neq 0, \mathrm{Re}\left(\lambda_\kappa\right) = 0 \text{ at } \kappa = 0. \tag{3.9}$$

As $\mathrm{tr}\left(\mathcal{A}^{E^*}\right) < 0$ is always satisfied, there is no occurrence of Hopf bifurcation. This occurs when at least one of the following conditions is not satisfied.

$$\varrho_1\left(\kappa^2\right) < 0;\ \varrho_2\left(\kappa^2\right) < 0. \tag{3.10}$$

We can easily see that $\varrho_1\left(\kappa^2\right) < 0$ is not violated when the requirement $\mathrm{tr}\left(\mathcal{A}^{E^*}\right) < 0$ is met because we assume $d_{11} > 0$ and $d_{22} > 0$. Hence, only a violation of the condition $\varrho_2\left(\kappa^2\right) > 0$ will occur. Then, the condition for diffusive instability is given by

$$\Psi = d_{22}\mathcal{A}_{11} + d_{11}\mathcal{A}_{22} - d_{21}\mathcal{A}_{12}^{E^*} - d_{12}\mathcal{A}_{21} > 0,$$

where $\varrho_2\left(\kappa^2\right) > 0$ for all $\kappa > 0$ since $\det\left(\mathcal{D}\right) > 0$ and $\det\left(\mathcal{A}^{E^*}\right) > 0$. For Turing's instability, one needs $\varrho_2\left(\kappa^2\right) < 0$ for some κ, and the function $\varrho_2\left(\kappa^2\right)$ achieves its minimum

$$\begin{aligned}\min \varrho_2\left(\kappa^2\right) &= \frac{4\Phi\Theta - \Psi^2}{4\Phi}\\ &= \frac{4\det\left(\mathcal{D}\right)\det\left(\mathcal{A}^{E^*}\right) - \left(d_{22}\mathcal{A}_{11} + d_{11}\mathcal{A}_{22} - d_{21}\mathcal{A}_{12} - d_{12}\mathcal{A}_{21}\right)^2}{4\det\left(\mathcal{D}\right)},\end{aligned}$$

at the critical value of $\kappa^2_{\min} > 0$ when

$$\begin{aligned}\kappa^2_{\min} &= \frac{\Psi}{2\Phi}\\ &= \frac{d_{22}\mathcal{A}_{11} + d_{11}\mathcal{A}_{22} - d_{21}\mathcal{A}_{12} - d_{12}\mathcal{A}_{21}}{2\ \det(D)}.\end{aligned}$$

Therefore, if $\Psi > 0$ and $\varrho_2\left(\kappa^2\right) < 0$ satisfied, the E^* is an equilibrium with respect to Model (1.1). In this case, $\varrho_2\left(\kappa^2\right) = 0$ has two positive roots κ_1^2 and κ_2^2, which are

$$\kappa^2_{1,2} = \frac{\Psi \pm \sqrt{\Lambda}}{2\Phi},$$

where $\Lambda = \Psi^2 - 4\Phi\Theta$. Then, if we can find some κ^2 such that $\kappa_1^2 < \kappa^2 < \kappa_2^2$, then $\varrho_2\left(\kappa^2\right) < 0$. From the above analysis, we obtain the following result.

Theorem 3.1 *Assuming that the UIE positive equilibrium E^* exists if the following conditions are true*
(i)

$$d_{22}\mathcal{A}_{11} + d_{11}\mathcal{A}_{22} > d_{21}\mathcal{A}_{12} + d_{12}\mathcal{A}_{21},$$

(ii)

$$\Psi = d_{22}\mathcal{A}_{11} + d_{11}\mathcal{A}_{22} - d_{21}\mathcal{A}_{12} - d_{12}\mathcal{A}_{21} > 2\sqrt{\Phi\Theta} = 2\sqrt{\det(\mathcal{D})\ \det\left(\mathcal{A}^{E^*}\right)},$$

then the positive equilibrium E^* of Model (1.1) is if $\kappa_1^2 < \kappa^2 < \kappa_2^2$ for some κ.

For some fixed parameter values, this gives a d_{12}^T as the root of the following equation

$$\begin{aligned}&(d_{11}\mathcal{A}_{22} + d_{22}\mathcal{A}_{11} - d_{12}\mathcal{A}_{21} - d_{21}\mathcal{A}_{12})^2\\ &\quad - 4\left(\mathcal{A}_{11}\mathcal{A}_{22} - \mathcal{A}_{21}\mathcal{A}_{21}\right)\left(d_{11}d_{22} - d_{12}d_{21}\right) = 0,\end{aligned} \tag{3.11}$$

and we have

$$d_{12}^T = \frac{-\mathcal{J} \pm \sqrt{\mathcal{J}^2 - 4\mathcal{R}\mathcal{Q}}}{2\mathcal{R}}, \tag{3.12}$$

where

$$\mathcal{J} = 2\mathcal{A}_{21}d_{21}\mathcal{A}_{12} - 2\mathcal{A}_{21}\left(d_{11}\mathcal{A}_{22} + d_{22}\mathcal{A}_{11}\right) + 4d_{21}\left(\mathcal{A}_{11}\mathcal{A}_{22} - \mathcal{A}_{12}\mathcal{A}_{21}\right),$$

$$\mathcal{R} = (d_{11}\mathcal{A}_{22} + d_{22}\mathcal{A}_{11})^2 + (d_{21}\mathcal{A}_{12})^2 - 2d_{21}\mathcal{A}_{12}(d_{11}\mathcal{A}_{22} + d_{22}\mathcal{A}_{11})$$
$$- 4d_{11}d_{22}(\mathcal{A}_{11}\mathcal{A}_{22} - \mathcal{A}_{12}\mathcal{A}_{21}),$$

$$\mathcal{Q} = (\mathcal{A}_{21})^2,$$

Then, the critical κ_T is given by

$$\kappa_T = \sqrt{\frac{\mathcal{A}_{11}\mathcal{A}_{22} - \mathcal{A}_{12}\mathcal{A}_{21}}{d_{11}d_{22} - d_{12}d_{21}}} = \sqrt{\frac{\det(\mathcal{A})}{\det(\mathcal{D})}}. \tag{3.13}$$

13.4 NON-LINEAR ANALYSIS

The Turing structures described by three pairs of modes $(\kappa_i, -\kappa_i)$, $i = 1, 2, 3$, make an angle $\frac{2\pi}{3}$ between each pair, with k_i, $i = 1, 2, 3$ equal in length, and magnitude equal to κ_T (i.e. $|\kappa_i| = \kappa_T$, $i = 1, 2, 3$), such that

$$\kappa_1 + \kappa_2 = -\kappa_3,\ \kappa_2 + \kappa_3 = -\kappa_1,\ \kappa_1 + \kappa_3 = -\kappa_2 \tag{4.1}$$

and

$$\kappa_1.\kappa_2 = \kappa_2.\kappa_3 = \kappa_1.\kappa_3 = \kappa_T^2 \cos\left(\frac{2\pi}{3}\right) \tag{4.2}$$

and

$$\sum_{i=1}^{3} \kappa_i = 1 \tag{4.3}$$

We now use the standard technique to derive the coefficients of Model (1.1) near the onset of $d_{12} = d_{12}^T$. The solution of Model (1.1) can be expanded as

$$\begin{pmatrix} u \\ v \end{pmatrix} = \begin{pmatrix} u^* \\ v^* \end{pmatrix} + \sum_{j=1}^{3} \mathcal{W}_0 \left[\mathcal{A}_j \exp(i\kappa_j.r) + \overline{\mathcal{A}_j} \exp(-i\kappa_j.r)\right]. \tag{4.4}$$

Here, $\mathcal{W}_0$ defines the direction of the eigenmodes, and $\mathcal{A}_j = \left(\mathcal{A}_j^u, \mathcal{A}_j^v\right)^t$, $\overline{\mathcal{A}_j} = \left(\overline{\mathcal{A}_j^u}, \overline{\mathcal{A}_j^v}\right)^t$, respectively denote the amplitudes associated with modes

$\kappa_j, -\kappa_j$. This leads to $u = u - u^*$ and $v = v - v^*$. Then, the Taylor expansion of Model (1.1) is

$$\begin{cases} \frac{\partial u}{\partial t} = \mathcal{A}_{11}u + \mathcal{A}_{12}v + d_{11}\Delta u + d_{12}\Delta v + \frac{1}{2}\left(F_{20}u^2 + 2F_{11}uv\right) + o\left(|(u,v)|^2\right), \\ \frac{\partial v}{\partial t} = \mathcal{A}_{21}u + \mathcal{A}_{22}v + d_{21}\Delta u + d_{22}\Delta v + \frac{1}{2}\left(2G_{11}uv + G_{02}v^2\right) + o\left(|(u,v)|^2\right), \end{cases} \tag{4.5}$$

where

$$\begin{aligned} F_{20} &= -2\frac{r}{k}, \quad G_{20} = 0, \\ F_{11} &= -\frac{r}{k} - \beta, \quad G_{11} = \beta - \frac{s}{k}, \\ F_{02} &= 0, \quad G_{02} = -2\frac{s}{k}, \\ F_{ij} &= 0, \quad G_{ij} = 0, \ \text{for } i + j > 2. \end{aligned}$$

Then, System (4.5) can be written in the following form:

$$\frac{\partial}{\partial t}\begin{pmatrix} u \\ v \end{pmatrix} = \mathcal{L}\begin{pmatrix} u \\ v \end{pmatrix} + \mathcal{N}(u,v), \tag{4.6}$$

where $\mathcal{L}$ is the linear operator given by

$$\mathcal{L} = \begin{pmatrix} \mathcal{A}_{11} + d_{11}\Delta & \mathcal{A}_{12} + d_{12}\Delta \\ \mathcal{A}_{21} + d_{22}\Delta & \mathcal{A}_{21} + d_{22}\Delta \end{pmatrix}, \tag{4.7}$$

and $\mathcal{N}(u,v)$ is the nonlinear vector defined as

$$\mathcal{N}(u,v) = \begin{pmatrix} \frac{1}{2}\left(F_{20}u^2 + 2F_{11}uv\right) \\ \frac{1}{2}\left(2G_{11}uv + G_{02}v^2\right) \end{pmatrix}. \tag{4.8}$$

Near the threshold $d_{12} = d_{12}^T$, we use the following perturbation of d_{12}, u, v, and t with respect to ϵ

$$\begin{cases} u = \epsilon u_1 + \epsilon^2 u_2 + \epsilon^3 u_3 + o\left(\epsilon^3\right), \\ v = \epsilon v_1 + \epsilon^2 v_2 + \epsilon^3 v_3 + o\left(\epsilon^3\right), \\ t = t_0 + \epsilon t_1 + \epsilon^2 t_2 + o\left(\epsilon^2\right), \\ d_{12} - d_{12}^T = \epsilon d_{12}^1 + \epsilon^2 d_{12}^2 + \epsilon^3 d_{12}^3 + o\left(\epsilon^3\right). \end{cases} \tag{4.9}$$

Here, $|\epsilon|$ is sufficiently small. By differentiating $\mathcal{L}$ with respect to ϵ, we get

$$\mathcal{L} = \mathcal{L}^T + \left(d_{12} - d_{12}^T\right)\mathcal{H}, \tag{4.10}$$

where

$$\mathcal{L}^T = \begin{pmatrix} \mathcal{A}_{11} + d_{11}\Delta & A_{12} + d_{12}^T\Delta \\ \mathcal{A}_{21} + d_{21}\Delta & \mathcal{A}_{22} + d_{22}\Delta \end{pmatrix} \tag{4.11}$$

and

$$\mathcal{H} = \begin{pmatrix} 0 & \Delta \\ 0 & 0 \end{pmatrix}. \tag{4.12}$$

Remark 4.1 Considering $t \sim t_0$ and assuming that $\frac{\partial A_j}{\partial t}\Big|_{t=t_0} = 0 j = 1, 2, 3$ means that $A_j (j = 1, 2, 3)$ changes slowly with respect to time.

Substituting u and v defined in (4.9) in the nonlinear part $\mathcal{N}(u, v)$ of System (4.5), we get

$$\mathcal{N} = \epsilon^2 \mathcal{N}_2 + \epsilon^3 \mathcal{N}_3 + o\left(\epsilon^3\right), \tag{4.13}$$

where

$$\mathcal{N}_2 = \begin{pmatrix} \frac{1}{2}F_{20}u_1^2 + F_{11}u_1v_1 \\ G_{11}u_1y_1 + \frac{1}{2}G_{02}v_1^2 \end{pmatrix} \tag{4.14}$$

and

$$\mathcal{N}_3 = \begin{pmatrix} F_{20}u_1u_2 + F_{11}\left(u_1v_2 + u_2v_1\right) \\ G_{11}\left(u_1v_2 + u_2v_1\right) + G_{02}v_1v_2 \end{pmatrix}. \tag{4.15}$$

Substituting (4.9) into (4.5), then we get different orders of ϵ as follows:

First order of ϵ:

$$\mathcal{L}^T \begin{pmatrix} u_1 \\ v_1 \end{pmatrix} = \begin{pmatrix} 0 \\ 0 \end{pmatrix}, \tag{4.16}$$

where $(u_1, v_1)^t$ is the linear combination corresponding to the eigenvalue zero. The solution of the linear problem (4.16) satisfying the conditions is given by

$$\begin{pmatrix} u_1 \\ v_1 \end{pmatrix} = \begin{pmatrix} \varphi \\ 1 \end{pmatrix} \left(\sum_{j=1}^{3} W_j \exp\left(i\kappa_j.r\right) + \overline{W_j} \exp\left(-i\kappa_j.r\right) \right), \tag{4.17}$$

where $\varphi = -\frac{A_{12}-d_{12}^T\kappa_T^2}{A_{11}-d_{11}\kappa_T^2}$ or $\varphi = -\frac{A_{22}-d_{22}\kappa_T^2}{A_{21}-d_{21}\kappa_T^2}$, $|\kappa_j| = \kappa_T, W_j$ is corresponding to the mode of $\exp(i\kappa_j.r)$, $(j = 1, 2, 3)$. Similarly, for the order ϵ^2, we obtain

$$\mathcal{L}^T\begin{pmatrix} u_2 \\ v_2 \end{pmatrix} = \frac{\partial}{\partial t_1}\begin{pmatrix} u_1 \\ v_1 \end{pmatrix} - d_{12}^T\mathcal{H}\begin{pmatrix} u_1 \\ v_1 \end{pmatrix} - \mathcal{N}_2 \triangleq \begin{pmatrix} \mathcal{F}_u \\ \mathcal{F}_v \end{pmatrix}. \tag{4.18}$$

Furthermore, we have

$$\mathcal{H}\begin{pmatrix} \varphi \\ 1 \end{pmatrix}\left(\sum_{j=1}^{3} W_j \exp(i\kappa_j.r) + \overline{W_j}\exp(-i\kappa_j.r)\right) = \begin{pmatrix} \Delta v_1 \\ 0 \end{pmatrix}. \tag{4.19}$$

From the conditions, the vector of the function on the right-hand side of Equation (4.18) must be orthogonal to the zero of operator $\mathcal{L}_*^T$ which is the zero of $\mathcal{L}^T$. Then, the zero of $\mathcal{L}_*^T$ is

$$\begin{pmatrix} 1 \\ \psi \end{pmatrix}\left(\exp(-i\kappa_j.r) + c.c.\right),\ j = 1, 2, 3, \tag{4.20}$$

where $\psi = -\frac{A_{11}-d_{11}\kappa_T^2}{A_{21}-d_{21}\kappa_T^2}$ or $\psi = -\frac{A_{12}-d_{12}^T\kappa_T^2}{A_{22}-d_{22}\kappa_T^2}$ and $c.c$ are the conjugate term of $\exp(-i\kappa_j.r)$. Based on orthogonal conditions, we have

$$\begin{pmatrix} 1, \psi \end{pmatrix}\begin{pmatrix} \mathcal{F}_u^j \\ \mathcal{F}_v^j \end{pmatrix} = 0,\ j = 1, 2, 3, \tag{4.21}$$

where $\mathcal{F}_u^j$, $\mathcal{F}_v^j$ denote the coefficients corresponding to $\exp(i\kappa_j.r)$ in $\mathcal{F}_u$, $\mathcal{F}_v$ respectively. From Equation (4.21), we get the following equations:

$$(\varphi + \psi)\frac{\partial W_1}{\partial t_1} = -d_{12}^1\kappa_T^2 W_1 + 2(\mathfrak{m}_2 + \psi\mathfrak{n}_2)\overline{W_2}\,\overline{W_3}, \tag{4.22}$$

$$(\varphi + \psi)\frac{\partial W_2}{\partial t_1} = -d_{12}^1\kappa_T^2 W_2 + 2(\mathfrak{m}_2 + \psi\mathfrak{n}_2)\overline{W_1}\,\overline{W_3}, \tag{4.23}$$

$$(\varphi + \psi)\frac{\partial W_3}{\partial t_1} = -d_{12}^1\kappa_T^2 W_3 + 2(\mathfrak{m}_2 + \psi\mathfrak{m}_2)\overline{W_1}\,\overline{W_2}, \tag{4.24}$$

where $\mathfrak{m}_2 = \frac{1}{2}F_{20}\varphi^2 + F_{11}\varphi$ and $\mathfrak{n}_2 = G_{11}\varphi + \frac{1}{2}G_{02}$. From (4.22) to (4.24), we cannot conclude with any result regarding the asymptotic behaviors of pattern amplitude. Therefore, we will continue to the higher order, and when we substitute (4.17) into (4.18), we get

$$\begin{pmatrix} u_2 \\ v_2 \end{pmatrix} = \begin{pmatrix} \mathcal{U}_0 \\ \mathcal{V}_0 \end{pmatrix} + \sum_{j=1}^{3} \begin{pmatrix} \mathcal{U}_j \\ \mathcal{V}_j \end{pmatrix} \exp\left(ik_j.r\right) + \sum_{j=1}^{3} \begin{pmatrix} \mathcal{U}_{jj} \\ \mathcal{V}_{jj} \end{pmatrix} \exp\left(2i\kappa_j.r\right)$$

$$+ \begin{pmatrix} \mathcal{U}_{12} \\ \mathcal{V}_{12} \end{pmatrix} \exp\left(i\left(\kappa_1 - \kappa_2\right).r\right) + \begin{pmatrix} \mathcal{U}_{23} \\ \mathcal{V}_{23} \end{pmatrix} \exp\left(i\left(\kappa_2 - \kappa_3\right).r\right)$$
$$+ \begin{pmatrix} \mathcal{U}_{31} \\ \mathcal{V}_{31} \end{pmatrix} \exp\left(i\left(\kappa_3 - \kappa_1\right).r\right) + c.c. \tag{4.25}$$

The coefficients introduced in Equation (4.25) can be calculated by solving equations of $\exp(0)$, $\exp\left(i\kappa_j.r\right)$, $\exp\left(2i\kappa_j.r\right)$, and $\exp(i\left(\kappa_j - \kappa_l\right)$ with $j \neq l$ and $j, l = 1, 2, 3$. Then, we obtain $\mathcal{U}_j = \varphi \mathcal{V}_j$, $\begin{pmatrix} \mathcal{U}_0 \\ \mathcal{V}_0 \end{pmatrix} = \begin{pmatrix} u_{00} \\ v_{00} \end{pmatrix} \left(|\mathcal{W}_1|^2 + |\mathcal{W}_2|^2 + |\mathcal{W}_3|^2\right)$, $\begin{pmatrix} \mathcal{U}_{jj} \\ \mathcal{V}_{jj} \end{pmatrix} = \begin{pmatrix} u_{11} \\ v_{11} \end{pmatrix} \mathcal{W}_j^2$, $\begin{pmatrix} \mathcal{U}_{12} \\ \mathcal{V}_{12} \end{pmatrix} = \begin{pmatrix} u_{22} \\ v_{22} \end{pmatrix} \mathcal{W}_1 \overline{\mathcal{W}_2}$, $\begin{pmatrix} \mathcal{U}_{23} \\ \mathcal{V}_{23} \end{pmatrix} = \begin{pmatrix} u_{22} \\ v_{22} \end{pmatrix} \mathcal{W}_2 \overline{\mathcal{W}_3}$, and $\begin{pmatrix} \mathcal{U}_{31} \\ \mathcal{V}_{31} \end{pmatrix} = \begin{pmatrix} u_{22} \\ v_{22} \end{pmatrix} \mathcal{W}_3 \overline{\mathcal{W}_1}$. Furthermore,

$$\begin{pmatrix} u_{00} \\ v_{00} \end{pmatrix} = \frac{-2}{det\left(\mathcal{A}^{E^*}\right)} \begin{pmatrix} \mathcal{A}_{22}\mathfrak{m}_2 - \mathcal{A}_{12}\mathfrak{n}_2 \\ -\mathcal{A}_{21}\mathfrak{m}_2 + \mathcal{A}_{11}\mathfrak{n}_2 \end{pmatrix}$$

$$\begin{pmatrix} u_{11} \\ v_{11} \end{pmatrix} = \begin{pmatrix} \dfrac{-(\mathcal{A}_{22}-4d_{22}\kappa_T^2)\mathfrak{m}_2+(\mathcal{A}_{12}-4d_{12}^T\kappa_T^2)\mathfrak{n}_2}{(\mathcal{A}_{11}-4d_{11}\kappa_T^2)(\mathcal{A}_{22}-4d_{22}\kappa_T^2)-(\mathcal{A}_{12}-4d_{12}^T\kappa_T^2)(\mathcal{A}_{21}-4d_{21}\kappa_T^2)} \\[2ex] \dfrac{-(\mathcal{A}_{11}-4d_{11}\kappa_T^2)\mathfrak{n}_2+(\mathcal{A}_{21}-4d_{21}\kappa_T^2)\mathfrak{m}_2}{(\mathcal{A}_{11}-4d_{11}\kappa_T^2)(\mathcal{A}_{22}-4d_{22}\kappa_T^2)-(\mathcal{A}_{12}-4d_{12}^T\kappa_T^2)(\mathcal{A}_{21}-4d_{21}\kappa_T^2)} \end{pmatrix},$$

and

$$\begin{pmatrix} u_{22} \\ v_{22} \end{pmatrix} = 2 \begin{pmatrix} \dfrac{-(\mathcal{A}_{22}-3d_{22}\kappa_T^2)\mathfrak{m}_2+(\mathcal{A}_{12}-3d_{12}^T\kappa_T^2)\mathfrak{n}_2}{(\mathcal{A}_{11}-4d_{11}\kappa_T^2)(\mathcal{A}_{22}-3d_{22}\kappa_T^2)-(\mathcal{A}_{12}-3d_{12}^T\kappa_T^2)(\mathcal{A}_{21}-3d_{21}\kappa_T^2)} \\[2ex] \dfrac{-(\mathcal{A}_{11}-3d_{11}\kappa_T^2)\mathfrak{n}_2+(\mathcal{A}_{21}-3d_{21}\kappa_T^2)\mathfrak{m}_2}{(\mathcal{A}_{11}-3d_{11}\kappa_T^2)(\mathcal{A}_{22}-3d_{22}\kappa_T^2)-(\mathcal{A}_{12}-3d_{12}^T\kappa_T^2)(\mathcal{A}_{21}-3d_{21}\kappa_T^2)} \end{pmatrix}.$$

At ϵ^3, we have

$$\mathcal{L}^T\begin{pmatrix} u_3 \\ v_3 \end{pmatrix} = \frac{\partial}{\partial t_1}\begin{pmatrix} u_2 \\ v_2 \end{pmatrix} + \frac{\partial}{\partial t_2}\begin{pmatrix} u_1 \\ v_1 \end{pmatrix} - d_{12}^1\mathcal{H}\begin{pmatrix} u_2 \\ v_2 \end{pmatrix} - d_{12}^2\mathcal{H}\begin{pmatrix} u_1 \\ v_1 \end{pmatrix} - \mathcal{N}_3 \triangleq \begin{pmatrix} \mathcal{J}_u \\ \mathcal{J}_v \end{pmatrix}. \tag{4.26}$$

From the condition applied to (4.26), we have the following equations

$$(\varphi+\psi)\left(\frac{\partial \mathcal{V}_1}{\partial t_1} + \frac{\partial \mathcal{W}_1}{\partial t_2}\right) = -\kappa_T^2\left(d_{12}^1\mathcal{V}_1 + d_{12}^2\mathcal{W}_1\right) + 2\left(\mathfrak{m}_2 + \psi\mathfrak{n}_2\right)\left(\overline{\mathcal{W}_2\mathcal{V}_3} + \overline{\mathcal{W}_3\mathcal{V}_2}\right) - \left[\mathcal{G}_1|\mathcal{W}_1|^2 + \mathcal{G}_2\left(|\mathcal{W}_2|^2 + |\mathcal{W}_3|^2\right)\right]\mathcal{W}_1, \tag{4.27}$$

$$(\varphi+\psi)\left(\frac{\partial \mathcal{V}_2}{\partial t_1} + \frac{\partial \mathcal{W}_2}{\partial t_2}\right) = -\kappa_T^2\left(d_{12}^1\mathcal{V}_2 + d_{12}^2\mathcal{W}_2\right) + 2\left(\mathfrak{m}_2 + \psi\mathfrak{n}_2\right)\left(\overline{\mathcal{W}_3\mathcal{V}_1} + \overline{\mathcal{W}_1\mathcal{V}_3}\right) - \left[\mathcal{G}_1|\mathcal{W}_2|^2 + \mathcal{G}_2\left(|\mathcal{W}_1|^2 + |\mathcal{W}_3|^2\right)\right]\mathcal{W}_2, \tag{4.28}$$

$$(\varphi+\psi)\left(\frac{\partial \mathcal{V}_3}{\partial t_1} + \frac{\partial \mathcal{W}_3}{\partial t_2}\right) = -\kappa_T^2\left(d_{12}^1\mathcal{V}_3 + d_{12}^2\mathcal{W}_3\right) + 2\left(\mathfrak{m}_2 + \psi\mathfrak{n}_2\right)\left(\overline{\mathcal{W}_1\mathcal{V}_2} + \overline{\mathcal{W}_2\mathcal{V}_1}\right) - \left[\mathcal{G}_1|\mathcal{W}_3|^2 + \mathcal{G}_2\left(|\mathcal{W}_1|^2 + |\mathcal{W}_2|^2\right)\right]\mathcal{W}_3, \tag{4.29}$$

where

$$-\mathcal{G}_1 = (\varphi F_{20} + F_{11})(u_{00} + u_{11}) + (F_{02} + \varphi F_{11})(v_{00} + v_{11}) + \psi\left[(\varphi G_{20} + G_{11})(u_{00} + u_{11}) + (G_{02} + \varphi G_{11})(v_{00} + v_{11})\right] \tag{4.30}$$

and

$$-\mathcal{G}_2 = (\varphi F_{20} + F_{11})(u_{00} + u_{22}) + (F_{02} + \varphi F_{11})(v_{00} + v_{22}) + \psi\left[(2\varphi G_{20} + G_{11})(u_{00} + u_{22}) + (2G_{02} + \varphi G_{11})(v_{00} + v_{22})\right]. \tag{4.31}$$

13.4.1 Amplitude Equations

Now, we derive an amplitude equation, which is very important to describe the dynamics of Model (1.1) for some values near the Turing bifurcation curve. From Remark 4.1, the equation of the $\mathcal{A}_j$, $(j = 1, 2, 3)$ is given by

$$\frac{\partial \mathcal{A}_j}{\partial t} = \epsilon \frac{\partial \mathcal{A}_j}{\partial t_1} + \epsilon^2 \frac{\partial \mathcal{A}_j}{\partial t_2} + o\left(\epsilon^2\right), \tag{4.32}$$

and

$$\mathcal{A}_j = \epsilon \mathcal{W}_j + \epsilon^2 \mathcal{V}_j + o\left(\epsilon^2\right). \tag{4.33}$$

Therefore, from the above analysis and by neglecting terms containing ϵ, including terms of order ϵ^3 and higher, the a Amplitude equations of $\mathcal{A}_j$, $(j = 1, 2, 3)$ are given by

$$\begin{cases} \zeta_0 \frac{\partial \mathcal{A}_1}{\partial t} = \xi \mathcal{A}_1 + \chi \overline{\mathcal{A}_2 \mathcal{A}_3} - (h_1 |\mathcal{A}_1|^2 + h_2 (|\mathcal{A}_2|^2 + |+|\mathcal{A}_3|^2))\mathcal{A}_1, \\ \zeta_0 \frac{\partial \mathcal{A}_2}{\partial t} = \xi \mathcal{A}_2 + \chi \overline{\mathcal{A}_1 \mathcal{A}_3} - (h_1 |\mathcal{A}_2|^2 + h_2 (|\mathcal{A}_1|^2 + |+|\mathcal{A}_3|^2))\mathcal{A}_2, \\ \zeta_0 \frac{\partial \mathcal{A}_3}{\partial t} = \xi \mathcal{A}_3 + \chi \overline{\mathcal{A}_1 \mathcal{A}_2} - (h_1 |\mathcal{A}_3|^2 + h_2 (|\mathcal{A}_1|^2 + |+|\mathcal{A}_2|^2))\mathcal{A}_3, \end{cases} \tag{4.34}$$

where

$$\xi = \frac{d_{12} - d_{12}^T}{d_{12}^T}, \quad \zeta_0 = -\frac{\varphi + \psi}{d_{12}^T \kappa_T^2}, \tag{4.35}$$

$$h_1 = -\frac{\mathcal{G}_1}{d_{12}^T \kappa_T^2}, \quad h_2 = -\frac{\mathcal{G}_2}{d_{12}^T \kappa_T^2}, \quad \chi = -\frac{2(\mathfrak{m}_2 + \psi \mathfrak{n}_2)}{d_{12}^T \kappa_T^2}. \tag{4.36}$$

13.4.2 Analysis of the Amplitude Equations

Next, suppose that each term in (4.34) can be decomposed to

$$\mathcal{A}_j = \rho_j \exp\left(i\theta_j\right), \; j = 1, 2, 3, \tag{4.37}$$

where ρ_j and $\theta_j, j = 1, 2, 3$ represent the mode and the corresponding phase angle, respectively, and $\theta = \sum_{j=1}^{3} \theta_j$. Therefore, we get

$$\begin{aligned}\frac{\partial\theta}{\partial t} &= \frac{1}{i}\sum_j \frac{1}{\mathcal{A}_j}\frac{\delta \mathcal{A}_j}{\delta t}\\ &= -\frac{\chi}{\zeta_0}\left(\rho_1^2\rho_3^2 + \rho_2^2\rho_3^2 + \rho_1^2\rho_2^2\right)\left(\rho_1^2\rho_2^2\rho_3^2\right)^{-1}(\sin(\theta) + i\cos(\theta))\\ &\quad + (i\zeta_0)^{-1}\left(3\mu - (h_1 + 2h_2)\left(|\mathcal{A}_1|^2 + |\mathcal{A}_2|^2 + |\mathcal{A}_3|^2\right)\right)\end{aligned}$$

The real part is obtained as follows:

$$\zeta_0\frac{\partial\theta}{\partial t} = -\chi\left(\rho_1^2\rho_3^2 + \rho_2^2\rho_3^2 + \rho_1^2\rho_2^2\right)\left(\rho_1^2\rho_2^2\rho_3^2\right)^{-1}\sin(\theta) \tag{4.38}$$

By substituting Equation (4.37) into System (4.34) and by separating the real and imaginary parts, we get the following system

$$\begin{cases}\zeta_0\frac{\partial\theta}{\partial t} = -\chi\frac{\rho_1^2\rho_2^2+\rho_2^2\rho_3^2+\rho_3^2\rho_1^2}{\rho_1\rho_2\rho_3}\sin(\theta)\\ \zeta_0\frac{\partial\rho_1}{\partial t} = \xi\rho_1 + \chi\rho_2\rho_3\cos(\theta) - h_1\rho_1^3 - h_2\left(\rho_2^2 + \rho_3^2\right)\rho_1,\\ \zeta_0\frac{\partial\rho_2}{\partial t} = \xi\rho_2 + \chi\rho_1\rho_3\cos(\theta) - h_1\rho_2^3 - h_2\left(\rho_1^2 + \rho_3^2\right)\rho_2,\\ \zeta_0\frac{\partial\rho_3}{\partial t} = \xi\rho_3 + \chi\rho_1\rho_2\cos(\theta) - h_1\rho_3^3 - h_2\left(\rho_1^2 + \rho_2^2\right)\rho_3,\end{cases} \tag{4.39}$$

where ζ_0, ξ, h_1, h_2, and χ are defined in (4.35) and (4.36). Then, System (4.39) has four types of stationary states defined by the following four expressions seen in Table 13.1.

- The equation is given by $\rho_1 = \rho_2 = \rho_3 = 0$.
- The representation is given by $\rho_1 = \rho_u, \rho_2 = \rho_3 = 0$.
- The representation by $\rho_1 = \rho_2 = \rho_3 = \rho_h$, with $\theta = 0$ or π (H_0 or H_π).
- The are given by $\rho_1 = \rho_{m1}, \rho_2 = \rho_3 = \rho_{m2}$.

13.5 NUMERICAL SIMULATIONS AND CONCLUSIONS

In this section, we give some numerical simulations illustrating the use of two sets of parameter values:

TABLE 13.1 Stability Analysis of Pattern Solution

Steady state	Formula	Existence	Stability
Stationary state	$\rho_1 = \rho_2 = \rho_3 = 0,$	Always	$\xi < \xi_2 = 0$; stable $\xi > \xi_2 = 0$; unstable
Stripe pattern	$\rho_1 = \sqrt{\frac{\xi}{h_1}} \neq 0, \rho_2 = \rho_3 = 0$	$\xi > 0$	$\xi > \xi_c = \chi^2 \frac{h_1}{(h_2 - h_1)^2}$; stable $\xi < \xi_c = \chi^2 \frac{h_1}{(h_2 - h_1)^2}$; unstable
Hexagonal pattern	$\rho_h^+ = \frac{\chi + \sqrt{\chi^2 + 4(h_1 + 2h_2)\xi}}{2(h_1 + 2h_2)}$	$\xi > \xi_1$	$\xi < \xi_4$; stable
	$\rho_h^- = \frac{\chi - \sqrt{\chi^2 + 4(h_1 + 2h_2)\xi}}{2(h_1 + 2h_2)}$		Always unstable
Mixed structure	$\rho_{m1} = \frac{\lvert\chi\rvert}{h_2 - h_1}$	$h_2 > h_1$	Always unstable
	$\rho_{m2} = \rho_{m3} = \sqrt{\frac{\xi - h_1 \rho_{m1}^2}{h_1 + h_2}}$	$\xi > h_1 \rho_{m1}^2$	

Set 1: The first set we will use was given in a book by Wodarz et al. $r = 4; s = 0.1; d = 0.5; k = 10; \beta = 0.8; a = 0.6$, and $\Omega = [0, l] \times [0, l] \subset \mathbb{R}^2$ with $l = 300$, space step $h = 2$, time step $\Delta t = 0.1$. The initial data is around the $E^* = (u^*, v^*)$ as follows,

$$u = u^* + \vartheta_1 (X); \; v = v^* + \vartheta_2 (X) \tag{5.1}$$

with $\vartheta_i \in [-1, 1], i = 1, 2; X \in \Omega$; see Figures 13.1 and 13.2.

Set 2: This second parameter set was introduced in [8, 12, 1, 2, 7] and is summarized in Table 13.2 with time step $\Delta t = 0.1$ and space step $h = 1$; see Figure 13.3.

In this work, we have studied the spatio-temporal dynamics of an interaction model under Neumann boundary conditions. We have shown that cross-diffusion can be induced by the presence of cross-diffusion by considering the infected coefficient as a parameter, and in the potential behavior ofthere is novia thegave an idea about the possiblefor the i near somevalue. One of our goals is to study the influence of these coefficients on the model. We can consider the coefficients of positive, negative, or zero values. A positive coefficient means that the spatial movement of the population moves toward a lower concentration of another population, while a negative coefficient means that one population moves toward a higher concentration (see Gui-Quan, 2012 [3]). Numerical simulations are carried out to illustrate different spatial distributions of concentrations of uninfected and infected

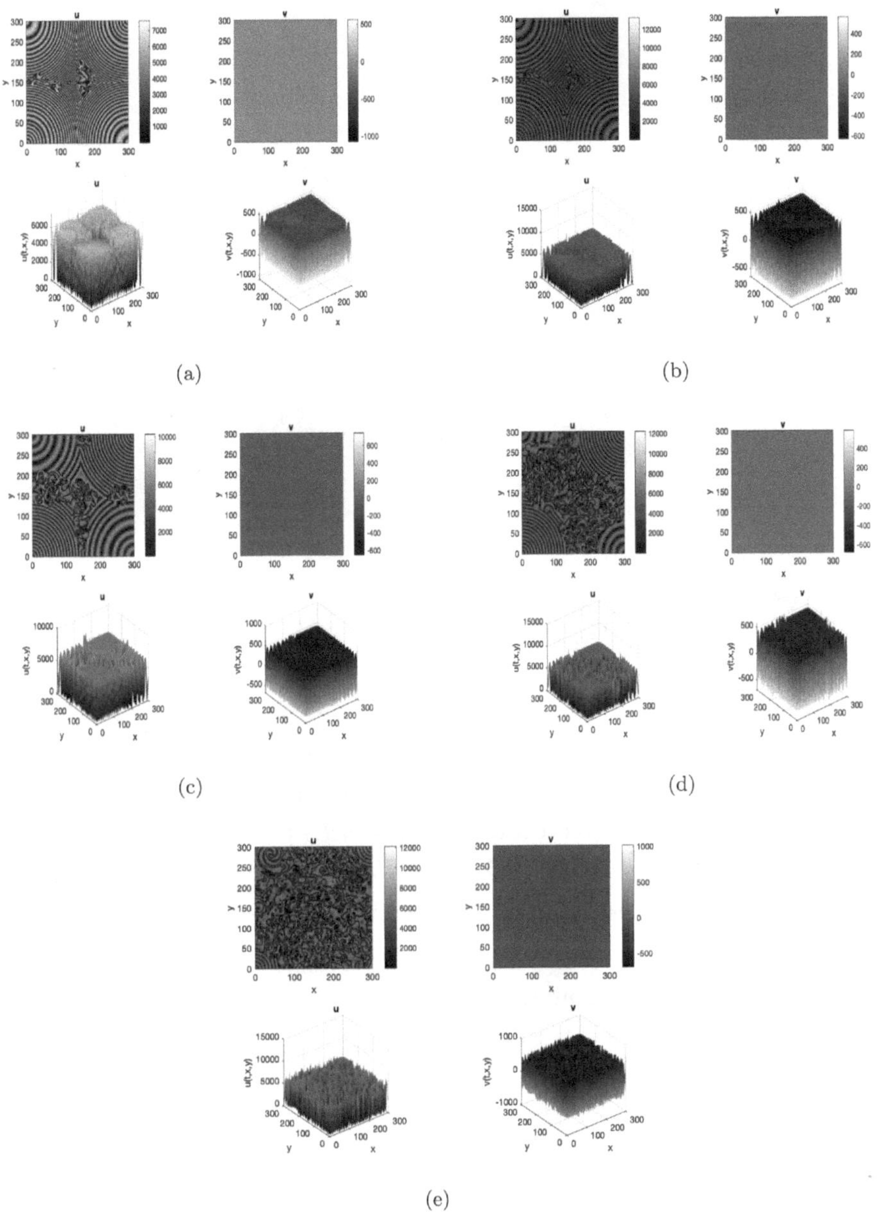

FIGURE 13.1 Simulated spatial patterns in uninfected and infected tumor cell concentrations with diffusion coefficients $d_{11} = 2; d_{12} = -120; d_{21} = -0.5; d_{22} = 0.05$ and time (a) $t = 200$, (b) $t = 300$, (c) $t = 500$, (d) $t = 800$, (e) $t = 3000$. Stripes are no longer stable.

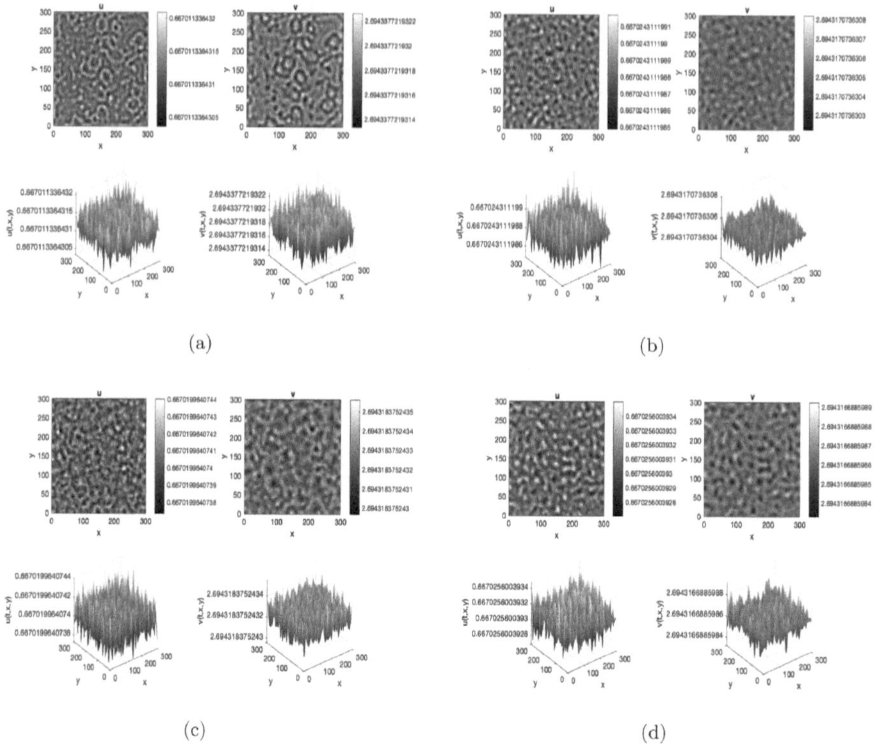

FIGURE 13.2 Simulated spatial patterns in uninfected and infected tumor cell concentrations with diffusion coefficients $d_{11} = 6; d_{12} = -140; d_{21} = -0.9; d_{22} = 0.3$ and time (a) $t = 200$, (b) $t = 300$, (c) $t = 500$, (d) $t = 800$. Spots are no longer stable.

tumor cells, which vary from spots to stripes, or a mixture of both, and their changes over time.

TABLE 13.2 Parameters and Initial Conditions

Symbol	Value (2D)	Reference
k	66 cells mm^{-1}	[12]
r	$0.0002 - 0.02\ h^{-1}$	[1]
s	$0.0002 - 0.02\ h^{-1}$	[1]
d_{11}	$3.6 \times 10^{-6} mm^2 h^{-1}$	[1]
d_{22}	$3.6 \times 10^{-3} mm^2 h^{-1}$	[13]
d_{12}	$- - - mm^2 h^{-1}$	estimated
d_{21}	$- - - mm^2 h^{-1}$	estimated
a	$0.056\ h^{-1}$	[2]
β	$7 \times 10^{-9} mm^2$	[7]
u_0	$1 - 3600$ cells mm^{-2}	[12]
v_0	3.2×10^8 viruses mm^{-2}	[8]

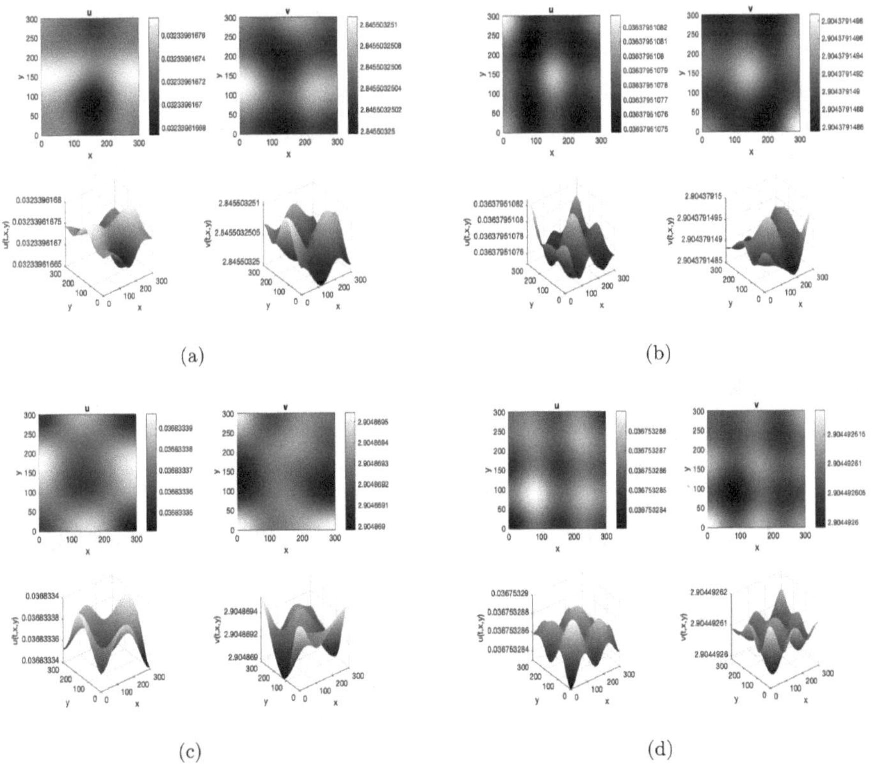

FIGURE 13.3 Simulated spatial patterns in uninfected and infected tumor cell concentrations with parameter values $r = 4; s = 0.1; d = 0.5; k = 10; \beta = 0.8; a = 0.1$, and with diffusion coefficients $d_{11} = 1; d_{12} = -0.5; d_{22} = 5; d_{21} = -70$, and time (a) $t = 200$, (b) $t = 400$, (c) $t = 500$, (d) $t = 600$.

REFERENCES

1. A. Friedman and W. Hao, The role of exosomes in pancreatic cancer microenvironment. *Bulletin of Mathematical Biology,* Vol. 80, No. 5, pp. 1111–1133 (2018).
2. A. Friedman , J. L. Tian, G. Fulci, E. A. Chiocca and J. Wang, Glioma virotherapy: Effects of innate immune suppression and increased viral replication capacity. *Cancer Research,* Vol. 66, No. 4, pp. 2314–2319 (2006).
3. S. Gui-Quan, J. Zhen, L. Li, H. Mainul and L. Bai-Lian, Spatial patterns of a predator-prey model with cross diffusion. *Nonlinear Dynamics,* Vol. 69, pp. 1631–1638 (2012).
4. H. A. Hoster, R. P. Zanes Jr, and E. Von Haam, Studies in Hodgkin's syndrome; The association of viral hepatitis and Hodgkin's disease; a preliminary report. *Cancer Research*, Vol. 9, pp. 473–480 (1949).
5. M. M. Hulou, C. F. Cho, E. A. Chiocca and R. Bjerkvig, 11 - Experimental therapies: Gene therapies and oncolytic viruses, in Gliomas. *Handbook of Clinical Neurology*, vol. 134. Elsevier, Amsterdam, 2016, pp. 183–197.
6. A. L. Jenner, A. C. F. Coster, P. S. Kim and F. Frascoli, Treating cancerous cells with viruses: Insights from a minimal model for oncolytic virotherapy. *Letters in Biomathematics*, Vol. 5 (Sup 1), pp. S117–S136 (2018).
7. M. S. Kathleen, E. L. Sean and T. L. Jackson, Modeling oncolytic viral therapy, immune checkpoint inhibition, and the complex dynamics of innate and adaptive immunity in glioblastoma treatment. *Frontiers in Physiology,* Vol. 11, p. 151 (2020).
8. J. H. Kim, Y. S. Lee, H. Kim, J. H. Huang, A. R. Yoon and C. O. Yun, Relaxin expression from tumor-targeting adenoviruses and its intratumoral spread, apoptosis induction, and efficacy. *Journal of the National Cancer Institute,* Vol. 98, No. 20, pp. 1482–1493 (2006).
9. Q. Li and Y. Xi, Modeling the virus-induced tumor-specific immune response with delay in tumor virotherapy. *Communications in Nonlinear Science and Numerical Simulation*, Vol. 10, p. 106196 (2022).
10. F. Najm, R. Yafia and M. A. Aziz-Alaoui, Hopf bifurcation in oncolytic therapeutic modeling: Viruses as anti-tumor means with viral lytic cycle. *International Journal of Bifurcation and Chaos,* Vol. 32, No. 11, p. 2250171 (2022).

11. Najm, F. , Yafia, R., Aziz Alaoui, M. A.: Turing ifurcation induced by cross-diffusion and amplitude equation in oncolytic therapeutic model: Viruses as anti-tumor means. *International Journal of Bifurcation and Chaos,* Vol. 33, No. 05, p. 2350062 (2023).
12. A. V. Nguyen, K. D. Nyberg, M. B. Scott, A. M. Welsh, A. H. Nguyen, N. Wu, S. V. Hohlbauch, N. A. Geisse, E. A. Gibb, A. G. Robertson et al., Stiffness of pancreatic cancer cells is associated with increased invasive potential. *Integrative Biology,* Vol. 8, No. 12, pp. 1232–1245 (2016).
13. A. Pluen, Y. Boucher, S. Ramanujan, T. D. McKee, T. Gohongi, E. di Tomaso, E. B. Brown, Y. Izumi, R. B. Campbell, D. A. Berk et al., Role of tumor–host interactions in interstitial diffusion of macromolecules: Cranial vs. subcutaneous tumors. *Proceedings of the National Academy of Sciences,* Vol. 98, No. 8, pp. 4628–4633 (2001).
14. J. Pol, G. Kroemer and L. Galluzzi, First oncolytic virus approved for melanoma immunotherapy. *Oncoimmunology,* Vol. 5, No. 1 (December 8, 2015) Article no. e1115641.
15. M. H. Protter and H. F. Weinberger, *Maximum Principles in Differential Equations.* Prentice Hall, Englewood Cliffs, NJ, 1967.
16. L. Russell and K. Peng, The emerging role of oncolytic virus therapy against cancer. *Chinese Clinical Oncology*, Vol. 7, No. 2, p. 16 (2018).
17. J. G. Sinkovics and J. C. Horvath, Natural and genetically engineered viral agents for oncolysis and gene therapy of human cancers. *Archivum Immunologiae et Therapiae Experimentalis (Warsz),* Vol. 56 (Suppl 1), pp. 3s–59s (2008).

CHAPTER 14

From Metric Spaces to O-Metric Spaces: Generalizing the Metrical Triangle Inequality

Hallowed O. Olaoluwa, Aminat O. Ige, and Johnson O. Olaleru

14.1 INTRODUCTION

The concept of metric spaces constitute one of the most general and suitable frameworks to model mathematical and physical problems, with axioms made to replicate the physical (at most three-dimensional) properties of distances. The Euclidean distance $d(x,y) = \sqrt{|x_1 - y_1|^2 + |x_2 - y_2|^2 + |x_3 - y_3|^2}$ between any two points $x = (x_1, x_2, x_3)$ and $y = (y_1, y_2, y_3)$ in the space $\mathbb{R}^3$ can be shown to satisfy the classical triangle inequality which is a special case of Minkowski's inequality. The space $C([a,b], \mathbb{R})$ of continuous real functions defined on a closed interval $[a,b]$ is also a metric space when endowed with the metric space $d(f,g) = \int_a^b |f(t) - g(t)|\ dt$, in what is an example of how the abstraction of metric spaces fits into a solution space for differential equations. In recent years, several generalizations of metric spaces have been considered, some of which focus on weakening or modifying altogether the triangle inequality for practical or theoretical purposes. In the sequel, we list only a few for reference.

 DOI: 10.1201/9781003530602-14

14.1.1 The Class of b-Metric Spaces and Some Extensions

Definition 1.1 [1] Let X be a nonempty set, and $s \geq 1$ a given positive real number. If the mapping $d : X \times X \longrightarrow [0, \infty)$ satisfies: $(d_1) d(x, y) = 0$ if and only if $x = y$, $(d_2) d(x, y) = d(y, x)$ for all $x, y \in X$, and $(d_3) d(x, y) \leq s[d(x, z) + d(z, y)]$ for all $x, y, z \in X$ (s-relaxed triangle inequality), then, d is called a b-metric on X and the triple (X, d, s) is called a b-metric space.

Besides the plurality of its applicability, the framework of b-metric spaces is remarkable due to the simplicity in the modification of the metrical triangle inequality: the inequality is weakened by the introduction of a multiplicative factor $s \geq 1$. A notable example of a b-metric space is the space l_p of p-summable sequences (where $p \geq 1$) with the distance-like function $d(x, y) = \sum_{n=1}^{\infty} |x_n - y_n|^p$ for $x = (x_n)_{n \in \mathbb{N}}$ and $y = (y_n)_{n \in \mathbb{N}}$ in l_p. Such function is a b-metric with multiplying factor $s = 2^p$, and generates the same topology as the classical metric $d_p = d^{\frac{1}{p}}$. Parvaneh and Ghoncheh in [2] went further to replace the multiplicative factor in the triangle inequality with an increasing function as follows.

Definition 1.2 [2] Let X be a nonempty set. A function $\tilde{d} : X \times X \rightarrow [0, \infty)$ is a p-metric if there exists a strictly increasing continuous function $\Omega : [0, \infty) \rightarrow [0, \infty)$ with $t \leq \Omega(t)$ for all $t \geq 0$ such that for all $x, y, z \in X$, the following conditions hold: $(p_1) \tilde{d}(x, y) = 0$ if $x = y$, $(p_2) \tilde{d}(x, y) = \tilde{d}(y, x)$, $(p_3) \tilde{d}(x, z) \leq \Omega\left(\tilde{d}(x, y) + \tilde{d}(y, z)\right)$. The pair $\left(X, \tilde{d}\right)$ is called a p-metric space, or, an extended b-metric space.

The class of p-metric spaces is larger than the class of b-metric spaces since any b-metric with multiplicative factor s is a p-metric with Ω such that $\Omega(t) = st$ for all $t \geq 0$. Mlaiki, Aydi, Souayah, and Abdeljawad [3] include a control function η, which, unlike in the cases of b-metrics and p-metrics, changes in function of the pair of elements whose distance it multiplies:

Definition 1.3 [3] Let X be a nonempty set and $\eta : X \times X \rightarrow [1, \infty)$ a given map. The function $d : X \times X \rightarrow [0, \infty)$ is called a controlled metric type if for all $x, y, z \in X$, the following conditions hold: $(\Omega_1) d(x, y) = 0$ if $x = y$, $(\Omega_2) d(x, y) = d(y, x)$, $(\Omega_3) d(x, y) \leq \eta(x, z) d(x, z) + \eta(z, y) d(z, y)$. The pair (X, d) is called a controlled metric type space.

It should be noted that, in the case where $\eta(x,z) = \eta(z,y) = \eta(x,y)$ for all $x,y,z \in X$ in inequality (Ω_3), the pair (X,d) is called an extended b-metric space by the authors in [4, 5]. By definition, b-metric spaces are controlled metric type spaces when the control function η is taken to be the constant function $\eta(x,y) = s$ for all $x,y \in X$.

14.1.2 θ-Metric Spaces

Earlier in 2013, Khojasteh, Karapinar, and Radenovic [6] introduced the concept of θ-metric spaces, where θ is defined with the following axioms:

Definition 1.4 [6] A function $\theta : [0,\infty)\times[0,\infty) \to [0,\infty)$ is said to be a B-action if it is continuous with respect to each of its arguments, and such that: (i) $\theta(0,0) = 0$ and $\theta(u,v) = \theta(v,u)$ for all $u,v \geq 0$, (ii) $\theta(w,t) < \theta(u,v)$ if either $w < u$ and $t \leq v$, or $w \leq u$ and $t < v$, (iii) for each $r \in \mathrm{Im}(\theta)$ and $v \in [0,r]$, there exists $u \in [0,r]$ such that $\theta(u,v) = r$,(recall that, $\mathrm{Im}(\theta) = \{\theta(u,v) : u \geq 0, v \geq 0\}$), and (iv) $\theta(u,0) \leq u$, for all $u > 0$.

In other terms, a non-negative function of two non-negative variables is called a B-action if it is zero at the origin, symmetric, continuous, and strictly increasing with respect to each of its arguments, such that for any $a,b,v \geq 0$, the equation $\theta(u,v) = \theta(a,b)$ has a solution $u \in [0,\theta(a,b)]$ whenever $v \in [0,\theta(a,b)]$, and $\theta(v,0) \leq v$. The following are examples of B-actions:

1. $\theta(u,v) = k(u+v+uv)$ for all $u,v \geq 0$, where k is a constant in $(0,1]$;
2. $\theta(u,v) = \frac{uv}{1+uv}$ for all $u,v \geq 0$;
3. $\theta(u,v) = (u^n + v^n)^{\frac{1}{n}}$ for all $u,v \geq 0$, where $n \in \mathbb{N}$.

It should be noted that the conditions on a B-action θ guarantees for each $a,b,s \geq 0$ such that $s \in [0,\theta(a,b)]$ the existence of a function $\rho : [0,\infty)\times[0,\infty) \to [0,\infty)$ (called B-inverse action of θ) and a non-negative number $t \in [0,\theta(a,b)]$ such that $\rho(r,s) = t$. Such function ρ has the following properties as enumerated in Lemma 7 of [6]:

- $\rho(0,0) = 0$;
- $\theta(\rho(r,s),s) = r$ and $\theta(r,\rho(s,r)) = s$;
- ρ is continuous with respect to the first variable;
- $\rho(r,s) \geq 0$ implies that $0 \leq s \leq r$.

For example, if θ is the B-action defined by $\theta(u,v) = (u^n + v^n)^{\frac{1}{n}}$ for all $u, v \geq 0$, where $n \in \mathbb{N}$, then its B-inverse action is simply defined by $\rho(u,v) = (u^n - v^n)^{\frac{1}{n}}$ If θ is a B-action, a θ-metric space is defined as follows.

Definition 1.5 [6] Let X be a nonempty set. A mapping $d_\Theta : X \times X \longrightarrow [0,\infty)$ is called a θ-metric on X with respect to B-action θ if d_Θ satisfies the following: (A_1) $d_\Theta(x,y) = 0$ if and only if $x = y$; (A_2) $d_\Theta(x,y) = d_\Theta(y,x)$, for all $x, y \in X$; (A_3) $d_\Theta(x,y) \leq \theta(d_\Theta(x,z), d_\Theta(z,y))$ for all $x, y, z \in X$. The pair (X, d_Θ) is called a θ-metric space.

Every metric space is a θ-metric space with $\theta(u,v) = u + v$, for all $u, v \geq 0$. However, b-metrics, and by extension, p-metrics and controlled metric type spaces, are not necessarily θ-metrics. Indeed, a b-metric with optimal scalar constant $s > 1$ cannot be a θ-metric because the condition (iv) in Definition 1.4 fails.

14.1.3 Metrics of Multiplicative Type

In 2008, Bashirov, Kurpinar, and Ozyapici [7] introduced the concept of multiplicative metric spaces by replacing the addition of distances in the right hand side of the triangle inequality in metric spaces with multiplication, and by adjusting the "self distance" to the unit. This concept arose out of their presentation of multiplicative calculus.

Definition 1.6 [7] Let X be a nonempty set. A function $d^* : X \times X \to [1,\infty)$ is called a multiplicative metric on X if for any $x, y, z \in X$, the following conditions hold: (i) $d^*(x,y) = 1 \Longleftrightarrow x = y$; (ii) $d^*(x,y) = d^*(y,x)$; (iii) $d^*(x,z) \leq d^*(x,y)\, d^*(y,z)$ (multiplicative triangle inequality). The pair (X, d^*) is called a multiplicative metric space.

The authors in [8] then introduced the concept of b-multiplicative metric spaces by combining the notions of b-metrics and multiplicative metrics.

Definition 1.7 [8] Let X be a non-empty set and let $s \geq 1$ be a given real number. A function $d^* : X \times X \to [1,\infty)$ is a b-multiplicative metric with coefficient s on X if it satisfies for all $x, y, z \in X$ the following: (i) $d^*(x,y) =$

$1 \Longleftrightarrow x = y$; (ii) $d^*(x, y) = d^*(y, x)$; (iii) $d^*(x, z) \le [d^*(x, y)\, d^*(y, z)]^s$. The triple (X, d^*, s) is called a b-multiplicative metric space.

It has been observed that multiplicative metric spaces are essentially metric spaces, and hence b-multiplicative metric spaces are b-metric spaces (see [9, 10]). The passage from metrics (respectively, b-metrics) to multiplicative metric spaces (respectively, b-multiplicative metrics) is achieved by the taking the exponential of the metric (or b-metric), or the logarithm of the metric of multiplicative type. The concept of multiplicative metrics is therefore an alternative to metric spaces rather than a proper generalization. The purpose of this chapter is to introduce a generalized metric-type space, with practical applications, that will unify all the metric-type spaces mentioned. Concepts defined in this chapter will be used in proving a fixed point result (see [11–13] for more), and applications to optimization, topology, and inequalities will be discussed as well. Many results presented in the chapter are extracted from the authors' manuscripts [14, 15], the first in the literature on O-metric spaces.

14.2 O-METRIC SPACES AND CLASSIFICATIONS

In the remaining part of the chapter, the following notations are adopted:

1. $\mathbb{R}_+$ and $\mathbb{R}$ denote the interval of real numbers $[0, \infty)$ and the set of all real numbers, respectively;
2. the floor function and the ceiling function are denoted by $\lfloor . \rfloor$ and $\lceil . \rceil$ respectively: for any $x \in \mathbb{R}$, $\lfloor x \rfloor := sup\{n \in \mathbb{Z} : n \le x\}$ and $\lceil x \rceil := inf\{n \in \mathbb{Z} : n \ge x\}$, where $\mathbb{Z}$ is the set of all integers;
3. given non-negative real numbers a and b, I_a and J_b will denote some interval of non-negative real numbers containing a and b respectively;
4. functions $\mathbf{o}_a : I_a \times I_a \to \mathbb{R}_+$ and $\mathbf{o}_b : J_b \times J_b \to \mathbb{R}_+$ will often be considered as binary operations on $\mathbb{R}_+$, with values $\mathbf{o}_a(u, v)$ (when $u, v \in I_a$) and $\mathbf{o}_b(u, v)$ (when $u, v \in J_b$) also denoted $u\ \mathbf{o}_a v$ and $u\ \mathbf{o}_b v$ respectively. When no confusion arises, we simply write o instead of $\mathbf{o}_a$.

14.2.1 Definition and Examples of O-Metric Spaces

Definition 2.1 Let X be a nonempty set. A function $d_{\mathbf{o}}: X \times X \to I_a$ is said to be an O-metric on X, and $(X, d_{\mathbf{o}}, a)$ an O-metric space, if for all $x, y, z \in X$ the following conditions hold:

- $d_{\mathbf{o}}(x, y) = a$ if and only if $x = y$;
- $d_{\mathbf{o}}(x, y) = d_{\mathbf{o}}(y, x)$;
- $d_{\mathbf{o}}(x, z) \leq \mathbf{o}(d_{\mathbf{o}}(x, y), d_{\mathbf{o}}(y, z))$ (triangle $\mathbf{o}$-inequality).[1]

The class of O-metric spaces is defined as the class of all O-metric spaces, for any operation $\mathbf{o}: I_a \times I_a \to \mathbb{R}_+$.

We distinguish two main classes of O-metric spaces based on the nature of I_a.

Definition 2.2 An O-metric space $(X, d_{\mathbf{o}}, a)$ is said to be:

- an a-upward (or upward) O-metric space if $I_a \subset [a, \infty)$;
- *an* a-downward (or downward) O-metric space if $I_a \subset [0, a]$.

Therefore, an a-upward (respectively, an a-downward) O-metric space $(X, d_{\mathbf{o}}, a)$ is an O-metric space for which the value a of self-metrics $d_{\mathbf{o}}(x, x)$ is the least (respectively, greatest) possible value of the O-metric $d_{\mathbf{o}}$. Without any loss of generality, we may take $I_a = [a, \infty)$ for a-upward O-metric spaces and $I_a = [0, a]$ for a-downward O-metric spaces.

Remark 2.3 The class of upward O-metric spaces include many known metric-type spaces when $a = 0$ or $a = 1$ (for metrics of multiplicative type). Indeed:

1. any metric space (X, d) is a 0-upward O-metric space $(X, d_{\mathbf{o}}, 0)$ (or more specifically, a +-metric space $(X, d_+, 0)$) if the function o is the addition operation (+) defined by $\mathbf{o}(u, v) = u + v$ for all $u, v \in I_0$, where $I_0 = [0, \infty)$;
2. a b-metric space (X, d, s) where $s \geq 1$, is a 0-upward O-metric space $(X, d_{\mathbf{o}}, 0)$ if the function o is defined as the scaled addition $\mathbf{o}(u, v) = s(u + v)$ for all $u, v \in I_0$, where $I_0 = [0, \infty)$;

[1] One can also write $d_{\mathbf{o}}(x, z) \leq d_{\mathbf{o}}(x, y)\ \mathbf{o}\ d_{\mathbf{o}}(y, z)$ for the triangle $\mathbf{o}$-inequality.

3. a p-metric space $(X, \tilde{d})$ is a 0-upward O-metric space $(X, d_{\mathbf{o}}, 0)$ if the function o is defined by $\mathbf{o}(u, v) = \Omega(u + v)$ for all $u, v \in I_0$, where $I_0 = [0, \infty)$ and Ω is the strictly increasing continuous function on $[0, \infty)$ as in Definition 1.2;
4. a θ-metric space, where θ is a B-action (as defined in Definition 1.4), is an a-upward O-metric space when $a = 0$, and $\mathbf{o} = \theta$;
5. an ultra metric space $(X, d_\wedge)$ (see [16] for definition) is a 0-upward O-metric space $(X, d_{\mathbf{o}}, 0)$ when $\mathbf{o}(u, v) = \max\{u, v\}$ for all $u, v \geq 0$, and $d_\wedge = d_{\mathbf{o}}$;
6. a multiplicative metric space (X, d^*) is a 1-upward O-metric space $(X, d_{\mathbf{o}}, 1)$ (or more specifically, a $\times$-metric space $(X, d_\times, 1)$) if o is the multiplication operation ($\times$) defined by $\mathbf{o}(u, v) = uv$ for all $u, v \in I_1$, where $I_1 = [1, \infty)$.

The following are examples of O-metrics which are new in the existing literature on metric-type spaces:

Example 2.4 Let $X = \mathbb{R}$. Consider the function o such that $\mathbf{o}(u, v) = \ln(e^u + e^v - 1)$ for all $u, v \in I_0$ where $I_0 = [0, \infty)$, and the map $d_{\mathbf{o}}(x, y) = \ln(1 + |x - y|)$ for all $x, y \in X$. Then $(X, d_{\mathbf{o}}, 0)$ defines a 0-upward O-metric space. To arrive at the triangle o-inequality, we consider the usual inequality $|x - y| \leq |x - z| + |z - y|$ for any $x, y, z \in X$, then proceed as follows:

$$\begin{aligned} d_{\mathbf{o}}(x, y) = \ln(1 + |x - y|) &\leq \ln(1 + |x - z| + |z - y|) \\ &= \ln[(1 + |x - z|) + (1 + |z - y|) - 1] \\ &= \ln\left[e^{d(x,z)} + e^{d(z,y)} - 1\right]. \end{aligned} \tag{1}$$

Thus $(X, d_{\mathbf{o}}, 0)$ is a 0-upward $\mathbf{o}$-metric space, with $\mathbf{o}(u, v) = \ln(e^u + e^v - 1)$ for all $u, v \geq 0$. In fact, $d_{\mathbf{o}}$ is also a metric in the usual sense since for all $x, y, z \in X$, $d_{\mathbf{o}}(x, y) \leq d_{\mathbf{o}}(x, z) + d_{\mathbf{o}}(z, y)$. However, the inequality (1) is more precise than than the metrical triangle inequality, since $\mathbf{o}(u, v) < u+v$ for all $u, v > 0$.

Example 2.5 Let $X = (0, \infty)$ and $d : X \times X \to \mathbb{R}_+$ be defined by:

$$d(x, y) = \begin{cases} \frac{2xy}{x^2+y^2}, & \text{if } x \neq y \\ 0, & \text{if } x = y. \end{cases}$$

Given that $2xy \leq x^2 + y^2$ with equality if and only if $x = y$, we have that $d(x, y) \in [0, 1)$ for any $x, y > 0$. In fact, for distinct positive real numbers x, y, z, $d(x, y) \leq \frac{2d(x,z)}{d(z,y)}$, since

$$\begin{aligned} d(x, z) &= \frac{2xz}{x^2 + z^2} \\ &= \frac{2xy}{x^2 + y^2} \cdot \frac{2yz}{y^2 + z^2} \cdot \frac{(x^2 + y^2)(y^2 + z^2)}{2y^2(x^2 + z^2)} \\ &= d(x, y)\, d(y, z) \frac{x^2y^2 + x^2z^2 + y^4 + y^2z^2}{2(x^2y^2 + y^2z^2)} \\ &\geq \frac{1}{2} d(x, y)\, d(y, z). \end{aligned}$$

In the case where x, y, z are not all distinct, $d(x, y) \leq d(x, z) + d(z, y)$. Therefore,

$$d(x, y) \leq \mathbf{o}(d(x, z), d(z, y))$$

for all $x, y, z \in X$, where $\mathbf{o} : I_0 \times I_0 \to \mathbb{R}_+$, with $I_0 = [0, 1)$, is defined by

$$\mathbf{o}(u, v) = \begin{cases} \frac{2u}{v}, & \text{if } uv \neq 0 \\ u + v, & \text{otherwise.} \end{cases}$$

Therefore $(X, d, 0)$ is an O-metric space.

Example 2.6 For any function $\mathbf{o}$: $[a, \infty) \times [a, \infty) \to [a, \infty)$ increasing in each of its two variables and such that $\mathbf{o}(a, a) = a$, $u \leq \mathbf{o}(u, a)$ and $\mathbf{o}(u, v) = \mathbf{o}(v, u)$ for all $u, v \in [a, \infty)$, the set $\mathbb{R}$ of real numbers is an O-metric space equipped with the O-metric $d_{\mathbf{o}}(x, y) = \begin{cases} a & \text{if } x = y, \\ \mathbf{o}(f(x), f(y)) & \text{if } x \neq y, \end{cases}$ where $f(u) = a + |u - a|$ for $u \in \mathbb{R}$.

The concept of downward O-metrics is not common. One of the usefulness of the concept is to accommodate binary operations that are not commutative (or symmetric) such as division. In addition, downward O-metrics can be seen in real-life applications when considering weights of pairs other than distance types. The following is an example of a downward O-metric:

Example 2.7 Define the operation $\div$ by $\div(u, v) = \frac{u}{v}$ for all $u, v \in I_1$, where $I_1 = (0, 1]$. The set $X = \mathbb{R}^n$ equipped with $d_{\div}(x, y) = e^{-\|x-y\|}$ for all

$x, y \in X$, where $\|\cdot\|$ is the Euclidean norm in $\mathbb{R}^n$, is a 1-downward $\div$-metric space $(X, d_{\div}, 1)$. The 1-downward metric $d_{\div}(x, y)$ between x and y may be considered as the degree of interaction between x and y, with 1 being the highest interaction level (only obtainable between any x and itself). In such model, a pair (x, y) will always interact (i.e., $d(x, y) > 0$) and the level of interaction diminishes as the Euclidean distance increases.

It should be noted that not all O-metrics are necessarily "upward" or "downward". The following is an example of an O-metric space which is neither an upward or a downward O-metric space.

Example 2.8 Let $X = \mathbb{R}$. Let I_1 be the interval $(0, \infty)$, and $\mathbf{o}: I_1 \times I_1 \to I_1$ the function defined by:

$$\mathbf{o}(u, v) = \begin{cases} \max\left\{\min\left\{\frac{u}{v}, \frac{v}{u}\right\}, -\ln(uv)\right\} & \text{if } 0 < u, v \le 1; \\ \max\left\{\min\left\{ue^v, \frac{e^{-v}}{u}\right\}, \ln\left(\frac{e^v}{u}\right)\right\} & \text{if } 0 < u \le 1 < v; \\ \max\left\{\min\left\{ve^u, \frac{e^{-u}}{v}\right\}, \ln\left(\frac{e^u}{v}\right)\right\} & \text{if } 0 < v \le 1 < u; \\ \max\left\{\min\left\{e^{u-v}, e^{v-u}\right\}, u + v\right\} & \text{if } u, v > 1. \end{cases}$$

Consider the function $d_{\mathbf{o}}: X \times X \to (0, \infty)$ defined by:

$$d_{\mathbf{o}}(x, y) = \begin{cases} e^{-|x-y|} & \text{if } |x - y| \le 1; \\ |x - y| & \text{if } |x - y| > 1. \end{cases}$$

Then $(X, d_{\mathbf{o}}, 1)$ is an O-metric space that is neither a 1-upward O-metric space nor a 1-downward O-metric space, since $d_{\mathbf{o}}$ takes values less than the self-metric 1 and also values greater than 1 at the same time.

14.2.2 Constructing New O-Metrics from Existing Ones

An uncountable number of O-metrics can be generated from an existing one on a given set X. The following proposition holds:

Proposition 2.9 *Let X be a nonempty set. Let $\theta: \mathbb{R}_+ \to \mathbb{R}_+$ be a nondecreasing function, bijective on I_a, with $\theta(a) = b$ and $\theta(I_a) = J_b$, where I_a and J_b are intervals of non-negative real numbers such that $a \in I_a$ and $b \in J_b$. Suppose $\mathbf{o}_a$ and $\mathbf{o}_b$ are two functions such that the following diagram commutes:*[2]

$$\begin{array}{ccc} J_b \times J_b \overset{\mathbf{o}_b}{\to} & & \mathbb{R}_+ \\ \uparrow \theta \times \theta & & \uparrow \theta \\ I_a \times I_a \overset{\mathbf{o}_a}{\to} & & \mathbb{R}_+ \end{array} \tag{2}$$

Let $d_{\mathbf{o}_a} : X \times X \to I_a$ and $d_{\mathbf{o}_b} : X \times X \to J_b$ be maps such that $d_{\mathbf{o}_b} = \theta \circ d_{\mathbf{o}_a}$. Then:

- $(X, d_{\mathbf{o}_b}, b)$ is an O-metric space if $(X, d_{\mathbf{o}_a}, a)$ is an O-metric space.
- $d_{\mathbf{o}_b}$ is b-upward (resp. b-downward) if $d_{\mathbf{o}_a}$ is a-upward (resp. a-downward).
- The converses of (i) and (ii) hold if θ is surjective.

Proof (i) Suppose $(X, d_{\mathbf{o}_a}, a)$ is an O-metric space. Note that $d_{\mathbf{o}_b}$ satisfies the first two axioms of an O-metric. Since the diagram (2) is commutative and θ is non-decreasing, for all $x, y, z \in X$, we have

$$\begin{aligned} d_{\mathbf{o}_b}(x,z) = \theta\left(d_{\mathbf{o}_a}(x,z)\right) &\le \theta\left(\mathbf{o}_a\left(d_{\mathbf{o}_a}(x,y), d_{\mathbf{o}_a}(y,z)\right)\right) \\ &= \mathbf{o}_b\left(\theta\left(d_{\mathbf{o}_a}(x,y)\right), \theta\left(d_{\mathbf{o}_a}(y,z)\right)\right) \\ &= \mathbf{o}_b\left(d_{\mathbf{o}_b}(x,y), d_{\mathbf{o}_b}(y,z)\right). \end{aligned}$$

Therefore $d_{\mathbf{o}_b}$ is an O-metric as well.

(ii) Suppose that $d_{\mathbf{o}_a}$ is an a-upward O-metric, that is, $I_a \subset [a, \infty)$, and let $t \in J_b$. Given that $\theta(I_a) = J_b$, there is $r \in I_a$ such that $t = \theta(r)$, and $t = \theta(r) \ge \theta(a) = b$. Therefore, $J_b \subset [b, \infty)$, hence $d_{\mathbf{o}_b}$ is a b-upward O-metric. Similarly, if $d_{\mathbf{o}_a}$ is a-downward O-metric, then $d_{\mathbf{o}_b}$ is a b-downward O-metric.

(iii) Suppose that θ is surjective. Then, one can construct a nondecreasing function $\theta^{-1} : \mathbb{R}_+ \to \mathbb{R}_+$, bijective on J_b, such that $\theta^{-1}(b) = a$, $\theta^{-1}(J_b) = I_a$, and the diagram obtained by replacing θ with θ^{-1} in (2) commutes. Since $d_{\mathbf{o}_a} = \theta^{-1} \circ d_{\mathbf{o}_b}$, it follows from (i) that $(X, d_{\mathbf{o}_a}, a)$ is an O-metric space if $(X, d_{\mathbf{o}_b}, b)$ is an O-metric space, and it follows from (ii) that $d_{\mathbf{o}_a}$ is a-upward (a-downward) if $d_{\mathbf{o}_b}$ is b-upward (b-downward). □

[2] The function $\theta \times \theta$ is defined by $(\theta \times \theta)(u, v) = (\theta(u), \theta(v))$ for all $u, v \ge 0$. The diagram (2) therefore commutes if and only if $\mathbf{o}_b(\theta(u), \theta(v)) = \theta(\mathbf{o}_a(u, v)) \quad \forall u, v \in I_a$.

As an application of Proposition 2.9, we prove that positive powers of O-metrics are O-metrics.

Corollary 2.10 *Let* $(X, d_{\mathbf{o}}, a)$ *be an O-metric space and* $r > 0$. *Then* (X, d_{φ}, a^r) *is an O-metric space with* $d_{\varphi} = (d_{\mathbf{o}})^r$ *and* $\varphi(u, v) = \left(\mathbf{o}\left(u^{\frac{1}{r}}, v^{\frac{1}{r}}\right)\right)^r$ *for all* $u, v \in I_a$.

Proof. It follows from Proposition 2.9, if we take $\theta(t) = t^r$ for all $t \in \mathbb{R}_+$.

□

The commutativity of the diagram (2) requirement can be weakened when constructing an upward O-metric from a metric (in the usual sense) or vice-versa. The proofs of the following are obvious and therefore omitted.

Proposition 2.11 *Let* (X, d) *be a metric space. Consider two functions* $\theta\colon [0, \infty) \to [a, \infty)$ *and* $\mathbf{o}\colon [a, \infty) \times [a, \infty) \to [a, \infty)$ *such that:*

- $\theta(t) = a \Longleftrightarrow t = 0$;
- θ *is monotone nondecreasing;*
- $\theta(t_1 + t_2) \le \theta(t_1)\ \mathbf{o}\ \theta(t_2)$ *for all* $t_1, t_2 \ge 0$.

Then $(X, d_{\mathbf{o}}, a)$ *with* $d_{\mathbf{o}} = \theta \circ d$ *is an a-upward O-metric space.*

Corollary 2.12 *If* (X, d) *is a metric, and* $\theta\colon [0, \infty) \to [0, \infty)$ *is nondecreasing and sub-additive, then* $(X, \theta \circ d)$ *is also a metric space (or, in the terminology of the article, a +-metric space).*

Proof. Take $a = 0$ and o is the addition on $[0, \infty)$. □

For example, if $\mathbb{R}$ is endowed with the metric $d(x, y) = |x - y|$ for all $x, y \in \mathbb{R}$, then d' defined by $d'(x, y) = \ln(1 + |x - y|)$ for all $x, y \ge 0$, is a metric on $\mathbb{R}$, since $d' = \theta \circ d$ and $\theta : [0, \infty) \to [0, \infty)$ such that $\theta(t) = \ln(1 + t)$ is nondecreasing and sub-additive. Similarly, an upward O-metric generates a metric in the usual sense if the conditions of the following proposition holds:

Proposition 2.13 *Let* $(X, d_{\mathbf{o}}, a)$ *be an O-metric space, where* $\mathbf{o}\colon [a, \infty) \times [a, \infty) \to [a, \infty)$. *Then, X is a metric space equipped with a metric* $d\colon X \times X \to [0, \infty)$ *if there exists a monotone nondecreasing function* $\lambda\colon [a, \infty) \to [0, \infty)$ *such that the following conditions hold:*

- $\lambda(t) = 0 \Longleftrightarrow t = a$;

- $\lambda(u \ \mathbf{o} \ v) \leq \lambda(u) + \lambda(v)$ *for all* $u, v \geq a$;
- $d = \lambda \circ d_{\mathbf{o}}$.

The following establishes a necessary and sufficient condition for a metric space to be an upward O-metric space.

Theorem 2.14 *Let* X *be a nonempty set,* $d_{\mathbf{o}}: X \times X \to [a, \infty)$, $\lambda: [a, \infty) \to [0, \infty)$ *and* $\mathbf{o}: [a, \infty) \times [a, \infty) \to [a, \infty)$, *functions such that:*

- λ *is increasing, with* $\lambda(a) = 0$;
- $\lambda(\mathbf{o}(u, v)) = \lambda(u) + \lambda(v) \quad \forall u, v \geq a.$

Then $(X, d_{\mathbf{o}}, a)$ *is an upward O-metric space if and only if* $(X, \lambda \circ d_{\mathbf{o}})$ *is a metric space (equivalently,* (X, d) *is a metric space if and only if* $(X, d_{\mathbf{o}}, a)$ *is an upward O-metric space with* $d_{\mathbf{o}} = \lambda^{-1} \circ d$*).*

Proof. From the condition (E_1), the function $\lambda : [a, \infty) \to \operatorname{Im}(\lambda)$ is bijective and hence the following holds: $\lambda^{-1}(\lambda(u)) = u$ for all $u \geq a$. Suppose that $d_{\mathbf{o}}$ is an O-metric on X. Then from Proposition 2.13, $\lambda \circ d_{\mathbf{o}}$ is a metric on X as the conditions $(\lambda_1) - (\lambda_3)$ are satisfied (particularly, (λ_1) is satisfied since λ is strictly increasing and $\lambda(a) = 0$). Conversely, suppose that $\lambda \circ d_{\mathbf{o}}$ is a metric on X. The function $\theta = \lambda^{-1}$ satisfies the conditions $(\theta_1) - (\theta_3)$ of Proposition 2.11. Thus, $\theta \circ (\lambda \circ d_{\mathbf{o}}) = d_{\mathbf{o}}$ is an upward O-metric. □

It should be noted that Theorem 2.14 can also be retrieved from Proposition 2.9 by taking $\theta = \lambda$, $b = 0$, and $J_b = [0, \infty)$.

From Theorem 2.14, one obtains the following corollary which establishes an equivalence between metric spaces and multiplicative metric spaces.

Corollary 2.15 [9] $(X, d_{\times}, 1)$ *is a multiplicative metric space if and only if* $(X, \ln \circ d_{\times})$ *is a metric space.*

Proof. Take $a = 1$, $\mathbf{o}(u, v) = uv$ for all $u, v \geq 1$, and $\lambda : [1, \infty) \to [0, \infty)$ defined by $\lambda(t) = \ln(t)$ for all $t \geq 1$. Note that, λ is increasing, $\lambda(1) = 0$ and $\lambda(\mathbf{o}(u, v)) = \ln(uv) = \ln(u) + \ln(v) = \lambda(u) + \lambda(v)$ for all $u, v \in [1, \infty)$. Thus, conditions (E_1) and (E_2) of Theorem 2.14 are satisfied. □

Example 2.16 Given any metric space (X, d) and a bijective mapping $\phi: \mathbb{R}_+ \to \mathbb{R}_+$ such that $\phi(0) = 0$, the composition $\phi \circ d$ is an $\mathbf{o}$-metric on X, with $u \ \mathbf{o} \ v = \phi(\phi^{-1}(u) + \phi^{-1}(v))$, satisfying (E_1) and (E_2) with $\lambda = \phi^{-1}$.

An O-metric can be constructed from two O-metrics on the same set, if o is associative, commutative, and monotone, as seen in the following proposition:

Proposition 2.17 *Suppose $d_{\mathbf{o}}$ and $d'_{\mathbf{o}}$ are two* $\mathbf{o}$*-metrics on a set* X*, where* $\mathbf{o}: I_a \times I_a \to \mathbb{R}_+$ *is strictly increasing in both variables, associative, commutative, and* $\mathbf{o}(a,a) = a$*. Then* $d_{\mathbf{o}}\ \mathbf{o}\ d'_{\mathbf{o}}$ *is also an O-metric on* X*.*

Proof. Define $D : X \times X \to \mathbb{R}_+$ by $D(x,y) = d_{\mathbf{o}}(x,y)\ \ \mathbf{o}\ \ d'_{\mathbf{o}}(x,y)$ for any $x, y \in X$. Then, for $x, y, z \in X$,

$$\begin{aligned} D(x,y) = a &\Longleftrightarrow \mathbf{o}\left(d_{\mathbf{o}}(x,y), d'_{\mathbf{o}}(x,y)\right) = a \\ &\Longleftrightarrow d_{\mathbf{o}}(x,y) = d'_{\mathbf{o}}(x,y) = a \\ &\Longleftrightarrow x = y; \end{aligned}$$

$$\begin{aligned} D(x,y) &= \mathbf{o}\left(d_{\mathbf{o}}(x,y), d'_{\mathbf{o}}(x,y)\right) \\ &= \mathbf{o}\left(d_{\mathbf{o}}(y,x), d'_{\mathbf{o}}(y,x)\right) \\ &= D(y,x); \end{aligned}$$

$$\begin{aligned} D(x,z) &= d_{\mathbf{o}}(x,z)\ \ \mathbf{o}\ \ d'_{\mathbf{o}}(x,z) \\ &\le \left(d_{\mathbf{o}}(x,y)\ \ \mathbf{o}\ \ d_{\mathbf{o}}(y,z)\right)\ \ \mathbf{o}\ \ \left(d'_{\mathbf{o}}(x,y)\ \ \mathbf{o}\ \ d'_{\mathbf{o}}(y,z)\right) \\ &= \left(d_{\mathbf{o}}(x,y)\ \ \mathbf{o}\ \ d'_{\mathbf{o}}(x,y)\right) \mathbf{o} \left(d_{\mathbf{o}}(y,z)\ \ \mathbf{o}\ \ d'_{\mathbf{o}}(y,z)\right) \\ &= D(x,y)\ \ \mathbf{o}\ \ D(y,z). \end{aligned}$$

Thus, D defines an O-metric on X. □

To end this subsection, we state the following proposition that the Cartesian product of two O-metric spaces is an O-metric space.

Proposition 2.18 *Let* $\left(X_1, d_{\mathbf{o}_1}, a\right)$ *and* $\left(X_2, d_{\mathbf{o}_2}, a\right)$ *be two O-metric spaces. Let* $X_1 \times X_2$ *be the Cartesian product of* X_1 *and* X_2*. Take* $\mathbf{o} = \max\{\mathbf{o}_1, \mathbf{o}_2\}$*, and define the function* $d_{\mathbf{o}}$ *on* $X_1 \times X_2$ *by:*

$$d_{\mathbf{o}}\left((x,y),(u,v)\right) = \varphi\left(d_{\mathbf{o}_1}(x,u), d_{\mathbf{o}_2}(y,v)\right), \forall (x,y),(u,v) \in X_1 \times X_2,$$

where $\varphi: I_a \times I_a \to \mathbb{R}_+$ *is such that:*

$$\begin{cases} \varphi(u_1, v_1) = a \iff u_1 = v_1 = a, \\ \varphi \text{is nondecreasing in both variables,} \\ \varphi(u_1 \mathbf{o} u_2, v_1 \mathbf{o} v_2) \le \varphi(u_1, v_1)\, \mathbf{o} \varphi(u_2, v_2), \forall u_1, u_2, v_1, v_2 \in I_a. \end{cases}$$

Then $(X_1 \times X_2, d_{\mathbf{o}}, a)$ *is an O-metric space.*

Proof. Let $(x, y), (u, v), (z, t) \in X_1 \times X_2$.

$$\begin{aligned} d_{\mathbf{o}}((x, y), (u, v)) = a &\iff \varphi\left(d_{\mathbf{o}_1}(x, u), d_{\mathbf{o}_2}(y, v)\right) = a \\ &\iff d_{\mathbf{o}_1}(x, u) = d_{\mathbf{o}_2}(y, v)) = a \\ &\iff x = u \wedge y = v \\ &\iff (x, y) = (u, v). \end{aligned}$$

$$\begin{aligned} d_{\mathbf{o}}((x, y), (u, v)) &= \varphi\left(d_{\mathbf{o}_1}(x, u), d_{\mathbf{o}_2}(y, v)\right) \\ &= \varphi\left(d_{\mathbf{o}_1}(u, x), d_{\mathbf{o}_2}(v, y)\right) \\ &= d_{\mathbf{o}}((u, v), (x, y)). \end{aligned}$$

$$\begin{aligned} d_{\mathbf{o}}((x, y), (z, t)) &\leq \varphi\left(d_{\mathbf{o}_1}(x, z), d_{\mathbf{o}_2}(y, t)\right) \\ &\leq \varphi\left(d_{\mathbf{o}_1}(x, u) \;\; \mathbf{o}_1 d_{\mathbf{o}_1}(u, z), d_{\mathbf{o}_2}(y, v) \;\; \mathbf{o}_2 d_{\mathbf{o}_2}(v, t)\right) \\ &\leq \varphi\left(d_{\mathbf{o}_1}(x, u) \;\; \mathbf{o} \;\; d_{\mathbf{o}}(u, z), d_{\mathbf{o}_2}(y, v) \;\; \mathbf{o} \;\; d_{\mathbf{o}}(v, t)\right) \\ &\leq \varphi\left(d_{\mathbf{o}_1}(x, u), d_{\mathbf{o}_2}(y, v)\right) \;\; \mathbf{o} \;\; \varphi\left(d_{\mathbf{o}_1}(u, z), d_{\mathbf{o}_2}(v, t)\right) \\ &= d_{\mathbf{o}}((x, y), (u, v)) \;\; \mathbf{o} \;\; d_{\mathbf{o}}((u, v), (z, t)). \end{aligned}$$

□

Note that, if $\mathbf{o}_1$ and $\mathbf{o}_2$ are nondecreasing in both variables, and $d_{\mathbf{o}_1}$ and $d_{\mathbf{o}_2}$ are a-upward O-metrics, then φ can be taken as the maximum function. The proposition above can be generalized to a finite number of O-metric spaces as follows:

Proposition 2.19 *Let $\left\{\left(X_i, d_{\mathbf{o}_i}, a\right)\right\}_{1 \leq i \leq n}$ be a family of O-metric spaces. The Cartesian product $X_1 \times X_2 \times \cdots \times X_n$ of the X_i's is an O-metric space, where* $\mathbf{o} = \max\{\mathbf{o}_1, \mathbf{o}_2, ..., \mathbf{o}_n\}$ *and*

$$d_{\mathbf{o}}((x_1, x_2, ..., x_n), (y_1, y_2, ..., y_n)) = \varphi\left(d_{\mathbf{o}_1}(x_1, y_1), d_{\mathbf{o}_2}(x_2, y_2), ..., d_{\mathbf{o}_n}(x_n, y_n)\right),$$

with $\varphi : I_a \times I_a \times \cdots \times I_a \to \mathbb{R}_+$ such that:

$$\begin{cases} \varphi(u_1, u_2, ..., u_n) = a \iff u_1 = u_2 = ... = u_n = a, \\ \varphi \text{is non-decreasing on all variables,} \\ \varphi(\mathbf{o}(u_1, v_1), \mathbf{o}(u_2, v_2), ..., \mathbf{o}(u_n, v_n)) \leq \mathbf{o}(\varphi(u_1, u_2, ..., u_n), \varphi(v_1, v_2, ..., v_n)), \end{cases}$$

$\forall u_i, v_i \in I_a \, \forall i \in \{1, 2, ..., n\}$.

14.2.3 Linking Upward and Downward O-Metric Spaces

Under some conditions on the binary operation o, an upward O-metric space may be associated to a particular downward O-metric. The following proposition gives such conditions:

Proposition 2.20 *Let $(X, d_{\mathbf{o}}, a)$ be an O-metric space and suppose there is a function $\varphi : I_a \times I_a \to \mathbb{R}$ such that*

$$w \leq \mathbf{o}(u, v) \Longleftrightarrow \varphi(w, v) \leq u, \forall u, v, w \in I_a. \tag{3}$$

Then, the reverse triangle ψ-inequality

$$\psi(d_{\mathbf{o}}(x, z), d_{\mathbf{o}}(z, y)) \leq d_{\mathbf{o}}(x, y)$$

holds for all $x, y, z \in X$, where $\psi(w, v) := \max\{\varphi(w, v), \varphi(v, w)\}$. If $\theta : I_a \cup \mathrm{Im}(\varphi) \to \mathbb{R}_+$ is a continuous function, decreasing on I_a, then $(X, d_\xi, \theta(a))$ is an O-metric space with $d_\xi = \theta \circ d_{\mathbf{o}}$ and for all $u, v \in \mathrm{Im}(\theta)$, $\xi(u, v) = \theta(\varphi(\theta^{-1}(u), \theta^{-1}(v)))$. If, in addition, $\theta([a, \infty)) \subset [0, a]$ with $\theta(a) = a \neq 0$, $(X, d_{\mathbf{o}}, a)$ is an upward O-metric space if and only if (X, d_ξ, a) is a downward O-metric space.

Proof. From equation (3), the symmetry of $d_{\mathbf{o}}$, and the triangle o-inequality, the reverse triangle ψ-inequality holds.

Suppose that $\theta : I_a \cup Im(\varphi) \to \mathbb{R}_+$ is a continuous mapping, decreasing on I_a. Since X is an O-metric space, for all $x, y, z \in X$,

$$d_\xi(x, y) = \theta(a) \Longleftrightarrow \theta(d_{\mathbf{o}}(x, y)) = \theta(a) \Longleftrightarrow d_{\mathbf{o}}(x, y) = a \Longleftrightarrow x = y,$$

and $d_\xi(x, y) = \theta(d_{\mathbf{o}}(x, y)) = \theta(d_{\mathbf{o}}(y, x)) = d_\xi(y, x)$. Taking the images by θ of the two sides of the inequality[3] in (a), for all $x, y \in X$,

$$\begin{aligned} \theta(d_{\mathbf{o}}(x, y)) &\leq \theta(\varphi(d_{\mathbf{o}}(x, z), d_{\mathbf{o}}(z, y))) \\ d_\xi(x, y) &\leq \xi(d_\xi(x, z), d_\xi(z, y)). \end{aligned}$$

The ξ-metric d_ξ has its values in $\theta(I_a)$ which is an interval (by virtue of the continuity of θ) containing $\theta(a)$. Thus $(X, d_\xi, \theta(a))$ is an O-metric space.

With the additional conditions $\theta([a, \infty)) \subset [0, a]$ and $\theta(a) = a \neq 0$, $(X, d_{\mathbf{o}}, a)$ is an upward O-metric space if and only if (X, d_ξ, a) is a downward O-metric space, since for all $x, y \in X$, $d_{\mathbf{o}}(x, y) \geq a$ if and only if $d_\xi(x, y) = \theta(d_{\mathbf{o}}(x, y)) \leq \theta(a) = a$. □

---

[3] The reverse triangle ψ-inequality remains true if ψ is substituted with φ since $\varphi(w, v) \leq \psi(w, v)$ for all $w, v \in I_a$.

We consider the reverse triangle inequalities in some known metric-types in literature:

Example 2.21 Suppose (X, d, s) is b-metric space (with $s \geq 1$); the condition (3) is satisfied when $a = 0$, $I_a = [0, \infty)$, $\mathbf{o}(u, v) = s(u + v)$, and $\varphi(u, v) = \frac{u}{s} - v$. Thus, $\psi(u, v) = \max\left\{\frac{u}{s} - v, \frac{v}{s} - u\right\} = \left(\frac{1}{2} + \frac{1}{2s}\right)|u - v| - \left(\frac{1}{2} - \frac{1}{2s}\right)(u + v)$, and the reverse triangle ψ-inequality is given by:

$$(s+1)\,|d(x,z) - d(z,y)| - (s-1)\,[d(x,z) + d(z,y)] \leq 2sd(x,y) \quad \forall x, y, z \in X. \tag{4}$$

In the special case when $s = 1$, (X, d) is a metric space and the reverse triangle inequality becomes:

$$|d(x,z) - d(z,y)| \leq d(x,y) \quad \forall x, y, z \in X. \tag{5}$$

In this case, $\varphi(u, v) = u - v$, and $\psi(u, v) = \max\{u - v, v - u\} = |u - v|$.

Example 2.22 If X is endowed with a multiplicative metric d^*, then (3) is satisfied when $\mathbf{o}(u, v) = uv$ and $\varphi(u, v) = \frac{u}{v}$, $a = 1$ and $I_a = [1, \infty)$. We have $\psi(u, v) = \max\left\{\frac{u}{v}, \frac{v}{u}\right\} = \frac{u^2+v^2}{2uv} + \frac{|u^2-v^2|}{2uv}$ with the following reverse "triangle multiplicative inequality" for all $x, y, z \in X$:

$$d^*(x,z)^2 + d^*(z,y)^2 + \left|d^*(x,z)^2 - d^*(z,y)^2\right| \leq 2d^*(x,z)\,d^*(z,y)\,d^*(y,x). \tag{6}$$

A reverse triangle ψ-inequality is possible if the condition (3) is not satisfied. Consider for example an a-upward O-metric space $(X, d_{\mathbf{o}}, a)$ such that o satisfies conditions (E_1) and (E_2) of Theorem 2.14 for some function $\lambda : [a, \infty) \to [0, \infty)$. Since $\mathbf{o}(u, v) = \lambda^{-1}(\lambda(u) + \lambda(v))$ for all $u, v \in I_a$, we obtain

$$\begin{aligned} w \leq \mathbf{o}(u,v) &\Longleftrightarrow w \leq \lambda^{-1}(\lambda(u) + \lambda(v)) \\ &\Longleftrightarrow \lambda(w) \leq \lambda(u) + \lambda(v) \\ &\Longleftrightarrow \lambda(w) - \lambda(v) \leq \lambda(u) \\ &\Longleftrightarrow \lambda^{-1}(\lambda(w) - \lambda(u)) \leq u, \end{aligned}$$

for all $u, v, w \in I_a$ such that $w \geq v$. A function φ such that $\varphi(w, u) = \lambda^{-1}(\lambda(w) - \lambda(v))$ may only be defined on the set $\mathbf{Dom}(\varphi) :=$

$\{(w, v) \in \mathbb{R}^2 : w \geq v \geq a\}$ because of the domain of definition of λ^{-1}. However, since $\lambda \circ d_{\mathbf{o}}$ is a metric on X, the reverse triangle ψ-inequality holds, with ψ defined by $\psi(w, v) = \lambda^{-1}(|\lambda(w) - \lambda(v)|)$ for all $w, v \geq a$:

$$\lambda^{-1}(|\lambda(d_{\mathbf{o}}(x, z)) - \lambda(d_{\mathbf{o}}(z, y))|) \leq d_{\mathbf{o}}(x, y) \quad \forall x, y, z \in X.$$

Remark 2.23 It is easy to see that inequality (3) in Proposition 2.20 holds if $\varphi : I_a \times I_a \to \mathbb{R}$, and $\mathbf{o}$ are increasing in the first variable and

$$\varphi(\mathbf{o}(u, v), v) = \mathbf{o}(\varphi(u, v), v) = u \text{ for all } u, v \in I_a. \tag{7}$$

14.3 TOPOLOGY INDUCED BY AN O-METRIC

Consistent with the literature of metric-type spaces and the topologies they define, "open" balls in an O-metric space are constructed. The term "open" is only employed to be consistent with relevant literature and does not indicate whether the "open" balls are open sets for the O-metric topology defined in the sequel.

Definition 3.1 Let $(X, d_{\mathbf{o}}, a)$ be an O-metric space. Given $x \in X$ and $r > 0$, the open ball centered on x and with radius r is given by

$$B(x, r) = \{y \in X : |d_{\mathbf{o}}(x, y) - a| < r\}.$$

If X is an a-upward O-metric, that is, if $I_a = [a, \infty)$, then for $r > 0$ and $x \in X$, $B(x, r) = \{y \in X | a \leq d_{\mathbf{o}}(x, y) < a + r\}$. This is compatible with the definition of open balls in metric spaces and b-metric spaces (when $a = 0$), and multiplicative metric spaces (when $a = 1$). If X is an a-downward O-metric, that is, if $I_a = [0, a]$, then $B(x, r) = \{y \in X | a - r < d_{\mathbf{o}}(x, y) \leq a\}$ for $r > 0$ and $x \in X$. A few topologies can be constructed from open balls:

Proposition 3.2 *An O-metric $d_{\mathbf{o}} : X \times X \to I_a$ generates the following topologies on X:*

$$\mathcal{T}_1 = \{A \subset X | \forall x \in A \exists r \in (Im\mathbf{o}) \setminus \{a\}, B(x, |r - a|) \subset A\}, \tag{8}$$

$$\mathcal{T}_2 = \{A \subset X : \forall x \in A \ \exists r > 0, B(x, r) \subset A\}, \tag{9}$$

$$\mathcal{T}_3 = \left\{ A \subset X \,\middle|\, \begin{array}{l} \text{for any sequence } \{x_n\} \text{ of points in } X \text{ such that } \lim_{n\to\infty} d_{\mathbf{o}}(x_n, x) = a \\ \text{for some } x \in A, \text{ there exists } n_0 \in \mathbb{N} \text{ such that } x_n \in A \text{ for all } n \geq n_0. \end{array} \right\}, \tag{10}$$

$$\mathcal{T}_4 = \left\{ A \subset X : A = \bigcup_{i\in I} A_i, \text{ where each} A_i \text{is a finite intersection of open balls.} \right\}. \tag{11}$$

We also have $\mathcal{T}_1 \subset \mathcal{T}_2 = \mathcal{T}_3 \subset \mathcal{T}_4$.

Proof . It has been omitted. □

The topology $\mathcal{T}_2$ (or $\mathcal{T}_3$) will be called the topology induced by the O-metric$d_{\mathbf{o}}$, or simply, the O-metric topology.

Definition 3.3 Let $(X, d_{\mathbf{o}}, a)$ be an O-metric space. A set $A \subset X$ is said to be open in the O-metric topology if for all $x \in A$, there is $r > 0$ such that $B(x, r) \subset A$, where $B(x, r)$ is defined in Definition 3.1.

It should be re-emphasized that the terminology “open” ball is employed for consistency with literature on metric-types: an “open” ball may not be an open set for the O-metric topology. To see this, we consider the following example, inspired by the authors in [17]:

Example 3.4 Let $X = [0, 1]$ and define d: $[0, 1] \times [0, 1] \to [0, \alpha]$ for all $x, y \in [0, 1]$ by

$$d(x, y) = \begin{cases} 0 & \text{if } x = y \\ 1 & \text{if } x \neq y \in \{0, 1\} \\ |x - y| & \text{if } x \neq y \in \{0\} \cup \{x_n : n = 1, 2, ...\} \\ \alpha & \text{otherwise,} \end{cases}$$

where $\{x_n\}$ is any sequence of points in the interval $(0, 1)$ such that $\lim_{n\to\infty} x_n = 0$ (in the usual topology on $\mathbb{R}$), and $\alpha > 1$ is a real number. Then (X, d) is a 0-upward O-metric space, with o defined by $\mathbf{o}(u, v) = \alpha(u + v)$ for all $u, v \geq 0$. For any β such that $1 < \beta < \alpha$, the open ball $B(1, \beta) = \{0, 1\}$ is not an open set (in the O-metric topology) since $0 \in B(1, \beta)$ but for any $r > 0$, $B(0, r)$ contains terms of the sequence $\{x_n\}$ less than r, none of which is in $B(1, \beta)$. The singleton $\{1\}$ is a member of the O-metric topology since one can find $r' > 0$ such that $B(1, r') = \{1\}$. Thus, the O-metric topology here is not metrizable.

It is therefore natural to impose some more conditions to have every open ball in an O-metric space as an an open set. We also investigate properties of the O-metric topology on an O-metric space $(X, d_{\mathbf{o}}, a)$ such as criteria for convergence of sequences of points in X, uniqueness of the limit of a convergent sequence, separation of disjoint open sets, and metrizability.

14.3.1 Convergence and O-Convergence

From the equality $\mathcal{T}_2 = \mathcal{T}_3$ established in Proposition 3.2, we obtain the following proposition:

Proposition 3.5 *If a sequence $\{x_n\}_{n\in\mathbb{N}}$ of points in an O-metric space $(X, d_{\mathbf{o}}, a)$ is such that $\lim\limits_{n\to\infty} d_{\mathbf{o}}(x_n, x) = a$ for some $x \in X$, then $\{x_n\}$ converges (in the topology induced by the O-metric) to $x \in X$.*

Proposition 3.5 gives a sufficient condition for convergence of a sequence of points in an O-metric space. It is not known whether the converse holds. Indeed, if $\{x_n\}$ is a convergent sequence in the O-metric topology, and x a limit of $\{x_n\}$, then the following statement is true:

$$\text{for any open set A, there is } n_0 \in \mathbb{N} \text{ such that } x_n \in A \text{ for all } n \geq n_0. \quad (12)$$

For any $\epsilon > 0$, $B(x, \epsilon)$ is not necessarily an open set, hence, the substitution $A = B(x, \epsilon)$ in (12) (which would then mean that $\lim_{n\to\infty} d_{\mathbf{o}}(x_n, x) = a$) is not allowed. However, if

$$\text{for all } \epsilon > 0, \text{there is an open set } A_\epsilon \text{ containing } x \text{ such that } A_\epsilon \subset B_{d_{\mathbf{o}}}(x, \epsilon), \quad (13)$$

then $\lim_{n\to\infty} d_{\mathbf{o}}(x_n, x) = a$. We therefore define the notion of *O-convergence* which, by Proposition 3.5, is a stronger notion than convergence in the O-metric topology.

Definition 3.6 Let $(X, d_{\mathbf{o}}, a)$ be an O-metric space, and x a point in X. A sequence $\{x_n\}_{n\in\mathbb{N}}$ of points in X is said to: (i) O-converge to x if $\lim_{n\to\infty} d_{\mathbf{o}}(x_n, x) = a$. In such case, x is called the O-limit of $\{x_n\}$, and we write $x_n \overset{O}{\to} x$. (ii) be a Cauchy sequence if $\lim_{n,m\to\infty} d_{\mathbf{o}}(x_n, x_m) = a$. The space $(X, d_{\mathbf{o}}, a)$ is said to be O-complete if the Cauchy sequences of X are the sequences that O-converge in X. In other terms, in an O-complete O-metric space, every O-convergent sequence is a Cauchy sequence, and every Cauchy sequence is O-convergent.

Note that the O-limit of an O-convergent sequence may not be unique:

Example 3.7 Consider the O-metric space $(X, d_{\mathbf{o}}, a)$ where $X = [-1, 1]$,

$$\mathbf{o}(u, v) = \begin{cases} \frac{1}{uv} & \text{if } u, v \neq 0, \\ 1 & \text{if } u = 0 \text{ or } v = 0, \end{cases}$$

$a = 1$, and

$$d_{\mathbf{o}}(x, y) = \begin{cases} 1 & \text{if } x = y \\ |xy| & \text{if } x \neq y. \end{cases}$$

Consider the sequence $\{x_n\}_{n \in \mathbb{N}}$ defined by $x_n = 1 - \frac{1}{n}$, for all $n \in \mathbb{N}$. For $x \in \{-1, 1\}$, $d_{\mathbf{o}}(x_n, x) = \left(1 - \frac{1}{n}\right)|x| = 1 - \frac{1}{n} \to 1$ as $n \to \infty$. Hence $\{x_n\}$ has two O-limits, ± 1.

It should also be noted that an O-convergent sequence may not be a Cauchy sequence. The following example is given as illustration:

Example 3.8 Consider the O-metric space $(X, d_{\mathbf{o}}, 0)$ considered in Example 2.5, where $X = (0, \infty)$, $\mathbf{o} : [0, 1) \times [0, 1) \to \mathbb{R}_+$ is the function defined by

$$\mathbf{o}(u, v) = \begin{cases} \frac{2u}{v}, & \text{if } uv \neq 0 \\ u + v, & \text{otherwise,} \end{cases}$$

and $d_{\mathbf{o}} : X \times X \to [0, 1)$ is the O-metric defined by:

$$d(x, y) = \begin{cases} \frac{2xy}{x^2+y^2}, & \text{if } x \neq y \\ 0, & \text{if } x = y. \end{cases}$$

Let $\{x_n\}$ be the sequence in X such that $x_n = \frac{1}{n}$ for all $n \in \mathbb{N}$. Then $x_n \overset{O}{\to} x$ for any $x > 0$ such that $1 \notin x\mathbb{N}$, since

$$d(x_n, x) = \frac{\frac{2x}{n}}{\frac{1}{n^2} + x^2} = \frac{2x}{n} \cdot \frac{n^2}{1 + n^2x^2} = \frac{2nx}{1 + n^2x^2} \to 0 \text{ as } n \to \infty.$$

The sequence $\{x_n\}$ has therefore infinitely many O-limits. However, $\{x_n\}$ is not a Cauchy sequence since for any $n, m \in \mathbb{N}$ such that $n \neq m$,

$$d(x_n, x_m) = \frac{\frac{2}{nm}}{\frac{1}{n^2} + \frac{1}{m^2}} = \frac{2nm}{n^2 + m^2} \nrightarrow 0 \text{ as } n, m \to \infty.$$

We specify sufficient conditions under which the O-limit of a sequence, if it exists, is unique.

Proposition 3.9 *Let $(X, d_{\mathbf{o}}, a)$ be an O-metric space. The O-limit of an O-convergent sequence is unique if the conditions (U_1) and (U_2), or (U_1) and (U_2') are satisfied, where:*

- $\mathbf{o}$ *is continuous at points (u, v) such that $u = a$ or $v = a$;*
- $\mathbf{o}$ *is nondecreasing in both variables and either $\mathbf{o}(u, a) = a \iff u = a$ for all $u \in I_a$, or $\mathbf{o}(a, u) = a \iff u = a$ for all $u \in I_a$.*
- $\mathbf{o}$ *is nondecreasing in one variable, say u_i, and for $u_j = a$ with $j \neq i$, $\mathbf{o}(u_1, u_2) = a \iff u_i = a$.*

Proof. Assume (U_1) and (U_2') hold (since (U_2) implies (U_2')). From condition (U_2'), $\mathbf{o}(a, a) = a$. Now, suppose $\lim_{n\to\infty} d_{\mathbf{o}}(x_n, x) = \lim_{n\to\infty} d_{\mathbf{o}}(x_n, y) = a$, where $x, y \in X$; then $d_{\mathbf{o}}(x, y) \leq \mathbf{o}(d_{\mathbf{o}}(x, x_n), d_{\mathbf{o}}(x_n, y))$. As $n \to \infty$, we obtain that $d_{\mathbf{o}}(x, y) \leq a$. Also, $d_{\mathbf{o}}(x_n, x) \leq \mathbf{o}(d_{\mathbf{o}}(x, y), d_{\mathbf{o}}(x_n, y))$ hence taking the limits, we have $a \leq \mathbf{o}(d_{\mathbf{o}}(x, y), a)$. Similarly, we obtain that $a \leq \mathbf{o}(a, d_{\mathbf{o}}(x, y))$. Thus, from condition (U_2'), $a \leq \mathbf{o}(u_1, u_2) \leq \mathbf{o}(a, a) = a$, with variable u_i in which $\mathbf{o}$ is monotone non-decreasing such that $u_i = d_{\mathbf{o}}(x, y)$ and $u_j = a$ for $j \neq i$. (U_2'), then this also implies that $d(x, y) = a$ so that $x = y$. □

From Remark 2.3, one can easily verify that metric spaces, b-metric spaces, multiplicative metric spaces, ultrametric spaces, θ-metric spaces, and p-metric spaces satisfy conditions (U_1) and (U_2), hence the limit of a "convergent"[4] sequence in any of these spaces converge with respect to the O-metric topology. Also, the O-metric space in Example 2.6 satisfies conditions (U_1) and (U_2). It turns out that conditions (U_1) and (U_2) can be weakened in the case of upward O-metric spaces, and that under those lesser conditions, every O-convergent sequence in an upward O-metric space is a Cauchy sequence.

Proposition 3.10 *Let $(X, d_{\mathbf{o}}, a)$ be an upward O-metric space such that:*

- $\mathbf{o}$ *is continuous at (a, a);*
- $\mathbf{o}(a, a) = a$.

[4] Here, "convergent" means "O-convergent", only valid when $d(x_n, x) \to 0$ as $n \to \infty$, where d is the metric-type considered. We use the term "convergent" since many authors in literature loosely employ it.

Then, every O-convergent sequence in X is a Cauchy sequence. Moreover, the O-limit of an O-convergent sequence is unique.

Proof. The result follows from the triangle o-inequality. Indeed, if $\{x_n\}$ is an O-convergent sequence with x an O-limit, then for all $n, m \in \mathbb{N}$,

$$a \leq d_{\mathbf{o}}(x_n, x_m) \leq \mathbf{o}\left(d_{\mathbf{o}}(x_n, x), d_{\mathbf{o}}(x, x_m)\right).$$

As $n, m \rightarrow \infty$, $d_{\mathbf{o}}(x_n, x_m) \rightarrow a$, hence $\{x_n\}$ is a Cauchy sequence. Suppose that $y \in X$ is also an O-limit of $\{x_n\}$: $\lim_{n\to\infty} d_{\mathbf{o}}(x_n, x) = \lim_{n\to\infty} d_{\mathbf{o}}(x_n, y) = a$. From the triangle o-inequality, for all $n \in \mathbb{N}$, $a \leq d_{\mathbf{o}}(x, y) \leq \mathbf{o}(d_{\mathbf{o}}(x, x_n), d_{\mathbf{o}}(x_n, y))$. By passage to limits, $d(x, y) = a$, hence $x = y$. □

The following proposition is a direct consequence of the definition of O-convergence.

Proposition 3.11 *Let $(X_1, d_{\mathbf{o}_1}, a)$ and $(X_2, d_{\mathbf{o}_2}, b)$ be two O-metric spaces, and $f: X_1 \rightarrow X_2$ a function. The following are equivalent:*

- f is sequentially continuous[5] *at* $\tilde{x} \in X_1$, that is, for any sequence $\{x_n\}_{n\in\mathbb{N}}$ of points in X_1, $x_n \overset{O}{\rightarrow} \tilde{x} \Longrightarrow f(x_n) \overset{O}{\rightarrow} f(\tilde{x})$;
- $\forall \epsilon > 0, \exists \delta > 0 : \left|d_{\mathbf{o}_1}(x, \tilde{x}) - a\right| < \delta \Longrightarrow \left|d_{\mathbf{o}_2}(f(x), f(\tilde{x})) - b\right| < \epsilon.$

14.3.2 Openness of Balls and Hausdorff Property

Example 3.4 was given earlier to show that open balls (defined in Definition 3.1) are not necessarily open sets (for the O-metric topology). The following theorem states sufficient conditions under which every open ball is a member of the O-metric topology. The proof is omitted.

Theorem 3.12 *Let $(X, d_{\mathbf{o}}, a)$ be an O-metric space. Suppose that:*

$$\forall r > 0 \quad \forall u \in I_a \; |u - a| < r \Longrightarrow \exists s > 0 \text{ such that } |w - a| < r \text{ whenever } w \in I_a \text{ with } w \leq \mathbf{o}(u, v) \text{ for some } v \in I_a \text{ such that } |v - a| < s. \quad (14)$$

Then every open ball in X is an open set.

[5] Sequential continuity here is relative to O-convergence.

In the case of upward O-metrics, we have the following:

Corollary 3.13 *Let $(X, d_{\mathbf{o}}, a)$ be an upward O-metric space. Then, every open ball is an open set if the following conditions simultaneously hold:*

- *There exists $\gamma : [a, \infty) \times [a, \infty) \to \mathbb{R}$ such that if $a \leq u < r$, then $\gamma(r, u) > a$ and $\mathbf{o}(u, \gamma(r, u)) \leq r$;*
- *$\mathbf{o}$ is increasing in both variables.*

Proof. If $r > 0$ and $u \in [a, \infty)$ are such that $|u - a| < r$, then $a \leq u < a + r$. Let $s = \gamma(a + r, u) - a$. From (C_1), $s > 0$. If $v \in [a, \infty)$ and $|v - a| < s$, then $a \leq v < a + \gamma(a + r, u)$; hence, given (C_2), if $w \in [a, \infty)$ and $w \leq \mathbf{o}(u, v)$, $w \leq \mathbf{o}(u, \gamma(a + r, u)) \leq a + r$ so $|w - a| < r$. Thus condition (14) is satisfied. □

In fact, one can show the following:

Proposition 3.14 *Let $(X, d_{\mathbf{o}}, a)$ be an upward O-metric space. Under conditions (C_1) and (C_2), the O-metric topology $\mathcal{T}$ is first countable and Hausdorff, and $\mathcal{T} = \mathcal{T}_1 = \mathcal{T}_2 = \mathcal{T}_3 = \mathcal{T}_4$.*

The conditions (C_1) and (C_2) also guarantee that O-convergence and convergence in the O-metric topology coincide, as shown in the following proposition:

Proposition 3.15 *Let $(X, d_{\mathbf{o}}, a)$ be an a-upward O-metric space such that o satisfies conditions (C_1) and (C_2). A sequence $\{x_n\}$ of points in X converges in the O-metric topology to some point $x \in X$ if and only if $x_n \overset{O}{\to} x$.*

Proof. We know from Proposition 3.5 that $x_n \overset{O}{\to} x$ implies that $\{x_n\}$ converges to x in the O-metric topology. Conversely, suppose that $\{x_n\}$ converges to x in the O-metric topology. Then, for any $\epsilon > 0$, given that $B(x, r)$ is an open set, there is $n_0 \in \mathbb{N}$ such that $x_n \in B(x, \epsilon)$, i.e., $|d_{\mathbf{o}}(x_n, x) - a| < \epsilon$, for all $n \geq n_0$. Therefore, $\lim_{n \to \infty} d_{\mathbf{o}}(x_n, x) = a$, and $x_n \overset{O}{\to} x$. □

It is interesting to distinguish metric-types in literature for which the underlining binary operation o in the triangle inequality satisfy conditions (C_1) and (C_2):

Remark 3.16

1. Metric spaces are upward *O*-metric spaces satisfying the conditions (C_1) and (C_2) with $a = 0$, $\mathbf{o}$ such that $\mathbf{o}(u, v) = u + v$ for all $u, v \geq 0$, and γ such that $\gamma(r, u) = r - u$ for all $r, u \geq 0$.
2. The conditions (C_1) and (C_2) hold in multiplicative metric spaces with $a = 1$, $\mathbf{o}$ being the product, and γ the right division defined by $\gamma(r, u) = \frac{r}{u}$ for all $r, u \geq 1$.
3. *Any* Θ-metric space (where Θ is a surjective B-action) satisfies the conditions (C_1) and (C_2), with $a = 0$, $\mathbf{o} = \Theta$, and γ taken as the B-inverse action of Θ.
4. Let (X, d, s) be a b-metric space where $s > 1$ is optimal, that is, for any $s' < s$, there are $x, y, z \in X$ such that $d(x, z) > s'(d(x, y) + d(y, z))$. Condition (C_1) does not hold. Indeed, if $\mathbf{o}(u, v) = s(u + v)$, and $u \in \left[\frac{r}{s}, r\right)$, then $\mathbf{o}(u, \gamma(r, u)) \leq r \Rightarrow \gamma(r, u) \leq \frac{r}{s} - u \leq 0$.

14.3.3 Metrizability and Topological Equivalence

Let $(X, d_{\mathbf{o}}, a)$ be an a-upward O-metric space for which there is a function $\lambda : [a, \infty) \to [0, \infty)$ satisfying the conditions:

- λ is increasing, with $\lambda(a) = 0$;
- $\lambda(\mathbf{o}(u, v)) = \lambda(u) + \lambda(v) \quad \forall u, v \geq a$.

By Theorem 2.14, $(X, \lambda \circ d_{\mathbf{o}})$ is a metric space. In this case, the O-metric topology on $(X, d_{\mathbf{o}}, a)$ and the metric topology on $(X, \lambda \circ d_{\mathbf{o}})$ coincide. Indeed, for $x \in X$ and $r > 0$, the open ball $B(x, r)$ is given by:

$$\begin{aligned} B(x, r) &= \{y \in X : |d_{\mathbf{o}}(x, y) - a| < r\} \\ &= \{y \in X : a \leq d_{\mathbf{o}}(x, y) < a + r\} \\ &= \{y \in X : 0 \leq \lambda(d_{\mathbf{o}}(x, y)) < \lambda(a + r)\} \\ &= B_{\lambda \circ d_{\mathbf{o}}}(x, \lambda(a + r)), \end{aligned}$$

where $B_{\lambda \circ d_{\mathbf{o}}}(x, \lambda(a + r))$ is the open ball centered on $x \in X$ with radius r, relative to the metric topology on $(X, \lambda \circ d_{\mathbf{o}})$. As λ is bijective, the open sets in the O-metric topology on $(X, d_{\mathbf{o}}, a)$ are exactly the open sets in the metric topology on $(X, \lambda \circ d_{\mathbf{o}})$. Thus, a sequence $\{x_n\}$ of points in X converges in the O-metric topology to some point $x \in X$ if and only if it O-converges to x ($x_n \overset{O}{\to} x$). Also, O-completeness in $(X, d_{\mathbf{o}}, a)$ is equivalent to completeness in the metric space $(X, \lambda \circ d_{\mathbf{o}})$. In the case of upward O-metric spaces, we can relate some of the conditions discussed in the chapter:

Proposition 3.17 *Consider an a-upward O-metric space* $(X, d_{\mathbf{o}}, a)$.

1. *If* $\mathbf{o}$ *satisfies* (E_1) *and* (E_2), *then it satisfies* (C_1) *and* (C_2);
2. *If* $\mathbf{o}$ *satisfies* (C_1) *and* (C_2), *then it satisfies* (U_1'), (U_2) *(and consequently,* (U_2')*).*

Multiplicative metric spaces satisfy the conditions (E_1) and (E_2) (see proof of Corollary 2.15), whereas b-metric spaces satisfy the conditions (U_1), (U_2) and (U_2') but not (C_1) in general. We end this section with the following theorem:

Theorem 3.18 *The topology of an O-metric space X is induced by an upward O-metric on X.*

Proof. Let $(X, d_{\mathbf{o}}, a)$ be an O-metric space, with $a \in \mathbb{R}_+$, and I_a an interval of non-negative numbers containing a. Let $J_a := I_a \cap [a, \infty)$, and define the maps $\xi : J_a \times J_a \to \mathbb{R}_+$ and $d_\xi : X \times X \to J_a$ by:

$$\xi(u, v) = \max\{\mathbf{o}(u, v), \mathbf{o}(u, 2a - v), \mathbf{o}(2a - u, v), \mathbf{o}(2a - u, 2a - v), 2a\} \quad \forall u, v \in J_a, \tag{15}$$

$$d_\xi(x, y) = a + |d_{\mathbf{o}}(x, y) - a| \quad \forall x, y \in X. \tag{16}$$

(X, d_ξ, a) is an a-upward O-metric space and the topology induced by $d_{\mathbf{o}}$ is the same as the topology induced by d_ξ as $B(x, r) = B_{d_\xi}(x, r)$ for all $x \in X$ and $r > 0$, where $B_{d_\xi}(x, r)$ is the open ball in the ξ-metric space (X, d_ξ, a).

□

It should be noted that the function ξ in (15) may not be easy to compute.

Example 3.19 Consider the O-metric space $(\mathbb{R}, d_{\mathbf{o}}, 1)$ in Example 2.8, where the function $\mathbf{o} : (0,\infty) \times (0,\infty) \to (0,\infty)$ is such that

$$\mathbf{o}(u,v) = \begin{cases} \max\left\{\min\left\{\frac{u}{v}, \frac{v}{u}\right\}, -\ln(uv)\right\} & \text{if } 0 < u, v \leq 1; \\ \max\left\{\min\left\{ue^{v}, \frac{e^{-v}}{u}\right\}, \ln\left(\frac{e^{v}}{u}\right)\right\} & \text{if } 0 < u \leq 1 < v; \\ \max\left\{\min\left\{ve^{u}, \frac{e^{-u}}{v}\right\}, \ln\left(\frac{e^{u}}{v}\right)\right\} & \text{if } 0 < v \leq 1 < u; \\ \max\left\{\min\left\{e^{u-v}, e^{v-u}\right\}, u+v\right\} & \text{if } u, v > 1, \end{cases}$$

and $d_{\mathbf{o}} : X \times X \to (0,\infty)$ is the map defined by:

$$d_{\mathbf{o}}(x,y) = \begin{cases} e^{-|x-y|} & \text{if } |x-y| \leq 1; \\ |x-y| & \text{if } |x-y| > 1. \end{cases}$$

By Theorem 3.18, the topology induced by $d_{\mathbf{o}}$ is also induced by the 1-upward O-metric given by (16):

$$d_{\xi}(x,y) = 1 + |d_{\mathbf{o}}(x,y) - 1| = \begin{cases} 2 - e^{-|x-y|} & \text{if } |x-y| \leq 1; \\ |x-y| & \text{if } |x-y| > 1. \end{cases}$$

14.4 POLYGON O-INEQUALITIES AND O-SERIES

The modification of the triangle inequality axiom of a metric space to accommodate some binary operation o which is not necessarily associative makes the resulting triangle $\mathbf{o}$-inequality interesting to study when applied to more than three points. Let $(X, d_{\mathbf{o}}, a)$ be an O-metric space. For $n \in \mathbb{N}$, let $x_0, x_1, x_2, \ldots, x_{n+1}$ be a finite sequence of $n+2$ points in X. The triangle $\mathbf{o}$-inequality provides many upper bounds for $d_{\mathbf{o}}(x_0, x_{n+1})$ as functions of exactly $d_{\mathbf{o}}(x_0,x_1), d_{\mathbf{o}}(x_1,x_2), \ldots, d_{\mathbf{o}}(x_n,x_{n+1})$ with each of $d_{\mathbf{o}}(x_0,x_1), d_{\mathbf{o}}(x_1,x_2), \ldots, d_{\mathbf{o}}(x_n,x_{n+1})$ occurring exactly once, without interchanging the order. In fact, if we view o as a binary operation, the upper bounds exactly correspond to the expressions

$$d_{\mathbf{o}}(x_0,x_1) \ \mathbf{o} \ d_{\mathbf{o}}(x_1,x_2) \ \mathbf{o} \cdots \mathbf{o} \ d_{\mathbf{o}}(x_n,x_{n+1}) \tag{17}$$

which depends on how parentheses are placed. The expression (17) has at most C_n values, where $C_n := \frac{1}{n+1}\binom{2n}{n}$ is the n-th Catalan number (see [18, 19]). The lemma below immediately follows:

Lemma 4.1 *Denote by $\Omega_{n,a}$ (or simply Ω_n when no confusion arises) the set (of order at most equal to C_n) of functions $\Delta_n : I_a^{n+1} \to I_a$ such that*

$\Delta_n(t_0, t_1, ..., t_n) = t_0\mathbf{o}\ t_1\mathbf{o}\cdots\mathbf{o}\ t_n$. *The triangle* **o**-*inequalities involving* $n+2$ *points* $x_0, x_1, ..., x_{n+1}$ *in an O-metric space* $(X, d_{\mathbf{o}}, a)$ *become:*

$$d_{\mathbf{o}}(x_0, x_{n+1}) \leq \Delta_n(d_{\mathbf{o}}(x_0, x_1), d_{\mathbf{o}}(x_1, x_2), ..., d_{\mathbf{o}}(x_n, x_{n+1}))\ \ \forall\Delta_n \in \Omega_n, \tag{18}$$

or simply

$$d_{\mathbf{o}}(x_0, x_{n+1}) \leq d_{\mathbf{o}}(x_0, x_1)\ \mathbf{o}\ d_{\mathbf{o}}(x_1, x_2)\ \mathbf{o}\cdots\mathbf{o}\ d_{\mathbf{o}}(x_n, x_{n+1}), \tag{19}$$

and are called polygon **o**-*inequalities.*

Example 4.2 Let o be a function defined for pairs of elements in the interval $I_0 = [0, \infty)$ by $\mathbf{o}(u, v) = u + 2v$. Then, for $0 \leq n \leq 3$, the set Ω_n is given by: $\Omega_0 = \{\Delta_0\}$, $\Omega_1 = \{\Delta_1\}$, $\Omega_2 = \{\Delta_2^1, \Delta_2^2\}$, and $\Omega_3 = \{\Delta_3^j : 1 \leq j \leq 5\}$, where for all $t_0, t_1, t_2, t_3 \geq 0$,

$$\begin{aligned}\Delta_0(t_0) &= t_0\\ \Delta_1(t_0, t_1) &= \mathbf{o}(t_0, t_1) = t_0 + 2t_1\end{aligned}$$

$$\begin{aligned}\Delta_2^1(t_0, t_1, t_2) &= t_0\mathbf{o}(t_1\mathbf{o}t_2) = t_0 + 2t_1 + 4t_3\\ \Delta_2^2(t_0, t_1, t_2) &= (t_0\mathbf{o}t_1)\mathbf{o}t_2 = t_0 + 2t_1 + 2t_2\end{aligned}$$

$$\begin{aligned}\Delta_3^1(t_0, t_1, t_2, t_3) &= t_0\mathbf{o}(t_1\mathbf{o}(t_2\mathbf{o}t_3)) = t_0 + 2t_1 + 4t_2 + 8t_3\\ \Delta_3^2(t_0, t_1, t_2, t_3) &= t_0\mathbf{o}((t_1\mathbf{o}t_2)\mathbf{o}t_3) = t_0 + 2t_1 + 4t_2 + 4t_3\\ \Delta_3^3(t_0, t_1, t_2, t_3) &= (t_0\mathbf{o}t_1)\mathbf{o}(t_2\mathbf{o}t_3) = t_0 + 2t_1 + 2t_2 + 4t_3\\ \Delta_3^4(t_0, t_1, t_2, t_3) &= (t_0\mathbf{o}(t_1\mathbf{o}t_2))\mathbf{o}t_3 = t_0 + 2t_1 + 4t_2 + 2t_3\\ \Delta_3^5(t_0, t_1, t_2, t_3) &= ((t_0\mathbf{o}t_1)\mathbf{o}t_2)\mathbf{o}t_3 = t_0 + 2t_1 + 2t_2 + 2t_3.\end{aligned}$$

It should be noted that if **o** is associative as a binary operation, then there is only one inequality (18–19). This is the case when **o** is the addition as in a metric space (X, d), with $d(x_0, x_{n+1}) \leq \sum_{i=0}^{n} d(x_i, x_{i+1})$, or when o is the multiplication as in multiplicative metric spaces $(X, d_\times)$, with $d_\times(x_0, x_{n+1}) \leq \prod_{i=0}^{n} d_\times(x_i, x_{i+1})$. In the non-associative case, it becomes necessary to define patterns that allow a function Δ_n to be expressed in function of some Δ_p and Δ_q, where $p + q = n$.

14.4.1 Patterns and Generalized Series

Definition 4.3 Let $\{\alpha_n\}_{n\in\mathbb{N}}$ be a sequence of integers such that $1 \leq \alpha_n \leq n-1$ for all $n \in \mathbb{N}$. A sequence $\{h_n\}_{n\in\mathbb{N}}$ of functions $h_n \in \Omega_{n-1}$ is said to follow the pattern of integers $\{\alpha_n\}_{n\in\mathbb{N}}$ if

$$h_n(t_1, t_2, ..., t_n) = h_{\alpha_n}(t_1, t_2, ..., t_{\alpha_n}) \ \mathbf{o} \ h_{n-\alpha_n}(t_{\alpha_n+1}, ..., t_n) \text{ for all } n \geq 2. \tag{20}$$

We consider the following example:

Example 4.4 Let $\{u_n\}$, $\{v_n\}$, $\{w_n\}$ and $\{z_n\}$ be sequences of functions defined as follow: If $n \in \mathbb{N}$ and $(t_1, t_2, ..., t_n) \in I_a^n$, then $u_1(t_1) = v_1(t_1) = w_1(t_1) = z_1(t_1) = t_1$, and for $n \geq 2$,

$$\begin{cases} u_n(t_1, t_2, ..., t_n) & = u_{n-1}(t_1, t_2, ..., t_{n-1}) \ \mathbf{o} \ t_n, \\ v_n(t_1, t_2, ..., t_n) & = v_p(t_1, t_2, ..., t_p) \ \mathbf{o} \ v_q(t_{p+1}, t_{p+2}, ..., t_n), \\ w_n(t_1, t_2, ..., t_n) & = w_{2^{l-1}}(t_1, ..., t_{2^{l-1}}) \ \mathbf{o} \ w_{n-2^{l-1}}(t_{2^{l-1}+1}, ..., t_n), \\ z_n(t_1, t_2, ..., t_n) & = t_1 \ \mathbf{o} \ z_{n-1}(t_2, ..., t_n), \end{cases} \tag{21}$$

where $p = \lceil \frac{n}{2} \rceil$, $q = \lfloor \frac{n}{2} \rfloor$, *and* $l = \lceil \log_2 n \rceil$. The sequence $\{u_n\}$ follows the pattern of integers $\alpha_n = n-1$. It can also be labelled **FIFO (First In, First Out)** in that the arguments $t_1, t_2, \cdots, t_n$ are composed by **o** in increasing order of indices. The sequence $\{z_n\}$ on the other hand can be labelled **LIFO (Last In First Out)** and follows the pattern of integers $\alpha_n = 1$. The sequence $\{v_n\}$ follows the pattern of integers $\alpha_n = \lceil \frac{n}{2} \rceil$ and is labelled **AISO (All In, Split Out)** while the sequence $\{w_n\}$ follows the pattern of integers $\alpha_n = 2^{\lceil \log_2 n \rceil - 1}$, with the arguments t_i split where the indice i is the highest power of 2.

In general, for an infinite sequence of points of real numbers, one can define **o**-series as a way of composing successively the terms following a given pattern. More precisely, we have the following definition:

Definition 4.5 (o-series) Consider a sequence $\{t_n\}_{n\geq 0}$ of points in I_a and an operation $\mathbf{o} : I_a \times I_a \to [0, \infty)$. The terms of the sequence can be composed successively starting from t_0 via the operation **o** as follows:

$$\begin{cases} \omega^0 & = t_0 \\ \omega^1 & = \mathbf{o}(t_0, t_1) \\ \omega^n & = h_{n+1}(t_0, t_1, t_2, ..., t_n), h_{n+1} \in \Omega_n, n \geq 2. \end{cases} \tag{22}$$

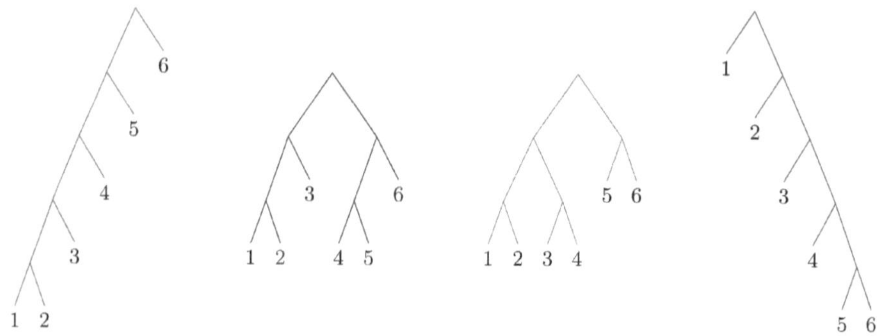

FIGURE 14.1 Binary trees for u_6, v_6, w_6 and z_6, from left to right.

The sequence $\{\omega^n\}_{n\geq 0}$ is defined to be the *sequence of partial compositions of* $\{t_n\}_{n\geq 0}$ following the pattern of functions $\{h_n\}_{n\in\mathbb{N}}$ and is denoted $\omega^n = \underset{i=0}{\overset{n}{\mathrm{O}}} t_i$. If $\{\omega^n\}_{n\in\mathbb{N}}$ converges, we say that $\{t_n\}_{n\geq 0}$ is *composable* and denote $\lim_{n\to\infty} \omega^n$ by $\underset{i=0}{\overset{\infty}{\mathrm{O}}} t_i$.

In general, the expression $\underset{i=0}{\overset{\infty}{\mathrm{O}}} t_i$ is called an infinite generalized series (or **o**-series) following the pattern of functions $\{h_n\}_{n\in\mathbb{N}}$, and $\{t_n\}_{n\geq 0}$ is called the sequence of terms, whether it is composable or not. If $\{t_n\}_{n\geq 0}$ is composable, the generalized series $\underset{i=1}{\overset{\infty}{\mathrm{O}}} t_i$ is said to converge; otherwise, it is said to diverge.

Given a sequence $\{t_n\}_{n\geq 0}$ of points in I_a, the n-th term ω^n of the sequence of partial compositions of $\{t_n\}_{n\geq 0}$ can be computed up to C_n ways (C_n being the n-th Catalan number). The following example serves as illustration:

Example 4.6 Consider $I_0 = [0, \infty)$ and o as in Example 4.2. If $t_n = n$ for any $n \geq 0$, then $\underset{i=0}{\overset{3}{\mathrm{O}}} t_i$ can be computed in $C_3 = 5$ ways:

$$\underset{i=0}{\overset{3}{\mathrm{O}}} t_i = \underset{i=0}{\overset{3}{\mathrm{O}}} i = \Delta_3^j(0,1,2,3) = \begin{cases} 34, & \text{for } j = 1 \\ 22, & \text{for } j = 2 \\ 18, & \text{for } j = 3 \\ 16, & \text{for } j = 4 \\ 12, & \text{for } j = 5. \end{cases}$$

It is therefore necessary to specify the pattern followed by the function $h_{n+1} \in \Omega_n$ (or Δ_n) used to compute the partial composition $\overset{n}{\underset{i=0}{\mathrm{O}}} t_i$. To this end, we adopt the following definition:

Definition 4.7 Let $\{\alpha_n\}_{n\in\mathbb{N}}$ be a sequence of integers such that $1 \le \alpha_n \le n-1$ for all $n \in \mathbb{N}$. If an **o**-series $\overset{\infty}{\underset{i=0}{\mathrm{O}}} t_i$ following the pattern of functions $\{h_n\}_{n\in\mathbb{N}}$ is such that $\{h_n\}_{n\in\mathbb{N}}$ follows the pattern of integers $\{\alpha_n\}_{n\in\mathbb{N}}$ in the sense of Definition 4.3, $\overset{\infty}{\underset{i=0}{\mathrm{O}}} t_i$ is also said to follow the pattern of integers $\{\alpha_n\}_{n\in\mathbb{N}}$.

Patterns are not necessary when the operation is **o** is associative as illustrated in the following example.

Example 4.8 Let $\{\omega^n\}_{n\ge 0}$ be the sequence of partial compositions of a sequence $\{t_n\}_{n\ge 0}$ of real numbers in the interval $[a,\infty)$, with $\mathbf{o} : [a,\infty) \times [a,\infty) \to [a,\infty)$ a function.

1. If $a = 0$ and $\mathbf{o}(u,v) = u+v$ for all $u, v \ge 0$, then whatever the pattern followed, **o**-series are series in the usual sense, i.e., $\overset{n}{\underset{i=0}{\mathrm{O}}} t_i = \sum_{i=0}^{n} t_i$ for $n \ge 0$;
2. If $\mathbf{o}(u,v) = \max\{u,v\}$ for all $u, v \ge a$, the sequence $\{\omega^n\}_{n\ge 0}$ of partial compositions of $\{t_n\}_{n\in\ge 0}$ following any pattern of functions is such that $\omega^n = \overset{n}{\underset{i=0}{\mathrm{O}}} t_i = \max\{t_0, t_2, \dots, t_n\}$, which means that $\overset{\infty}{\underset{i=0}{\mathrm{O}}} t_i = \lim_{n\to\infty} t_n$ if the sequence of terms $\{t_n\}_{n\in\mathbb{N}}$ is nondecreasing and $\overset{\infty}{\underset{i=0}{\mathrm{O}}} t_i = t_0$ if $\{t_n\}_{n\in\mathbb{N}}$ is non-increasing.
3. If $a = 1$ and $\mathbf{o}(u,v) = uv$ for all $u, v \ge 1$, then $\overset{n}{\underset{i=0}{\mathrm{O}}} t_i = \prod_{i=0}^{n} t_i$ for $n \ge 0$.

In general, when o satisfies conditions (E_1) and (E_2) with a function λ, then o is associative and given a sequence $\{t_n\}$ of numbers greater than or equal to a, we have for all $n \ge 0$,

$$\overset{n}{\underset{i=0}{\mathrm{O}}} t_i = \lambda^{-1}\left(\sum_{i=0}^{n} \lambda(t_i)\right). \tag{23}$$

In the next subsection, we consider a case when o is not associative.

14.4.2 Polygon O-Inequalities in b-Metric Spaces

Motivated by the relaxed triangle inequality in a b-metric space (X, d, s), let $\{\omega^n\}_{n\geq 0}$ be the sequence of partial compositions of a sequence $\{t_n\}_{n\geq 0} \subset [0, \infty)$, with o defined by $\mathbf{o}(u, v) = s(u + v)$ for all $u, v \geq 0$, where $s > 0$ is a constant. If the o-series follows the pattern of integer $\{1\}_{n\in\mathbb{N}}$, then for $n \geq 2$, $\underset{i=0}{\overset{n}{\mathrm{O}}} t_i = z_{n+1}(t_0, t_1, ..., t_n)$, where $\{z_n\}$ is defined as in (21):

$$\begin{aligned}
\underset{i=0}{\overset{n}{\mathrm{O}}} t_i = z_{n+1}(t_0, t_1, ..., t_n) = s\left[t_0 + z_n(t_1, ..., t_n)\right] \\
&= st_0 + sz_n(t_1, ..., t_n) \\
&= st_0 + s^2\left[t_1 + z_{n-1}(t_2, ..., t_n)\right] \\
&= st_0 + s^2 t_1 + s^2 z_{n-1}(t_2, ..., t_n) \\
&\vdots \\
&= \sum_{i=1}^{n-1} s^i t_{i-1} + s^{n-1} z_2(t_{n-1}, t_n) \\
&= \sum_{i=1}^{n-1} s^i t_{i-1} + s^n t_{n-1} + s^n t_n \\
&= \sum_{i=1}^{n} s^i t_{i-1} + s^n t_n. \qquad (24)
\end{aligned}$$

Therefore, for any $n \in \mathbb{N}$, $\underset{i=0}{\overset{n}{\mathrm{O}}} t_i = z_{n+1}(t_0, t_1, ..., t_n) = \sum_{i=1}^{n} s^i t_{i-1} + s^n t_n$. Similarly, if the o-series follows the pattern of integers $\{n-1\}_{n\in\mathbb{N}}$, then

$$\underset{i=0}{\overset{n}{\mathrm{O}}} t_i = u_{n+1}(t_0, t_1, ..., t_n) = s^n t_0 + \sum_{i=1}^{n} s^i t_{n-i+1} \ \forall n \in \mathbb{N}, \qquad (25)$$

where $\{u_n\}$ is defined as in (21). Now, consider the **o**-series following the pattern of integers $\{2^{\lceil \log_2 n\rceil - 1}\}_{n\in\mathbb{N}}$. Then $\underset{i=1}{\overset{n}{\mathrm{O}}} t_i = w_n(t_1, t_2, \cdots, t_n)$ for all $n \in \mathbb{N}$, where $\{w_n\}$ is defined as in (21). If $l = l(n) = \lceil \log_2(n)\rceil$, with $n \geq 2$, then $1 \leq 2^{l-1} < n \leq 2^l$ and

$$w_n(t_1, t_2, ..., t_n) = s\left(w_{2^{l-1}}(t_1, ..., t_{2^{l-1}}) + w_{n-2^{l-1}}(t_{2^{l-1}+1}, ..., t_n)\right). \qquad (26)$$

By a simple recursion, for any $r \geq 1$,

$$\begin{aligned} w_{2^r}(t_1, t_2, ..., t_{2^r}) &= s\left[w_{2^{r-1}}(t_1, t_2, ..., t_{2^{r-1}}) + w_{2^{r-1}}(t_{2^{r-1}+1}, t_{2^{r-1}+2}, ..., t_{2^r})\right] \\ &= s^2\left[w_{2^{r-2}}(t_1, t_2, ..., t_{2^{r-2}}) + w_{2^{r-2}}(t_{2^{r-2}+1}, t_{2^{r-2}+2}, ..., t_{2^{r-1}})\right. \\ &\quad \left. + w_{2^{r-2}}(t_{2^{r-1}+1}, ..., t_{3\times 2^{r-2}}) + w_{2^{r-2}}(t_{3\times 2^{r-2}+1}, ..., t_{2^r})\right] \\ &\ \vdots \\ &= s^r \sum_{i=1}^{2^r} w_1(t_i) = s^r \sum_{i=1}^{2^r} t_i. \end{aligned} \tag{27}$$

$$\text{Thus}\begin{cases} w_{2^{l-1}}(t_1, ..., t_{2^{l-1}}) = s^{l-1}\sum_{i=1}^{2^{l-1}} t_i \\ w_{n-2^{l-1}}(t_{2^{l-1}+1}, ..., t_n) \leq w_{2^{l-1}}(t_{2^{l-1}+1}, ..., t_n, 0, 0, ..., 0) = s^{l-1}\sum_{i=2^{l-1}+1}^{n} t_i. \end{cases}$$

Hence, for $n \geq 2$,

$$w_n(t_1, t_2, ..., t_n) \leq s\left[s^{l-1}\sum_{i=1}^{2^{l-1}} t_i + s^{l-1}\sum_{i=2^{l-1}+1}^{n} t_i\right] = s^l \sum_{i=1}^{n} t_i. \tag{28}$$

In fact, repeating the processes in (26) and (27), we have for n sufficiently large, align

$$\begin{aligned} & w_n(t_1, t_2, ..., t_n) \\ &= s\left(w_{2^{l(n)-1}}(t_1, ..., t_{2^{l(n)-1}}) + w_{n-2^{l(n)-1}}(t_{2^{l(n)-1}+1}, ..., t_n)\right) \\ &= s\left(s^{l(n)-1}\sum_{i=1}^{2^{l(n)-1}} t_i + w_{n-2^{l(n)-1}}(t_{2^{l(n)-1}+1}, ..., t_n)\right) \\ &= s\left(s^{l_0-1}\sum_{i=1}^{n_1} t_i + w_{n-n_1}(t_{n_1+1}, ..., t_n)\right), l_0 = l(n), n_1 = 2^{l_0-1}, \\ &= s^{l_0}\sum_{i=1}^{n_1} t_i + s w_{n-n_1}(t_{n_1+1}, ..., t_n) \\ &= s^{l_0}\sum_{i=1}^{n_1} t_i + s\left[s^{l_1}\sum_{i=n_1+1}^{n_2} t_i + s w_{n-n_2}(t_{n_2+1}, ..., t_n)\right] \end{aligned}$$

where $\quad l_1 = l(n - n_1), n_2 = n_1 + 2^{l_1 - 1},$

$$= s^{l_0} \sum_{i=1}^{n_1} t_i + s^{1+l_1} \sum_{i=n_1+1}^{n_2} t_i + s^2 w_{n-n_2} (t_{n_2+1}, \dots, t_n)$$

$$\vdots$$

$$= s^{l_0} \sum_{i=1}^{n_1} t_i + s^{1+l_1} \sum_{i=n_1+1}^{n_2} t_i + \cdots + s^{r-1+l_{r-1}} \sum_{i=n_{r-1}+1}^{n_r} t_i$$

$$+ s^r w_{n-n_r} (t_{n_r+1}, \dots, t_n), \tag{29}$$

where $r \in \mathbb{N}$ is such that $n - n_r \geq 1$, and $l_j = \lceil \log_2 (n - n_j) \rceil$ and $n_{j+1} = n_j + 2^{l_j - 1}$ for $j \in [0, r]$, with $n_0 = 0$. The sequence (n_j) of integers is strictly increasing and bounded above by n hence finite: there is $N \in \mathbb{N}$ such that $N - 1 = \max \{r \in \mathbb{N} : n - n_r \geq 2\}$. By definition of N, $n - n_N < 2$. Suppose $n = n_N$. Then $l_{N-1} = \lceil \log_2 (n - n_{N-1}) \rceil = \lceil \log_2 (n_N - n_{N-1}) \rceil = l_{N-1} - 1$, a contradiction. Thus $n - n_N = 1$ so $n_N = n - 1$. Therefore, from (29), taking $r = N$, the last term of the inequality becomes $s^N w_1 (t_n) = s^N t_n$ so that:

$$\bigcirc_{i=1}^{n} t_i = w_n (t_1, t_2, \cdots, t_n) = \sum_{r=0}^{N-1} \left(s^{r+l_r} \sum_{i=n_r+1}^{n_{r+1}} t_i \right) + s^N t_n$$

$$= \sum_{r=0}^{N} \left(s^{r+l_r} \sum_{i=n_r+1}^{n_{r+1}} t_i \right), \tag{30}$$

where $(n_r)_{0 \leq r \leq N+1}$ and $(l_r)_{0 \leq r \leq N}$ are such that:

$$\begin{cases} n_0 & = 0 \\ n_{j+1} & = n_j + 2^{l_j - 1} = \sum_{r=0}^{j} 2^{l_r - 1}, \ j \in \{0, 1, \dots, N-1\} \\ N - 1 & = \max \{r \in \mathbb{N} : n - n_r \geq 2\} \\ l_j & = \lceil \log_2 (n - n_j) \rceil, \ j \in \{0, 1, \dots, N\} \\ n_{N+1} & = n_N + 1 = n. \end{cases} \tag{31}$$

The sequence (n_j) in (31) provides the binary representation of $n \in \mathbb{N}$. Indeed, from (31),

$$n = n_N + 1 = \sum_{r=0}^{N-1} 2^{l_r - 1} + 1 = 2^{l_0 - 1} + 2^{l_1 - 1} + \cdots + 2^{l_{N-1} - 1} + 1.$$

Therefore, $n = \sum_{j=0}^{k} a_j 2^j$, where $k = l_0 - 1$, and

$$a_j = \begin{cases} 1, & \text{if } j \in \{0, l_0 - 1, l_1 - 1, \cdots, l_{N-1} - 1\} \\ 0, & \text{else where.} \end{cases}$$

As application, we determine some polygon (relaxed) inequalities that hold in b-metric spaces in the following proposition:

Proposition 4.9 (Polygon Inequalities in a b-Metric Space) *Let (X, d, s) be a b-metric space, with $s \geq 1$. Then given $n+2$ points $x_0, x_1, \cdots, x_{n+1}$, where $n \in \mathbb{N}$, the following polygon inequalities hold:*

$$d(x_0, x_{n+1}) \leq \sum_{i=1}^{n+1} a_i d(x_{i-1}, x_i) \tag{32}$$

with $(a_i)_{1 \leq i \leq n+1}$ a sequence of nonnegative real numbers such that either of the following hold:

1. *There exist $p \neq q \in \{1, 2, ..., n+1\}$ such that $a_p = a_q = s^n$, with the other a_i's distinct and equal to some power s^r of s, with $1 \leq r < n-1$.*
2. *The a_i's are constant, $a_i = K$, where $K = \frac{1}{n+1}\left(\frac{2s^{n+1} - s^n - s}{s-1}\right)$ or $K = s^{\lceil \log_2(n+1) \rceil}$.*

Proof. Under the conditions of the proposition, the polygon inequality (18) holds for any $\Delta_n \in \Omega_n$:

$$d_{\mathbf{o}}(x_0, x_{n+1}) \leq \Delta_n (d_{\mathbf{o}}(x_0, x_1), d_{\mathbf{o}}(x_1, x_2), ..., d_{\mathbf{o}}(x_n, x_{n+1})).$$

If we let $\Delta_n = z_{n+1}$, where (z_n) is the recursion defined in (21), then from (24), we obtain:

$$d(x_0, x_{n+1}) \leq \sum_{i=1}^{n} s^i d(x_{i-1}, x_i) + s^n d(x_n, x_{n+1}). \tag{33}$$

If we let $\Delta_n = u_{n+1}$, where (u_n) is the recursion defined in (21), then from (25), we obtain:

$$d(x_0, x_{n+1}) \leq s^n d(x_0, x_1) + \sum_{i=1}^{n} s^i d(x_{n-i+1}, x_{n-i+2}). \tag{34}$$

Rearranging the points x_i, $1 \leq i \leq n$, we obtain $d(x_0, x_{n+1}) \leq \sum_{i=1}^{n+1} a_i d(x_{i-1}, x_i)$, where $(a_i)_{1 \leq i \leq n+1}$ satisfies condition 1. In fact, if one sums $n+1$ of such inequalities $d(x_0, x_{n+1}) \leq \sum_{i=1}^{n+1} a_{\sigma_j(i)} d(x_{i-1}, x_i)$, $1 \leq j \leq n+1$, with permutations $\sigma_j \in S_{n+1}$ chosen such that for each i, the $\sigma_j(i)$ are distinct, then

$$(n+1)\, d(x_0, x_{n+1}) \leq \left(\sum_{i=1}^{n} s^i + s^n\right) \sum_{i=1}^{n+1} d(x_{i-1}, x_i).$$

Therefore,

$$\begin{cases} d(x_0, x_{n+1}) \leq \sum_{i=1}^{n+1} K d(x_{i-1}, x_i), \\ \text{where, } K = \frac{1}{n+1}\left(\sum_{i=1}^{n} s^i + s^n\right) = \frac{1}{n+1}\left(\frac{2s^{n+1} - s^n - s}{s-1}\right). \end{cases} \tag{35}$$

If in (18) we let $\Delta_n = w_{n+1}$, where (w_n) is the recursion defined in (21), then from (28), we obtain:

$$d(x_0, x_{n+1}) \leq \sum_{i=1}^{n+1} s^{\lceil \log_2 (n+1) \rceil} d(x_{i-1}, x_i). \tag{36}$$

From (35) and (36), condition 2 holds. □

It should be noted that polygon inequalities are used to prove that "contractive" sequences are Cauchy sequences. In a b-metric space (X, d, s), a sequence $\{x_n\}_{n \geq 0}$ is said to be contractive if $d(x_n, x_{n+1}) \leq k d(x_{n-1}, x_n)$ for all $n \in \mathbb{N}$, where $k \in [0, 1)$. Suzuki combined inequalities of type (32) to show that contractive sequences in b-metric spaces are Cauchy sequences.

14.4.3 s-Constrained Triangle Inequalities and Infinite Symmetric Matrices

Note that if the constant s in Definition 1.1 is taken less than 1 ($s < 1$), the set X in (X, d, s) is a singleton. On the other hand, the discrete metric on any nonempty set X satisfies (d_1), (d_2) and the s-constrained triangle inequality $d(x, z) \leq s[d(x, y) + d(y, z)]$ for all distinct $x, y, z \in X$ with $s = \frac{1}{2}$, although any infinite sequence $\{x_n\}_{n \in \mathbb{N}}$ of distinct points in the discrete metric is such that $\sum_{i=1}^{\infty} d(x_i, x_{i+1}) = \infty$. We obtain the following characterization of metric spaces satisfying the s-constrained triangle inequality on every non-degenerate triangle.

Proposition 4.10 *Let (X, d) be a metric space. There is no infinite sequence $\{x_n\}$ of distinct points satisfying the s-constrained triangle inequality for some $s \in [0, 1)$ such that $\sum_{i=1}^{\infty} d(x_i, x_{i+1})$ converges.*

Proof. Suppose $\{x_n\}$ is an infinite sequence of distinct points satisfying the s-constrained triangle inequality, and that $\sum_{i=1}^{\infty} d(x_i, x_{i+1})$ converges. Following (36), we obtain that for each $i \in \mathbb{N}$,

$$d(x_i, x_n) \leq s^{g(n-i)} \sum_{j=i}^{n-1} d(x_j, x_{j+1}) \leq s^{g(n-i)} \sum_{j=1}^{n-1} d(x_j, x_{j+1}), \forall n \in \mathbb{N}, \quad (37)$$

where $g(n) = [\log_2 (n+1)] + 1$ (with $[t]$ denoting the maximum integer not exceeding a real number t). Thus, $\lim_{n\to\infty} d(x_i, x_n) = 0$ for each $i \in \mathbb{N}$, which implies that each term is a limit of $\{x_n\}$, an absurd statement. □

From Proposition 4.10, we note the following:

Remark 4.11 (Characteristics of Infinite Metric Spaces Satisfying an s-Constrained Triangle Inequality in Non-Degenerate[6] Triangles) In an infinite metric space (X, d) (i.e., a metric space with an infinite number of points) satisfying the s-constrained triangle inequality on non-degenerate triangles (i.e., $d(x, y) \leq s[d(x, z) + d(z, y)]$ for x, y, z distinct, where $s < 1$):

1. Any infinite sequence $\{x_n\}$ of distinct points is such that $\sum_{i=1}^{\infty} d(x_i, x_{i+1}) = \infty$;
2. Any infinite sequence $\{x_n\}$ of distinct points cannot be a Cauchy sequence. Equivalently, any Cauchy sequence must be finite, and hence convergent;
3. X is a complete metric space;
4. Any non-trivial[7] curve is of infinite length: for any continuous map $\gamma : [0, T] \to X$,

$$\sup \left\{ \sum_{i=1}^{n} d(\gamma(t_{i-1}), \gamma(t_i)) : 0 = t_0 < t_1 < \dots < t_n = T \right\} = \infty.$$

[6] A non-degenerate triangle is a triple of distinct points.
[7] A non-trivial curve $\gamma : [0, T] \to X$ is such that $\gamma([0, T])$ is not a singleton.

It turns out that the proof of Proposition 4.10 can be adapted to obtain a parallel result on the maximization of "triangular quotients"[8] in some infinite symmetric matrices:

Proposition 4.12 *Let $A = \left(\alpha_{ij}\right)_{i,j\geq 1}$ be a symmetric infinite matrix with positive real numbers as non-diagonal entries, zeros as diagonal entries, and such that the sum $\sum_{i=1}^{\infty}\alpha_{i,i+1}$ of the super-diagonal entries is finite and the triangle inequality $\alpha_{ij} \leq \alpha_{ik} + \alpha_{kj}$ is satisfied for all i, j, k. Then*

$$\sup_{\substack{i,j,k \in \mathbb{N} \\ i \neq k \neq j}} \frac{\alpha_{ij}}{\alpha_{ik} + \alpha_{kj}} = 1.$$

Proof. The infinite set $X = \mathbb{N}$ can be equipped with the metric $d(i,j) = \alpha_{ij}$ for $i, j \geq 1$. The sequence $\{i\}_{i\in\mathbb{N}}$ is an infinite sequence of distinct points in X such that $\sum_{i=1}^{\infty} d(i, i+1) < \infty$. By Proposition 4.12, no s-constrained triangle inequality (where $s < 1$) can hold for terms of the sequence $\{i\}_{i\in\mathbb{N}}$: for any $s < 1$, there are $i, j, k \in \mathbb{N}$ such that $d(i,j) > s(d(i,k) + d(k,j))$. By the triangle inequality,

$$\sup_{\substack{i,j,k \in \mathbb{N} \\ i \neq k \neq j}} \frac{\alpha_{ij}}{\alpha_{ik} + \alpha_{kj}} = \sup_{\substack{i,j,k \in \mathbb{N} \\ i \neq k \neq j}} \frac{d(i,j)}{d(i,k) + d(k,j)} = 1.$$

□

The following example shows that the condition $\sum_{i=1}^{\infty}\alpha_{i,i+1} < \infty$ in Proposition 4.12 is necessary and cannot be omitted.

Example 4.13 The following matrix $B = \left(\beta_{ij}\right)$ is symmetric, infinite, with positive real numbers as non-diagonal entries, zeros as diagonal entries, and

[8] Triangular quotients refer to the quotients $\frac{\alpha_{ij}}{\alpha_{ik}+\alpha_{kj}}$, $i \neq k \neq j$, in Proposition 4.12.

such that the triangle inequality $\beta_{ij} \leq \beta_{ik} + \beta_{kj}$ is satisfied for all i, j, k:

$$B = \begin{pmatrix} 0 & 1 & 1 & 1 & \cdots \\ 1 & 0 & 1 & 1 & \cdots \\ 1 & 1 & 0 & 1 & \cdots \\ 1 & 1 & 1 & 0 & \cdots \\ \vdots & \vdots & \vdots & \vdots & \ddots \end{pmatrix}$$

However,

$$\sup_{\substack{i,j,k \in \mathbb{N} \\ i \neq k \neq j}} \frac{\beta_{ij}}{\beta_{ik} + \beta_{kj}} = \frac{1}{2} < 1.$$

14.5 FIXED POINT THEORY IN O-METRIC SPACES

The aim of this section is to state and prove the Banach Contraction Principle in the setting of O-metric spaces. More general fixed point theorems can be proved using similar methodology. We begin by constructing types of contractions that are somewhat compatible with the O-metric structure.

14.5.1 k-ϕ Lipschitz Maps and Contractions

Definition 5.1 Let $(X, d_{\mathbf{o}}, a)$ be an a-upward O-metric space, with o non-decreasing in both variables, continuous at (a, a) and $\mathbf{o}(a, a) = a$. Let $\phi : [0, \infty) \times [a, \infty) \to [a, \infty)$ be a function satisfying the following conditions:

- $\phi(0, t) = \phi(r, a) = a$ for all $t \geq a$ and $r \geq 0$;
- $\phi|_{(0,\infty)\times(a,\infty)}$ is increasing on both variables, and continuous in the second variable at a;
- $\forall r_1, r_2 \in [0, \infty) \forall t \in [a, \infty), \phi(r_1, \phi(r_2, t)) = \phi(r_1 r_2, t)$.

A map $T : X \to X$ is said to be k-ϕ *Lipschitz* on X, with $k \geq 0$, if

$$d_{\mathbf{o}}(Tx, Ty) \leq \phi(k, d_{\mathbf{o}}(x, y)) \quad \forall x, y \in X. \tag{38}$$

The k-ϕ Lipschitz map $T : X \to X$ is said to be a *contraction* if $k < 1$. A sequence $\{x_n\}_{n\geq 0}$ of points in X is said to be a k-ϕ *contractive* sequence if for all $n \in \mathbb{N}$,

$$d_{\mathbf{o}}(x_n, x_{n+1}) \leq \phi(k, d_{\mathbf{o}}(x_{n-1}, x_n)). \tag{39}$$

The following are examples of mappings satisfying conditions $(\phi_1) - (\phi_3)$.

Example 5.2 When $a = 0$, the following maps $\phi\colon [0,\infty)\times[0,\infty) \to [0,\infty)$ satisfying conditions $(\phi_1) - (\phi_3)$:

1. $\phi(t,u) = tu$ for all $t, u \geq 0$.
2. $\phi(t,u) = (1+u)^t - 1$ for all $t, u \geq 0$.
3. $\phi(t,u) = \ln(1 - t + te^u)$ for all $t, u \geq 0$.

Example 5.3 The map $\phi\colon [0,\infty)\times[1,\infty) \to [1,\infty)$ defined by $\phi(t,u) = u^t$ satisfies conditions $(\phi_1) - (\phi_3)$ when $a = 1$.

Example 5.4 Given an a-upward O-metric space $(X, d_{\mathbf{o}}, a)$, if $\lambda\colon [a,\infty) \to [0,\infty)$ is an increasing function satisfying $\lambda(a) = 0$ and $\lambda(u \mathbf{o} v) = \lambda(u) + \lambda(v)$ (as in conditions (E_1) and (E_2) of Theorem 2.14), the map $\phi\colon [0,\infty) \times [a,\infty) \to [a,\infty)$ defined by $\phi(t,u) = \lambda^{-1}(t\lambda(u))$ satisfies conditions $(\phi_1) - (\phi_3)$.

We note the following about maps satisfying condition (38):

Remark 5.5 Let $(X, d_{\mathbf{o}}, a)$ be an a-upward O-metric space.

1. The term *contraction* is justified. Indeed, let T be a contraction, with the inequality $d_{\mathbf{o}}(Tx, Ty) \leq \phi(k, d_{\mathbf{o}}(x,y))$ for all $x, y \in X$ and some $k < 1$; then T contracts the symmetric function D_r defined by $D_r(x,y) = \phi(r, d_{\mathbf{o}}(x,y))$ for some $r > 0$ and all $x, y \in X$. In fact, D_r is an a-upward $\mathbf{o}$-metric if $\phi(r, \mathbf{o}(t_1,t_2)) \leq \mathbf{o}(\phi(r,t_1), \phi(r,t_2))$ for all $t_1, t_2 \geq a$.
2. As expected of *Lipschitz* maps, a k-ϕ Lipschitz map $T : X \to X$ as defined in Definition 5.1 is continuous. Indeed, let G be an open set in X (for the O-metric topology $\mathcal{T}$). To show that $T^{-1}(G)$ is an open set for the topology, we let $x \in T^{-1}(G)$. Since $Tx \in G$, then there is $r > 0$ such that $B(Tx, r) \subset G$. Choose $\delta > 0$ such that

$\phi(k, a+\delta) - a = r$. For any $y \in B(x, \delta)$, since $d_{\mathbf{o}}(x, y) < a + \delta$, we have that $d_{\mathbf{o}}(Tx, Ty) - a \leq \phi(k, d_{\mathbf{o}}(x, y)) - a < \phi(k, a+\delta) - a = r$. Thus $Ty \in B(Tx, r) \subset G$ and $y \in T^{-1}(G)$.

3. A k-ϕ Lipschitz map $T : X \to X$ also preserves O-convergence (i.e., T is sequentially continuous): if a sequence $\{x_n\}$ of points in X is such that $x_n \overset{\mathrm{O}}{\to} x$, then, for each $n \in \mathbb{N}$, $a \leq d_{\mathbf{o}}(Tx_n, Tx) \leq \phi(k, d_{\mathbf{o}}(x_n, x))$, hence, as $n \to \infty$, $Tx_n \overset{\mathrm{O}}{\to} Tx$.

In order to determine values of k for which a k-ϕ contractive sequence is a Cauchy sequence, we introduce the set C_ϕ as in the proposition below.

Proposition 5.6 *Let o be nondecreasing in both variables, continuous at (a, a) and such that $\mathbf{o}(a, a) = a$. For a function $\phi : [0, \infty) \times [a, \infty) \to [a, \infty)$ satisfying $(\phi_1) - (\phi_3)$, define the set C_ϕ by:*

$$\begin{array}{l} C_\phi = \{r \geq 0 | \, \forall \epsilon \geq a \lim_{n,i\to\infty} h_{n,i}(r, \epsilon) = a\}, \\ \text{where } h_{n,i}(r, \epsilon) = h\left(\phi(r^n, \epsilon), \ldots, \phi\left(r^{n+i}, \epsilon\right)\right) \text{ for some } h \in \Omega_i. \end{array} \tag{40}$$

Then C_ϕ is an interval such that $0 \in C_\phi \subset [0, 1)$.

Proof. If $r = 0$, then $\phi(r^n, \epsilon) = \ldots = \phi\left(r^{n+i}, \epsilon\right) = a$ for all $\epsilon > a$ and $n, i \in \mathbb{N}_0$, hence $h_{n,i}(r, \epsilon) = h(a, a, \ldots, a) = a$ for all $h \in \Omega_i$. Therefore $0 \in C_\phi$. Suppose $r \in C_\phi$ and $s \in [0, r]$. Since ϕ is nondecreasing in the first variable, $\phi(s^m, \epsilon) \leq \phi(r^m, \epsilon)$ for $m \in \{n, n+1, \ldots, n+i\}$ with $n, i \in \mathbb{N}_0$. As o is nondecreasing in both variables, h is nondecreasing in all variables hence $h_{n,i}(s, \epsilon) \leq h_{n,i}(r, \epsilon)$ for all $h \in \Omega_i$. Thus $s \in C_\phi$ and C_ϕ is an interval. Let $r = 1$. For all $n, i \in \mathbb{N}_0$, $\epsilon > a$ and for any $h \in \Omega_i$, $h_{n,i}(1, \epsilon) = h(\phi(1, \epsilon), \ldots, \phi(1, \epsilon))$. $h_{n,i}(1, \epsilon) \nrightarrow a$ for $\phi(1, \epsilon) > a$ hence $1 \notin C_\phi$. Therefore C_ϕ is an interval, $0 \in C_\phi \subset [0, 1)$ and $sup C_\phi \leq 1$. □

In the next lemmas, we find C_ϕ for some maps ϕ satisfying conditions $(\phi_1) - (\phi_3)$.

Lemma 5.7 *Suppose* $\mathbf{o}$ *is nondecreasing in both variables, continuous at (a, a) with $\mathbf{o}(a, a) = a$, and satisfying conditions (E_1) and (E_2) of Theorem 2.14 for a function $\lambda\colon [a, \infty) \to [0, \infty)$. If $\phi\colon [0, \infty) \times [a, \infty) \to [a, \infty)$ is defined by $\phi(t, u) = \lambda^{-1}(t\lambda(u)) \forall t \geq 0 \;\; \forall u \geq a$, then $C_\phi = [0, 1)$.*

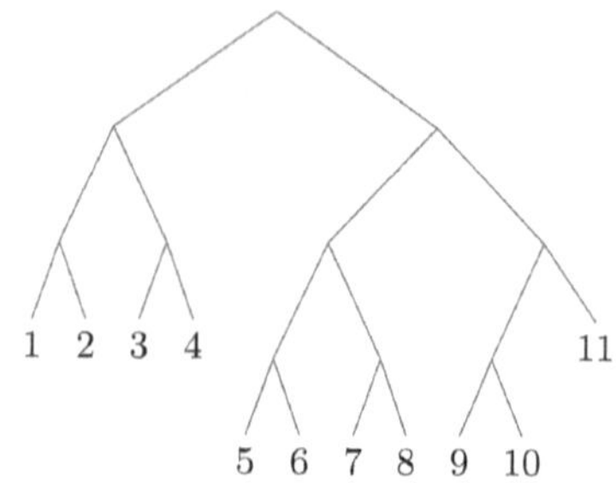

FIGURE 14.2 Binary tree for h defined in (42) for $i = 10, l = 2, \mu = 2$.

Proof. It is easy to check that ϕ so defined satisfies conditions $(\phi_1) - (\phi_3)$. Also, o is associative. Thus, if $r \geq 0, \epsilon > a, i \in \mathbb{N}_0$ and $h \in \Omega_i$, then by (23),

$$h(t_0, t_1, \cdots, t_n) = \underset{j=0}{\overset{n}{\mathrm{O}}} t_j = \lambda^{-1}\left(\sum_{j=0}^{n} \lambda(t_j)\right).$$

for $t_1, t_2, ..., t_{i+1} \geq a$. Therefore,

$$\begin{aligned}
\lambda(h_{n,i}(r, \epsilon)) &= \lambda(h(\phi(r^n, \epsilon), \phi(r^{n+1}, \epsilon), ..., \phi(r^{n+i}, \epsilon))) \\
&= \sum_{j=0}^{i} \lambda(\phi(r^{n+j}, \epsilon)) \\
&= \sum_{j=0}^{i} r^{n+j} \lambda(\epsilon) \\
&= \frac{1 - r^{i+1}}{1 - r} r^n \lambda(\epsilon) \to 0 \text{ as } n, i \to \infty \text{ if and only if } r < 1.
\end{aligned}$$

Thus $C_\phi = [0, 1)$ and $\sup C_\phi = 1$. □

Lemma 5.8 *If* **o** *is defined by* $\mathbf{o}(u, v) = s(u + v)$ *for all* $u, v \geq 0$, *where* s *is a constant greater than or equal to* 1, *then* $C_\phi = [0, 1)$ *for* $\phi\colon [0, \infty) \times [0, \infty) \to [0, \infty)$ *defined by* $\phi(t, u) = tu$ *for all* $t, u \geq 0$.

Proof. The function **o** so defined is nondecreasing in both variables, continuous at $(0, 0)$ and such that $\mathbf{o}(0, 0) = 0$. Furthermore, the map ϕ defined by $\phi(t, u) = tu$ for all $t, u \geq 0$ satisfies conditions $(\phi_1) - (\phi_3)$ for $a = 0$. Let $r \in (0, 1)$, $\epsilon > 0$, $n, i \in \mathbb{N}_0$. Let $l \in \mathbb{N}$ be such that $sr^{2^l} < 1$. If $i + 1 \leq 2^l$ then taking $h = w_{i+1} \in \Omega_i$, where $\{w_n\}$ is defined as in (21), we have from (28) that:

$$\begin{aligned} h_{n,i}(r,\epsilon) &= w_{i+1}\left(r^n\epsilon, ..., r^{n+i}\epsilon\right) \\ &\le s^{\lceil \log_2(i+1) \rceil} \sum_{j=0}^{i} r^{n+j}\epsilon \\ &\le s^l r^n \sum_{j=0}^{\infty} r^j \epsilon = s^l r^n \frac{\epsilon}{1-r} \to 0 \text{ as } n, i \to \infty. \end{aligned} \tag{41}$$

If $2^l < i+1$, then putting $\mu = \lfloor \frac{i+1}{2^l} \rfloor$, we take $h \in \Omega_i$ defined for all $t_1, ..., t_{i+1} \ge 0$ by

$$\begin{aligned} h\left(t_1, t_2, ..., t_{i+1}\right) =& z_{\mu+1}\left(w_{2^l}\left(T_1\right), w_{2^l}\left(T_2\right), ..., w_{2^l}\left(T_\mu\right), w_{i+1-\mu 2^l}\right. \\ & \times \left.\left(t_{\mu 2^l+1}, ..., t_{i+1}\right)\right) \end{aligned} \tag{42}$$

where $T_j = \left(t_{(j-1)2^l+1}, ..., t_{j2^l}\right)$ for $1 \le j \le \mu$ and $z_{\mu+1} \in \Omega_\mu$ as defined in (21). If we write $R_j = \left(r^{n+(j-1)2^l}\epsilon, ..., r^{n+j2^l-1}\epsilon\right)$, then

$$h_{n,i}(r,\epsilon) = z_{\mu+1}\left(w_{2^l}\left(R_1\right), w_{2^l}\left(R_2\right), ..., w_{2^l}\left(R_\mu\right), w_{i+1-\mu 2^l}\left(r^{n+\mu 2^l}\epsilon, ..., r^{n+i}\epsilon\right)\right).$$

From (41), $w_{2^l}\left(R_j\right) \le \frac{s^l r^{n+(j-1)2^l}\epsilon}{1-r}$ for all j. Since **o** is nondecreasing in both variables, $z_{\mu+1}$ is nondecreasing in all variables hence from (41) and (24),

$$\begin{aligned} h_{n,i}(r,\epsilon) &\le z_{\mu+1}\left(\frac{s^l r^n\epsilon}{1-r}, \frac{s^l r^{n+2^l}\epsilon}{1-r}, ..., \frac{s^l r^{n+(\mu-1)2^l}\epsilon}{1-r}, w_{i+1-\mu 2^l}\left(r^{n+\mu 2^l}\epsilon, ..., r^{n+i}\epsilon\right)\right) \\ &\le z_{\mu+1}\left(\frac{s^l r^n\epsilon}{1-r}, \frac{s^l r^{n+2^l}\epsilon}{1-r}, ..., \frac{s^l r^{n+(\mu-1)2^l}\epsilon}{1-r}, \frac{s^l r^{n+\mu 2^l}\epsilon}{1-r}\right) \\ &= \sum_{j=1}^{\mu}\left[\frac{s^{j+l} r^{n+(j-1)2^l}\epsilon}{1-r}\right] + \frac{s^{\mu+l} r^{n+\mu 2^l}\epsilon}{1-r} \\ &\le \frac{r^n s^l \epsilon}{1-r} \sum_{j=1}^{\mu+1} s^j r^{(j-1)2^l} \le \frac{r^n s^l \epsilon}{1-r} \sum_{j=1}^{\mu+1} \left(s r^{2^l}\right)^j \to 0 \text{ as } n \to \infty. \end{aligned}$$

Thus $r \in C_\phi$ and so $C_\phi = [0,1)$. □

14.5.2 The Banach Contraction Principle

Now, we prove the Banach Contraction Principle (see [20]) in the setting of O-metric spaces:

Theorem 5.9 *(Banach Contraction Principle). Let $(X, d_{\mathbf{o}}, a)$ be a complete a-upward O-metric space, with o nondecreasing in both variables, continuous at (a, a), and $\mathbf{o}(a, a) = a$. Let $T{:}X \to X$ be a k-ϕ contraction, where $\phi{:}\ [0, \infty) \times [a, \infty) \to [a, \infty)$ satisfies conditions $(\phi_1) - (\phi_3)$, and $k < \kappa$, with $\kappa{:}{=}\sup C_{\phi}$. Then T has a unique fixed point.*

Proof. Let $x_0 \in X$ and $\{x_n\}_{n\in\mathbb{N}}$ be a sequence[9] such that $x_{n+1} = Tx_n$ for $n \geq 0$. For $n \geq 0$,

$$\begin{aligned} d_{\mathbf{o}}(x_n, x_{n+1}) = d_{\mathbf{o}}(Tx_{n-1}, Tx_n) &\leq \phi(k, d_{\mathbf{o}}(x_{n-1}, x_n)) \\ &\leq \phi(k, \phi(k, d_{\mathbf{o}}(x_{n-2}, x_{n-1}))) \\ &= \phi(k^2, d_{\mathbf{o}}(x_{n-2}, x_{n-1})) \\ &\ \ \vdots \\ &\leq \phi(k^n, d_{\mathbf{o}}(x_0, x_1)). \end{aligned} \tag{43}$$

From Lemma 4.1, $d_{\mathbf{o}}(x_n, x_{n+i}) \leq h(d(x_n, x_{n+1}), d(x_{n+1}, x_{n+2}), \ldots, d(x_{n+i-1}, x_{n+i}))$ for all $h \in \Omega_{i-1}$, $n, i \in \mathbb{N}$. Since h is nondecreasing in all its variables, and for all $n \in \mathbb{N}$, $d_{\mathbf{o}}(x_n, x_{n+1}) \leq \phi(k^n, d_{\mathbf{o}}(x_0, x_1))$, we have:

$$d_{\mathbf{o}}(x_n, x_{n+i}) \leq h_{n,i-1}(k, d_{\mathbf{o}}(x_0, x_1)) \quad \forall n, i \in \mathbb{N}.$$

If $d_{\mathbf{o}}(x_0, x_1) = a$, then $x_0 = x_1 = Tx_0$ and x_0 is a fixed point of T. Suppose that $d_{\mathbf{o}}(x_0, x_1) > a$. Since $k < \kappa$, $d_{\mathbf{o}}(x_n, x_m) \to a$ as $n, m \to \infty$, hence $\{x_n\}$ is a Cauchy sequence,[10] and thus converges to some point $x^* \in X$. For all $n \in \mathbb{N}$, $d_{\mathbf{o}}(x_{n+1}, Tx^*) = d_{\mathbf{o}}(Tx_n, Tx^*) \leq \phi(k, d_{\mathbf{o}}(x_n, x^*)) \to \phi(k, a) = a$. Hence $x_{n+1} \overset{\mathrm{O}}{\to} Tx^*$ and $Tx^* = x^*$ since the O-limit is unique. Suppose that x_1^* and x_2^* are two fixed points of T such that $x_1^* \neq x_2^*$. Then $d_{\mathbf{o}}(x_1^*, x_2^*) = d_{\mathbf{o}}(Tx_1^*, Tx_2^*) \leq \phi(k, d_{\mathbf{o}}(x_1^*, x_2^*)) < \phi(1, d_{\mathbf{o}}(x_1^*, x_2^*)) = d_{\mathbf{o}}(x_1^*, x_2^*)$, a contradiction. Hence, the fixed point of T is unique. □

Corollary 5.10 *Let $(X, d_{\mathbf{o}}, a)$ be an a-upward O-metric space such that there is a function $\lambda{:}\ [a, \infty) \to [0, \infty)$ satisfying conditions (E_1) and (E_2). Any*

---

9 As seen in (43), the sequence $\{x_n\}$ of iterates of T is a contractive sequence.

10 C_{ϕ} can thus be considered as the *interval of Cauchyness of contractive sequences*: a k-ϕ contractive sequence is a Cauchy sequence if $k \in [0, \sup C_{\phi})$.

map $T: X \to X$ *such that for some* $k \in (0,1)$ *and for all* $x, y \in X$

$$d_{\mathbf{o}}(Tx, Ty) \leq \lambda^{-1}(k\lambda(d_{\mathbf{o}}(x,y))),$$

has a unique fixed point in X.

Proof. Under the conditions of the corollary, $\kappa = \sup C_\phi = 1$ for $\phi(r,u) = \lambda^{-1}(r\lambda(u))$ as seen in Proposition 5.7. The map T is a k-ϕ contraction, with $k < 1 = \kappa$. Thus, from Theorem 5.9, T has a unique fixed point in X. □

Corollary 5.11 *Let* (X, d, s) *be a b-metric space, with* $s \geq 1$ *and* $T: X \to X$ *a mapping such that for some* $k \in (0,1)$ *and for all* $x, y \in X$

$$d(Tx, Ty) \leq kd(x,y).$$

Then T has a unique fixed point in X.

Proof. The map T is a k-ϕ contraction with $\phi : [0,\infty) \times [0,\infty) \to [0,\infty)$ defined by $\phi(r,u) = ru$ for all $r, u \geq 0$. From Lemma 5.8, $\kappa = supC_\phi = 1$ hence from Theorem 5.9, T has a unique fixed point. □

The readers are referred to papers [14, 15] for more theorems on fixed points of self mappings of O-metric spaces.

REFERENCES

1. S. Czerwik, Contraction mapping in b-metric spaces, *Acta Mathematica et Informatica Universitasis Ostraviensis* 1 (1993) 5–11.
2. V. Parvaneh, S. J. H. Ghoncheh, Fixed points of $(\phi, \varphi)\Omega$-contractive mappings in ordered p-metric spaces, *Global Analysis and Discrete Mathematics* 4 (1) (2020) 15–29. doi:10.22128/gadm.2019.290.1019.
3. N. Mlaiki, H. Aydi, N. Souayah, T. Abdeljawad, Controlled metric type spaces and the related contraction principle, *Mathematics* 6 (10) (2018) 194. doi:10.3390/math6100194.
4. T. Kamran, M. Samreen, Q. UL. Ain, A generalization of b-metric space and some fixed point theorems, *Mathematics* 5 (19) (2017) 1–7. doi:10.3390/math5020019.
5. M. Samreen, T. Kamran, M. Postolache, Extended b-metric space, extended b-comparison function and nonlinear contractions, *University Politehnica of Bucharest Scientific Bulletin, Series A* 80 (2018) 21–28.

6. F. Khojasteh, E. Karapnar, S. Radenovic, θ-metric spaces: A generalization, *Mathematical Problems in Engineering* 2013 (5) (2013) 1–7. doi:10.1155/2013/504609.
7. A. Bashirov, E. Kurpinar, A.Ozyapici, Multiplicative calculus and its applications, *Journal of Mathematical Analysis and its Applications* 337 (1) (2008) 36–48. doi:10.1016/j.jmaa.2007.03.081.
8. M. U. Ali, T. Kamran, A. Kurdi, Fixed point theorems in b-multiplicative metric spaces, *University Politehnica of Bucharest Scientific Bulletin,* Series A 79 (3) (2017) 107–116.
9. T. Dosenović, M. Postolache, S. Radenović, On multiplicative metric spaces: Survey, *Fixed Point Theory and Applications* 2016 (92) (2016) 1–17. doi:10.1186/s13663-016-0584-6.
10. A. O. Ige, H. O. Olaoluwa, J. O. Olaleru, Some fixed points of multivalued maps in multiplicative metric spaces, *Helyon* 8 (2022) 1–8. doi:10.1016/j.heliyon.2022.e12453.
11. J. Olaleru, Some generalizations of fixed point theorems in cone metric spaces, *Fixed Point Theory and Applications* 2009 (2009) 1–10. doi:10.1155/2009/657914.
12. H. Olaoluwa, J. Olaleru, On common fixed points and multipled fixed points of contractive mappings in metric-type spaces, *Journal of the Nigerian Mathematical Society* 34 (3) (2015) 249–258. doi:10.1016/j.jnnms.2015.06.001.
13. H. Olaoluwa, J. Olaleru, A hybrid class of expansive-contractive mappings in cone b-metric spaces, *Afrika Matematika* 27 (5–6) (2016) 825–840. doi:10.1007/s13370-015-0381-0.
14. H. O. Olaoluwa, A. O. Ige, J. O. Olaleru, A generalized metric-type structure with some applications, arXiv:2208.12546.
15. H. O. Olaoluwa, A. O. Ige, J. O. Olaleru, Fixed points of generalized contractions on o-metric spaces, arXiv:2403.09655.
16. A. C. M. Van Rooij, *Non-archimedean functional analysis,* Marcel Dekker, New York, 1978.
17. T. V. An, L. Q. Tuyen, N. V. Dung, Stone-type theorem on b-metric spaces and applications, *Topology and its Applications* 185–186 (2015) 50–64. doi:10.1016/j.topol.2015.02.005.
18. H. W. Becker, Problem 4277 (solution), *The American Mathematical Monthly* 56 (1949) 697–699.
19. G. Szekeres, Problem 3954, *The American Mathematical Monthly* 48 (1941) 565.

20. S. Banach, Sur les operations dans les ensembles abstraits et leur applications aux equations integrales, *Fundamenta Mathematicae* 3 (1922) 133–181.

CHAPTER 15

Stability Analysis of a Diffusive SVIR Epidemic Model with Distributed Delay, Imperfect Vaccine, and General Incidence Rate

Achraf Zinihi, Mostafa Tahiri, and Moulay Rchid Sidi Ammi

15.1 INTRODUCTION

Global health security remains under the persistent threat of infectious diseases [14]. In the last few decades, a plethora of mathematical models have emerged to evaluate and address the transmission dynamics of infectious diseases worldwide [3, 5]. These modeling frameworks play a pivotal role in diverse domains, including policy formulation, public health readiness, risk assessment, and the evaluation of control programs [19, 24, 27]. Commonly, these models represent the infection status of individuals through the categorization of populations into various compartments, such as susceptible, infected, removed, and other relevant categories like vaccinated. These compartments are typically structured as ODEs, wherein the progression of infection takes place among these compartments. However,

DOI: 10.1201/9781003530602-15

such compartmental ODE systems often neglect the impact of population movement processes and spatial heterogeneity. Recent advancements in the field have introduced reaction–diffusion models utilizing PDEs [20, 21], integrating the spatial dispersal component into the ODE systems. This incorporation enables the exploration of disease spread across diverse geographical regions and the examination of spatial heterogeneity's influence on transmission patterns.

A prominent instance of the PDE model is the reaction–diffusion SVIR epidemiological system [23, 25], serving as a widely employed mathematical framework for understanding the dynamics of infectious diseases. This system encapsulates SVIR type kinetics, delineating the population's evolution through four compartments: Susceptible (S), Vaccinated (V), Infected (I), and Removed (R). The SVIR model's dynamics are subject to various influencing factors, including disease transmission rate [10] and the recovery (or treatment) rate of infected individuals [25]. Numerical simulations have become a frequent choice for studying this epidemiological model, enabling researchers to analyze disease dynamics. Stability analysis has also found application in comprehending the qualitative characteristics of dynamical systems, encompassing both ODE models [7] and PDE systems [1]. Through the examination of stability properties, valuable insights can be derived into a dynamical system's long-term behavior, elucidating aspects such as convergence to equilibrium, manifestation of damped oscillatory behavior, and other dynamic patterns [23, 25]. However, when extending stability analysis to PDE systems using numerical methods, challenges arise in terms of accuracy and complexity [16]. The traditional approach involves discretizing spatial derivatives in the PDE model and estimating eigenvalues from a considerably large discrete system. This method can be computationally demanding and susceptible to errors arising from spectral intricacies introduced during the discretization process [1].

The exploration of a reaction–diffusion SVIR model with distributed delay offers a compelling avenue for research with multifaceted motivations. Distributed delays in epidemic models are instrumental in capturing the intricate interplay between the incubation period of disease and the spatial dispersal of individuals. By incorporating distributed delays, the model accommodates variations in the time it takes for an infected individual to transmit the disease to others, thus reflecting more realistic scenarios in the context of infectious disease dynamics. This approach enables a nuanced examination of the spatial and temporal aspects of disease spread, providing insights into how delayed transmission influences the overall epidemiological landscape. Studying such a model is essential for gaining a deeper understanding of the impact of spatial heterogeneity, time delays, and diffusion processes on the persistence, propagation, and control of infectious diseases. The rate of disease incidence, a critical factor in epidemic modeling, denotes the number of new infections occurring per unit of time. Traditionally, this rate has been assumed to exhibit bilinear dependence on the number of susceptible (S) and infected (I) individuals, and sometimes vaccinated individuals (V) and infected individuals (I). However, alternative functions have been proposed to capture disease transmission dynamics. Generally, these functions can be expressed as $f(I)$. Various specific forms of these incidence functions have been explored in the field of mathematical epidemiology, including:

- βI (bilinear) [2];
- $\beta e^{-mI} I$ with $m > 0$ [6];
- $\beta \frac{I}{1+a_1 I}$ (saturated with respect to infectives) [11];
- $\beta \frac{I}{1+\omega_1 I+\omega_2 I^2}$ with $\omega_1, \omega_2 > 0$ [26];

where β is the transmission rate for the susceptible or the vaccinated individuals.

The subsequent sections of this manuscript are structured as follows. In Section 15.2, the SVIR model incorporating distributed delay is presented. Section 15.3 is dedicated to establishing the existence and uniqueness of a global solution for the proposed model. Section 15.4 encompasses the computation of the basic reproduction number, examining model persistence, and investigating the existence of endemic equilibria. The focus then shifts to the global stability analysis of both the disease-free equilibrium and the endemic equilibrium in Section 15.5. To complement the theoretical findings, Section 15.6 presents a series of numerical simulations. A closure for this work will be presented in Section 15.7.

15.2 MATHEMATICAL MODEL

Based on what we discussed earlier, our proposed model is defined by

$$\begin{cases} \frac{\partial S(t,x)}{\partial t} - d_S \Delta S(t,x) = & \Lambda - S(t,x) \int_0^k g(\tau) f(I(t-\tau,x))\, d\tau - (\mu+\alpha) S(t,x), \\ \frac{\partial V(t,x)}{\partial t} - d_V \Delta V(t,x) = & \alpha S(t,x) - V(t,x) \int_0^k g(\tau) h(I(t-\tau,x))\, d\tau - (\gamma_1+\mu) V(t,x), \\ \frac{\partial I(t,x)}{\partial t} - d_I \Delta I(t,x) = & V(t,x) \int_0^k g(\tau) h(I(t-\tau,x))\, d\tau \\ & + S(t,x) \int_0^k g(\tau) f(I(t-\tau,x))\, d\tau - (\gamma+\mu+c) I(t,x), \\ \frac{\partial R(t,x)}{\partial t} - d_R \Delta R(t,x) = & \gamma_1 V(t,x) + \gamma I(t,x) - \mu R(t,x), \end{cases} \tag{1}$$

with

$$\frac{\partial S(t,x)}{\partial \nu} = \frac{\partial V(t,x)}{\partial \nu} = \frac{\partial I(t,x)}{\partial \nu} = \frac{\partial R(t,x)}{\partial \nu} = 0, \; x \in \partial\Omega, \tag{2}$$

where $t > 0$, Ω is a bounded domain in $\mathbb{R}^n$ with smooth boundary $\partial\Omega$, ν is the outward normal to $\partial\Omega$, $d_S > 0$, $d_V > 0$, $d_I > 0$ and $d_R > 0$ stand for the diffusion rates, and $S(t,x), V(t,x), I(t,x), R(t,x)$ denote the number of susceptible, vaccinated, infected, and recovered individuals at time t and position x, respectively. Λ is the recruitment rate of new susceptibles, μ is the natural death rate of the population, γ represents the natural recovery rate of infective individuals, and α is the vaccination rate. $\frac{1}{\gamma_1}$ is the approximative time spent in V-class before getting immunity, and $\frac{1}{\gamma}$ the average time of the infectious period. Individuals leave the S-class at the following rate

$$S(t,x) \int_0^k g(\tau) f(I(t-\tau,x))\, d\tau,$$

and leave the V-class at the rate

$$V(t,x) \int_0^k g(\tau) h(I(t-\tau,x))\, d\tau,$$

where k represents the maximum time taken to become infectious and g is a non-negative function satisfying

$$\int_0^k g(\tau)\, d\tau = 1.$$

The initial condition for the above system (1) is given for $\theta \in [-k, 0]$ by

$$\begin{aligned}\Phi(\theta)(x) &= (\Phi_1(\theta, x), \Phi_2(\theta, x), \Phi_3(\theta, x), \Phi_4(\theta, x)) \\ &= (S(\theta, x), V(\theta, x), I(\theta, x), R(\theta, x)), \ x \in \overline{\Omega}.\end{aligned}$$

The function Φ belongs $C([-k, 0], \mathbb{X})$, where $C([-k, 0], \mathbb{X})$ denotes the space of continuous functions mapping from $[-k, 0]$ to $\mathbb{X}$ equipped with the sup-norm, and $\mathbb{X} = C(\Omega, \mathbb{R}_+^4)$ denotes the space of continuous functions mapping from Ω to $\mathbb{R}_+^4$. Our main objective is to discuss the global stability of the SVIR model (1). For that, we will construct suitable Lyapunov functions. Throughout this chapter, we assume that $f : \mathbb{R}_+ \longrightarrow \mathbb{R}_+$ and $h : \mathbb{R}_+ \longrightarrow \mathbb{R}_+$ are continuously differentiable in the interior of $\mathbb{R}_+$ with $f(I) = 0$ for $I = 0$ and $h(I) = 0$ for $I = 0$. The following hypotheses hold:

(H_1) For $I > 0$, we have $f(I) > 0$ and $h(I) > 0$.
(H_2) If $I \geq 0$, then $f'(I) > 0, h'(I) > 0, f''(I) \leq 0$ and $h''(I) \leq 0$.

Epidemiologically, (H_1) means that individuals are positive. (H_2) implies that the incidences of

$$V(t,x) \int_0^k g(\tau) h(I(t-\tau, x))\, d\tau,$$

and

$$S(t,x) \int_0^k g(\tau) f(I(t-\tau, x))\, d\tau,$$

become faster with an increase in the number of infectious individuals. However, the per capita infection rate will slow down because of a certain inhibition effect, since $f''(I) \leq 0$ and $h''(I) \leq 0$ imply that $\left(\frac{f(I)}{I}\right)', \left(\frac{h(I)}{I}\right)' < 0$.

15.3 BASIC PROPERTIES OF THE MODEL

Let A be the operator defined on $\mathbb{X}$ as follows

$$\begin{aligned} A : D(A) \subset \mathbb{X} &\longrightarrow \mathbb{X}, \\ u &\longmapsto Au = (A_1 u, A_2 u, A_3 u, A_4 u), \end{aligned}$$

where

$$D(A) := \left\{ u \in \mathbb{X} : \Delta u \in \mathbb{X}, \frac{\partial u}{\partial \nu} = 0 \text{ on } \partial\Omega \right\}, \tag{5}$$

and

$$\begin{aligned} A_1 u &= (d_S \Delta - \delta_1) u, \ \delta_1 = \mu + \alpha, \\ A_2 u &= (d_V \Delta - \delta_2) u, \ \delta_2 = \mu + \gamma_1, \\ A_3 u &= (d_I \Delta - \delta_3) u, \ \delta_3 = \mu + \gamma + c, \\ A_4 u &= (d_R \Delta - \delta_4) u, \ \delta_4 = \mu. \end{aligned}$$

According to the classical theory of semi-groups of partial differential equations [17], A_i is the infinitesimal generator of a strongly continuous semi-group $\exp(tA_i)$, where $i = 1, 2, 3, 4$.

Let $\mathcal{X}_i(t) : C(\Omega, \mathbb{R}) \longrightarrow C(\Omega, \mathbb{R})$ be the C_0 semi-group associated with A_i subject to (2). Then,

$$(\mathcal{X}_i(t) u)(x) = \int_\Omega \Gamma_i(t, x, y) u(y) \, dy,$$

where Γ_i represents the Green function associated with A_i subject to (2). By [22, Corollary 7.2.3], we establish that $\mathcal{X}(t) = (\mathcal{X}_1(t), \mathcal{X}_2(t), \mathcal{X}_3(t), \mathcal{X}_4(t))$ is compact and strongly positive for each $t > 0$. Following [13, 18], there exist constants $\xi_i > 0$ such that

$$\|\mathcal{X}_i(t)\| \leq \xi_i e^{\lambda_i t}, \ \forall t \geq 0, \tag{6}$$

where λ_i is the principal eigenvalue of A_i subject to (2).

For any function $u : [-k, \sigma) \longrightarrow \mathbb{X}$ with some $\sigma > 0$, we define $u_t \in C([-k, 0], \mathbb{X})$ by $u_t(\theta) = u(t + \theta), \theta \in [-k, 0]$. Let F be a function defined by

$$\begin{aligned} F : C([-k, 0], \mathbb{X}) &\longrightarrow C\left(\Omega, \mathbb{R}^4\right), \\ \varphi &\longmapsto F(\varphi) = (F_1(\varphi), F_2(\varphi), F_3(\varphi), F_4(\varphi)), \end{aligned}$$

where

$$F_1(\varphi) = \Lambda - \varphi_1(t, x) \int_0^k g(\tau) f(\varphi_3(t - \tau, x)) \, d\tau,$$

$$F_2(\varphi) = \alpha\varphi_1(t, x) - \varphi_2(t, x) \int_0^k g(\tau) h(\varphi_3(t - \tau, x)) \, d\tau,$$

$$
\begin{aligned}
F_3(\varphi) = & \varphi_2(t,x)\int_0^k g(\tau)h(\varphi_3(t-\tau,x))\,d\tau + \varphi_1(t,x) \\
& \times \int_0^k g(\tau)f(\varphi_3(t-\tau,x))\,d\tau, \\
F_4(\varphi) = & \gamma_1\varphi_2(t,x) + \gamma\varphi_3(t,x).
\end{aligned}
$$

Then, system (1)-(3) can be written in the following abstract form

$$
\begin{cases} \frac{d\vartheta(t)}{dt} = A\vartheta(t) + F(\vartheta_t), \; t \geq 0, \\ \vartheta_0 = \Phi, \end{cases} \tag{7}
$$

where $\vartheta(t) = (S(t), V(t), I(t), R(t))^T$ and $\Phi = (S_0, V_0, I_0, R_0)^T$. In addition, the proposed system (1) can be rewritten as the integral equation

$$
\vartheta(t) = \mathcal{X}(t)\Phi + \int_0^t \mathcal{X}(t-s)F(\vartheta_s)\,ds. \tag{8}
$$

Function F is locally Lipschitz on $C([-k,0],\mathbb{X})$, then problem (1) has a unique local non-negative solution $(S(t), V(t), I(t), R(t))^T$ for all $t \in [0,b)$.

Remark 1 The unique local positive solution becomes the global one if $b = +\infty$.

We will now show that the local solution can be extended to a global one. Suppose that $b < +\infty$. By [15, Theorem 2], we know that

$$
\|\vartheta(t)\| \longrightarrow +\infty \text{ as } t \longrightarrow b. \tag{9}
$$

Put

$$
N(t) = \int_\Omega \{S(t,x) + V(t,x) + I(t,x) + R(t,x)\}\,dx, \; \forall t \in [0,b).
$$

By the divergence theorem [8, Theorem 3.7] and (2), we get

$$
\begin{aligned}
& \int_\Omega d_S\Delta S(t,x)\,dx = 0, \; \int_\Omega d_V\Delta V(t,x)\,dx = 0, \\
& \int_\Omega d_I\Delta I(t,x)\,dx = 0, \; \int_\Omega d_R\Delta R(t,x)\,dx = 0.
\end{aligned}
$$

which gives

$$
\frac{dN(t)}{dt} \leq \Lambda|\Omega| - \mu N(t).
$$

By the comparison principle, there exist constants $M_1 > 0$ and $t_1 > 0$ such that $N(t) \leq M_1$ for all $t \in [t_1, b)$. Consequently,

$$\begin{array}{l} \int_\Omega S(t,x)\,dx \leq M_1, \ \int_\Omega V(t,x)\,dx \leq M_1, \\ \int_\Omega I(t,x)\,dx \leq M_1, \ \int_\Omega R(t,x)\,dx \leq M_1. \end{array} \forall t \in [t_1, b), \tag{10}$$

Let λ_i^j the eigenvalue of A_i subject to (2) corresponding to the eigenfunction ϕ_i^j, such that $0 \geq -\delta_i = \lambda_i^1 > \lambda_i^2 > \lambda_i^3 > \cdots$. From [9, Chapter 5], we have

$$\Gamma_i(t,x,y) = \sum_{j\geq 1} \exp\left(\lambda_i^j t\right) \phi_i^j(x)\, \phi_i^j(y).$$

Since ϕ_i^j is uniformly bounded, there exists $\beta > 0$ such that

$$\Gamma_i(t,x,y) \leq \beta \sum_{j\geq 1} \exp\left(\lambda_i^j t\right), \ \forall x > 0, x, y \in \Omega\, t > 0, x \in \Omega.$$

Adopting a similar methodology as outlined in [4, Page 5] and using (6) together with (8) and (10), we obtain

$$\begin{array}{l} S(t,x) < \infty, V(t,x) < \infty, \\ I(t,x) < \infty, R(t,x) < \infty. \end{array} \forall (t,x) \in [0,b) \times \Omega.$$

which contradicts (9). Thus, $b = +\infty$. These analyses lead to the following inevitable result.

Theorem 1 *For any initial value function in $\mathbb{X}$, system (1) has a unique non-negative solution $(S(t,x), V(t,x), I(t,x), R(t,x))$ for all $(t,x) \in [0,\infty) \times \Omega$.*

15.4 EQUILIBRIA AND THE BASIC REPRODUCTION NUMBER

Since the first three equations in (1) do not contain $R(t,x)$, it is sufficient to analyze the behavior of solutions to the following system

$$\left\{ \begin{array}{ll} \frac{\partial S(t,x)}{\partial t} - d_S \Delta S(t,x) = \Lambda - S(t,x) \int_0^k g(\tau) f(I(t-\tau,x))\,d\tau & \\ \qquad - (\mu + \alpha) S(t,x), & \\ \frac{\partial V(t,x)}{\partial t} - d_V \Delta V(t,x) = \alpha S(t,x) - V(t,x) \int_0^k g(\tau) h(I(t-\tau,x))\,d\tau & \\ \qquad - (\gamma_1 + \mu) V(t,x), & x \in \Omega, \\ \frac{\partial I(t,x)}{\partial t} - d_I \Delta I(t,x) = V(t,x) \int_0^k g(\tau) h(I(t-\tau,x))\,d\tau & \\ \qquad + S(t,x) \int_0^k g(\tau) f(I(t-\tau,x))\,d\tau & \\ \qquad - (\gamma + \mu + c) I(t,x), & \end{array} \right. \tag{11}$$

with

$$\frac{\partial S(t,x)}{\partial \nu} = \frac{\partial V(t,x)}{\partial \nu} = \frac{\partial I(t,x)}{\partial \nu} = 0, \quad x \in \partial\Omega. \tag{12}$$

System (11) always has a disease-free equilibrium $E_0 = (S_0, V_0, 0)^T$, where $S_0 = \frac{\Lambda}{\mu+\alpha}$ and $V_0 = \frac{\alpha\Lambda}{(\mu+\alpha)(\gamma_1+\mu)}$. Furthermore, by a simple and direct calculation, we conclude that the basic reproduction number for this proposed model (11) is given by

$$\mathcal{R}_0 = \frac{S_0}{(\gamma+\mu+c)} f'(0) + \frac{V_0}{(\gamma+\mu+c)} h'(0).$$

We have the following result.

Theorem 2 *If $\mathcal{R}_0 > 1$, then system (11) has a unique endemic equilibrium $E^* = (S^*, V^*, I^*)$ with $S^* = \frac{\Lambda}{f(I^*)+\mu+\alpha}$ and $V^* = \frac{\alpha\Lambda}{(h(I^*)+\gamma_1+\mu)(f(I^*)+\mu+\alpha)}$.*

Proof. When $\mathcal{R}_0 > 1$, the result is obvious. According to the first two equations of (11), one can get

$$S = \frac{\Lambda}{f(I)+\mu+\alpha}, V = \frac{\alpha\Lambda}{(h(I)+\gamma_1+\mu)(f(I)+\mu+\alpha)}.$$

Then, we have

$$\begin{aligned} H(I) = & \frac{\Lambda}{f(I)+\mu+\alpha} f(I) \\ & + \frac{\alpha\Lambda}{(h(I)+\gamma_1+\mu)(f(I)+\mu+\alpha)} h(I) - (\gamma+\mu+c) I. \end{aligned}$$

Obviously, $H(+\infty) = -\infty$ and $H(0) = 0$. It follows from $H'(0) > 0$ that $H(I) = 0$ has at least one positive solution denoted by I^*, where

$$\begin{aligned} H'(0) &= \frac{\Lambda}{\mu+\alpha} + \frac{\alpha\Lambda h'(0)}{(\gamma_1+\mu)(\mu+\alpha)} - (\gamma+\mu+c) \\ &= (\gamma+\mu+c)(\mathcal{R}_0 - 1) > 0. \end{aligned}$$

This is equivalent to $\mathcal{R}_0 > 1$. Thus, (11) has at least one positive solution with

$$S^* = \frac{\Lambda}{f(I^*)+\mu+\alpha}, \quad V^* = \frac{\alpha\Lambda}{(h(I^*)+\gamma_1+\mu)(f(I^*)+\mu+\alpha)}.$$

Now, we prove that the endemic equilibrium is unique. Note that

$$\begin{aligned}
H'(I) =& \frac{\Lambda f'(I)(f(I)+\mu+\alpha) - \Lambda f(I) f'(I)}{(f(I)+\mu+\alpha)^2} \\
&+ \frac{\alpha\Lambda h'(I)}{[(h(I)+\gamma_1+\mu)(f(I)+\mu+\alpha)]} \\
&- \frac{\alpha\Lambda h(I)[h'(I)(f(I)+\mu+\alpha)+(h(I)+\gamma_1+\mu)f'(I)]}{[(h(I)+\gamma_1+\mu)(f(I)+\mu+\alpha)]^2} \\
&- (\gamma+\mu+c)
\end{aligned}$$

and

$$\begin{aligned}
H''(I) =& \frac{\Lambda f''(I)\mu}{(f(I)+\mu+\alpha)^3} + \frac{\alpha\Lambda f''(I)(\gamma_1\mu)}{(f(I)+\mu+\alpha)^2(h(I)+\gamma_1+\mu)} \\
&+ \frac{\alpha\Lambda h''(I)(\gamma_1+\mu)}{(f(I)+\mu+\alpha)(h(I)+\gamma_1)+\mu)2} \\
&- [\frac{\alpha\Lambda h'(I) f'(I)(\gamma_1+\mu)}{(f(I)+\mu+\alpha)(h(I)+\gamma_1)+\mu)2} + \frac{2\alpha\Lambda(h'(I))^2(\gamma_1+\mu)}{(f(I)+\mu+\alpha)(h(I)+\gamma_1+\mu)^3} \\
&+ \frac{2\Lambda\mu(f'(I))^2}{(f(I)+\mu+\alpha)^3} + \frac{\alpha\Lambda h'(I) f'(I)}{(f(I)+\mu+\alpha)^2(h(I)+\gamma_1+\mu)} \\
&+ \frac{2\alpha\Lambda(f'(I))^2(\gamma_1+\mu)}{(f(I)+\mu+\alpha)^3(h(I)+\gamma_1+\mu)}]
\end{aligned}$$

By (H_2), we know that $h''(I) < 0$ for $I > 0$. If there exists more than one positive equilibrium, then there is a point $E^* = (S^*, V^*, I^*)$ such that $h(I^*) = 0$. We obtain a contradiction. □

15.5 GLOBAL STABILITY OF EQUILIBRIA

In this section, we show the global asymptotic stability of the disease-free equilibrium E_0 of the system (11) by constructing a Lyapunov functional. The following result holds.

Theorem 3 *Under hypotheses* (H_1) *and* (H_2)*, the disease-free equilibrium* E_0 *of system (11) is globally asymptotically stable if, and only if,* $\mathcal{R}_0 < 1$.

Proof. To prove our result, we consider the following Lyapunov functional

$$L(t) = \int_\Omega (L_1(t,x) + L_2(t,x))\,dx$$

where

$$L_1(t,x) = I(t,x) + S_0\varphi\left(\frac{S(t,x)}{S_0}\right) + V_0\varphi\left(\frac{V(t,x)}{V_0}\right)$$

with $\varphi(x) = x - 1 - \ln(x)$ and

$$\begin{aligned} L_2(t,x) = & \int_0^k g(\tau)\int_{t-\tau}^{\tau} f(I(u,x))\,S(u,x)\,du \\ & + \int_0^k g(\tau)\int_{t-\tau}^{\tau} h(I(u,x))\,V(u,x)\,dud\tau. \end{aligned}$$

According to

$$\ln x \le x - 1, S_0 = \frac{\Lambda}{\alpha + \mu} \text{ and } \alpha S_0 = (\gamma_1 + \mu),$$

we have

$$\begin{aligned} \frac{\partial L_1(t,x)}{\partial t} &= \frac{\partial I}{\partial t} + \left(1 - \frac{S_0}{S}\right)\frac{\partial S}{\partial t} + \left(1 - \frac{V_0}{V}\right)\frac{\partial V}{\partial t}, \\ \frac{\partial L_2(t,x)}{\partial t} &= \int_0^k g(\tau)(f(I)\,S - f(I(t-\tau,x)\,S(t-\tau,x))\,d\tau \\ & \quad + \int_0^k g(\tau)(h(I)\,V - h(I(t-\tau,x)\,V(t-\tau,x))\,d\tau. \end{aligned}$$

Thus, we get

$$\begin{aligned} \frac{\partial(L_1 + L_2)(t,x)}{\partial t} &= d_I\Delta I + S\int_0^k g(\tau)f(I(t-\tau,x))\,d\tau \\ & \quad + V\int_0^k g(\tau)h(I(t-\tau,x))\,d\tau - (\gamma + \mu + c)\,I \\ & \quad + \left(1 - \frac{S_0}{S}\right)(d_S\Delta S + \Lambda \end{aligned}$$

$$
\begin{aligned}
&- S\int_0^k g(\tau)f(I(t-\tau,x))\,d\tau - (\mu+\alpha)\,S)+\\
&\left(1-\frac{V_0}{V}\right)(d_V\Delta V+\alpha S\\
&- V\int_0^k g(\tau)\,h\,(I\,(t-\tau,x))\,d\tau - (\gamma_1+\mu)\,V)d\tau\\
&+\int_0^k g(\tau)f(I)\,Sd\tau - \int_0^k g(\tau)f(I(t-\tau,x))\,S\,(t-\tau,x)\,d\tau\\
&+\int_0^k g(\tau)\,h\,(I)\,Vd\tau - \int_0^k g(\tau)\,h\,(I(t-\tau,x))\,V\,(t-\tau,x)\,d\tau.
\end{aligned}
$$

That is to say

$$
\begin{aligned}
\frac{\partial\,(L_1+L_2)\,(t,x)}{\partial t} =& -(\mu+\alpha)\left(\frac{S}{S_0}+\frac{S_0}{S}-2\right)+\alpha S_0\left(\frac{S}{S_0}-\frac{V}{V_0}-\frac{V_0S}{VS_0}\right)\\
&+\int_0^k g(\tau)\,(S_0f(I\,(t-\tau,x))+f(I)\,S\\
&-f(I\,(t-\tau,x))\,S\,(t-\tau,x))d\tau\\
&+\int_0^k g(\tau)\,(V_0h\,(I\,(t-\tau,x))\\
&+h\,(I)\,V-h\,(I\,(t-\tau,x))\,V\,(t-\tau,x))d\tau\\
&-(\gamma+\mu+c)\,I\\
\le\,& \alpha S_0\left(\varphi\left(\frac{S}{S_0}\right)-\varphi\left(\frac{V}{V_0}\right)-\varphi\left(\frac{V_0S}{S_0V}\right)-2\,\ln\left(\frac{VS_0}{V_0S}\right)\right)\\
&-S_0\,(\mu+\alpha)\left(\varphi\left(\frac{S}{S_0}\right)+\varphi\left(\frac{S_0}{S}\right)\right)+S_0f'\,(0)\,I\\
&+V_0h'\,(0)\,I-(\gamma+\mu+c)\,I\\
=& -\alpha S_0\left(\varphi\left(\frac{S_0}{S}\right)+\varphi\left(\frac{V}{V_0}\right)+\varphi\left(\frac{V_0S}{S_0V}\right)+2\,\ln\left(\frac{VS_0}{V_0S}\right)\right)\\
&-\mu S_0\left(\varphi\left(\frac{S}{S_0}\right)+\varphi\left(\frac{S_0}{S}\right)\right)+(\gamma+\mu+c)\,(\mathcal{R}_0-1).
\end{aligned}
$$

From this, we conclude that

$$\frac{dL(t)}{dt} \le \int_{\Omega} \{-\alpha S_0 \left(\varphi\left(\frac{S_0}{S}\right) + \varphi\left(\frac{V}{V_0}\right) + \varphi\left(\frac{V_0 S}{S_0 V}\right) + 2\ln\left(\frac{V S_0}{V_0 S}\right)\right)$$
$$- \mu S_0 \left(\varphi\left(\frac{S}{S_0}\right) + \varphi\left(\frac{S_0}{S}\right)\right) + (\gamma + \mu + c)(\mathcal{R}_0 - 1) + d_I \Delta I$$
$$+ d_S \Delta S + d_V \Delta V - \frac{S_0}{S} d_S \Delta S - \frac{V_0}{V} d_V \Delta V\} dx.$$

By Green's formula and (2), we establish that

$$\int_{\Omega} (d_I \Delta I(t,x) + d_S \Delta S(t,x) + d_V \Delta V(t,x))\, dx = 0,$$

$$\int_{\Omega} \frac{\Delta S(t,x)}{S(t,x)} dx = \int_{\Omega} \frac{(\nabla S(t,x))^2}{S(t,x)^2} dx \ge 0,$$

and

$$\int_{\Omega} \frac{\Delta V(t,x)}{V(t,x)} dx = \int_{\Omega} \frac{(\nabla V(t,x))^2}{V(t,x)^2} dx \ge 0.$$

If $\mathcal{R}_0 < 1$, then

$$\frac{dL(t)}{dt} \le \int_{\Omega} \{-\alpha S_0 \left(\varphi\left(\frac{S_0}{S}\right) + \varphi\left(\frac{V}{V_0}\right) + \varphi\left(\frac{V_0 S}{S_0 V}\right) + 2\ln\left(\frac{V S_0}{V_0 S}\right)\right)$$
$$- \mu S_0 \left(\varphi\left(\frac{S}{S_0}\right) + \varphi\left(\frac{S_0}{S}\right)\right) + (\gamma + \mu + c)(\mathcal{R}_0 - 1)\} dx \le 0.$$

Thus, the disease-free equilibrium of (11) is globally asymptotically stable.
□

Theorem 4 *If $\mathcal{R}_0 \ge 1$, then E^* of system (11) is globally asymptotically stable.*

Proof. Define

$$H(t) = \int_{\Omega} (H_1(t,x) + H_2(t,x))\, dx,$$

where

$$H_1(t,x) = S^* \varphi\left(\frac{S}{S*}\right) + V^* \varphi\left(\frac{V}{V*}\right) + I^* \varphi\left(\frac{I}{I*}\right),$$

$$H_2(t,x) = S^* f(I^*) \int_{t-\tau}^{t} g(\tau) \int_0^{\tau} \varphi\left(\frac{S(\Theta) f(I(\Theta))}{S^* f(I^*)}\right) d\Theta d\tau$$
$$+ V^* h(I^*) \int_{t-\tau}^{t} g(\tau) \int_0^{\tau} \varphi\left(\frac{V(\Theta) h(I(\Theta))}{V^* h(I^*)}\right) d\Theta d\tau.$$

Then,

$$\frac{\partial H_1(t,x)}{\partial t} = \left(1 - \frac{S^*}{S}\right)\frac{\partial S}{\partial t} + \left(1 - \frac{V^*}{V}\right)\frac{\partial V}{\partial t} + \left(1 - \frac{I^*}{I}\right)\frac{\partial I}{\partial t},$$

and

$$\begin{aligned}
&\frac{\partial H_2(t,x)}{\partial t} \\
&= \int_0^k g(\tau)\left(Sf(I) - S(\tau)f(I(\tau)) + S^*f(I^*)\ln\left(\frac{S(\tau)f(I(\tau))}{Sf(I)}\right)\right)d\tau \\
&+ \int_0^k g(\tau)\left(Vh(I) - V(\tau)h(I(\tau)) + V^*h(I^*)\ln\left(\frac{V(\tau)h(I(\tau))}{Vh(I)}\right)\right)d\tau.
\end{aligned}$$

Then,

$$\begin{aligned}
&\frac{\partial (H_1 + H_2)(t,x)}{\partial t} \\
&= -S^*(\mu + \alpha)\left(\frac{S}{S^*} + \frac{S^*}{S} - 2\right) \\
&+ \left(1 - \frac{S^*}{S}\right)\left(S^*f(I^*) - \int_0^k g(\tau)\,Sf(I(\tau))\,d\tau\right) \\
&+ \left(1 - \frac{S^*}{S}\right)d_S\Delta S + (\gamma_1 + \mu)\,V^*\left(\frac{S}{S^*} - \frac{SV^*}{VS^*} - \frac{V}{V^*} + 1\right) \\
&+ \left(1 - \frac{V^*}{V}\right)\left(\frac{SV^*h(I^*)}{(\gamma_1 + \mu)\,S^*} - \int_0^k g(\tau)\,Vh(I(\tau))\,d\tau\right) \\
&+ \left(1 - \frac{V^*}{V}\right)d_V\Delta V + S^*f(I^*) + V^*h(I^*) - \frac{S^*If(I^*)}{I^*} \\
&- \frac{IV^*h(I^*)}{I^*} + \left(1 - \frac{I^*}{I}\right)\left(\int_0^k g(\tau)\,Sf(I(\tau))\right. \\
&\left. + \int_0^k g(\tau)\,Vh(I(\tau))\right).
\end{aligned}$$

Thus, align

$$\frac{\partial (H_1+H_2)(t,x)}{\partial t} = -S^*(\mu+\alpha)\varphi\left(\frac{S^*}{S}\right) + (\gamma_1+\mu)V^*$$
$$\left(-\varphi\left(\frac{SV^*}{VS^*}\right) - \varphi\left(\frac{V}{V^*}\right)\right) - S^*f(I^*)\int_0^k g(\tau)\left[\varphi\left(\frac{S^*}{S}\right)\right.$$
$$+\varphi\left(\frac{I^*Sf(I^*)}{IS^*f(I)}\right) + \varphi\left(\frac{S(\tau)f(I(\tau))}{S^*f(I^*)}\right) + 1$$
$$\left.-\frac{f(I(\tau))}{f(I^*)} - \frac{Sf(I)}{S^*f(I^*)} - \ln\left(\frac{IS(\tau)(f(I(\tau)))^3}{ISf(I)(f(I))^2}\right)\right]d\tau$$
$$- V^*h(I^*)\int_0^k g(\tau)\left[\varphi\left(\frac{SV^*}{(\gamma_1+\mu)S^*V}\right) + \varphi\left(\frac{I}{I^*}\right)\right.$$
$$+\varphi\left(\frac{I^*Vh(I(\tau))}{IV^*h(I^*)}\right) + \varphi\left(\frac{V(\tau)h((I(\tau))}{V^*h(I^*)}\right) - \frac{SV^*}{VS^*} - \frac{V}{V^*}$$
$$-\frac{S}{(\gamma_1+\mu)S^*} - \frac{h(I(\tau))}{h(I^*)} - 1 - \frac{Vh(I)}{V^*h(I^*)}$$
$$\left.-\ln\left(\frac{(V(\tau))^2(h(I(\tau)))^2}{(\gamma_1+\mu)V^*h(I)}\right)\right]d\tau + dS\Delta S - \frac{S^*}{S}dS\Delta S$$
$$+ d_V\Delta V - \frac{V^*}{V}d_V\Delta V + d_I\Delta I - \frac{I^*}{I}d_I\Delta I.$$

Note that

$$\int_\Omega d_S\Delta S(t,x) = \int_\Omega d_V\Delta V(t,x) = \int_\Omega d_I\Delta I(t,x) = 0,$$

and

$$\int_\Omega \frac{\Delta S}{S}dx = \int_\Omega \frac{\nabla S^2}{S^2}dx \geq 0, \int_\Omega \frac{\Delta V}{V}dx = \int_\Omega \frac{\nabla V^2}{V^2}dx \geq 0,$$
$$\int_\Omega \frac{\Delta I}{I}dx = \int_\Omega \frac{\nabla I^2}{I^2}dx \geq 0,$$

then, align

$$
\begin{aligned}
&\frac{dH(t)}{dt} \\
&= \int_{\Omega} \frac{\partial (H_1(t,x) + H_2(t,x))}{\partial t} dx \\
&\leq \int_{\Omega} -S^*(\mu+\alpha)\varphi\left(\frac{S^*}{S}\right) + (\gamma_1+\mu)V^*\left(-\varphi\left(\frac{SV^*}{VS^*}\right) - \varphi\left(\frac{V}{V^*}\right)\right) \\
&\quad - S^*f(I^*)\int_0^k g(\tau)\left[\varphi(\frac{S^*}{S} + \varphi\left(\frac{I^*Sf(I^*)}{IS^*f(I)}\right) + \varphi\left(\frac{S(\tau)f(I(\tau))}{S^*f(I^*)}\right) + 1 - \frac{f(I(\tau))}{f(I^*)}\right. \\
&\quad \left. -\frac{Sf(I)}{S^*f(I^*)} - \ln\left(\frac{IS(\tau)(f(I(\tau)))^3}{ISf(I)(f(I))^2}\right)\right] d\tau - V*h(I^*)\int_0^k g(\tau)\left[\varphi\left(\frac{SV^*}{(\gamma_1+\mu)S^*V}\right)\right. \\
&\quad + \varphi\left(\frac{I}{I^*}\right) + \varphi\left(\frac{I^*Vh(I(\tau))}{IV^*h(I^*)}\right) + \varphi\left(\frac{V(\tau)h((I(\tau))}{V^*h(I^*)}\right) - \frac{SV^*}{VS^*} - \frac{V}{V^*} - \frac{S}{(\gamma_1+\mu)S^*} \\
&\quad \left. -\frac{h(I(\tau))}{h(I^*)} - 1 - \frac{Vh(I)}{V^*h(I^*)} - \ln\left(\frac{(V(\tau))^2(h(I(\tau)))^2}{(\gamma_1+\mu)V^*h(I)}\right)\right] d\tau.
\end{aligned}
$$

By assumption (H_2) and $\ln(x) \leq x - 1$, we can get

$$
\begin{aligned}
&\frac{f(I(\tau))}{f(I^*)} + \frac{Sf(I)}{S^*f(I^*)} - \ln\left(\frac{IS(\tau)(f(I(\tau)))^3}{ISf(I)(f(I))^2}\right) \\
&\leq \frac{f(I(\tau))}{f(I^*)} + \frac{Sf(I)}{S^*f(I^*)} + \frac{IS(\tau)(f(I(\tau)))^3}{ISf(I)(f(I))^2} - 1 \leq 0,
\end{aligned}
$$

and

$$
\begin{aligned}
&\frac{SV^*}{VS^*} + \frac{V}{V^*} + \frac{S}{(\gamma_1+\mu)S^*} + \frac{h(I(\tau))}{h(I^*)} + 1 + \frac{Vh(I)}{V^*h(I^*)} \\
&+ \ln\left(\frac{(V(\tau))^2(h(I(\tau)))^2}{(\gamma_1+\mu)V^*h(I)}\right) \\
&\leq \frac{SV^*}{VS^*} + \frac{V}{V^*} + \frac{S}{(\gamma_1+\mu)S^*} + \frac{h(I(\tau))}{h(I^*)} \\
&+ \frac{Vh(I)}{V^*h(I^*)} + \frac{V(\tau)^2(h(I(\tau)))^2}{(\gamma_1+\mu)V^*h(I)}.
\end{aligned}
$$

Hence, $\frac{dH(t)}{dt} \leq 0$. By applying LaSalle's invariance principle (see [12]), we conclude that the endemic equilibrium point of system (11) is globally asymptotically stable. The proof is complete. □

TABLE 15.1 Parameter Values and Initial Conditions of (1).

Symbol	Description	Value
Λ	Birth (or demographic) rate of the population	0.392465
β_1	Transmission rate for the susceptible	0.002 (for $\mathcal{R}_0 > 1$) 0.0008 (for $\mathcal{R}_0 < 1$)
β_2	Transmission rate for the vaccinated	0.0016 (for $\mathcal{R}_0 > 1$) 0.00064 (for $\mathcal{R}_0 < 1$)
α	The rate of vaccination of susceptible	0.005
μ	Natural death rates	0.001
γ_1	Recovered rate of the vaccinated	0.005
γ	Recovered rate of the infected	0.009
c	Disease-induced death rate	0.09
$d_S = d_V = d_I = d_R$	Diffusion coefficients	0.1
S_0	Initial susceptible individuals	30
V_0	Initial vaccinated individuals	10
I_0	Initial infected individuals	5
R_0	Initial recovered individuals	0

15.6 NUMERICAL RESULTS

The numerical simulations for the proposed model (1) with (2) were solved using MATLAB. We employed the finite difference method (FDM) to approximate the solution of the PDE system in one spatial variable x and time t. In the FDM, the spatial domain $0 \leq x \leq 1$ was discretized into a grid with $N + 1$ equally spaced points $x_k = kh$ for $k = 0, 1, ..., N$, where $h = \frac{1}{N}$ denotes the uniform grid width. Moreover, the diffusion term in model (1) was discretized using a second-order central difference formula, as outlined by [16]

$$\Delta y \approx \frac{y_{k+1} - 2y_k + y_{k-1}}{h^2}.$$

This results in a system of $3(N + 1)$ ODEs, with one equation for each compartment discretized across uniformly spaced spatial nodes. The solution of this extensive ODE system is computed until a steady state is reached (i.e., until $t = 1500$). The grid spacing is set as $h = 10^{-2}$, and the parameter values are as indicated in Table 15.1 with $f(I) = \beta_1 I$ and $h(I) = \beta_2 I$.

In Figure 15.1, the condition $\mathcal{R}_0 = 0.8721 < 1$ is illustrated for the given parameters $\beta_1 = 0.0008$ and $\beta_2 = 0.00064$. In this scenario, the disease-free equilibrium is shown to be asymptotically stable, and the system's solutions converge toward a state with no disease.

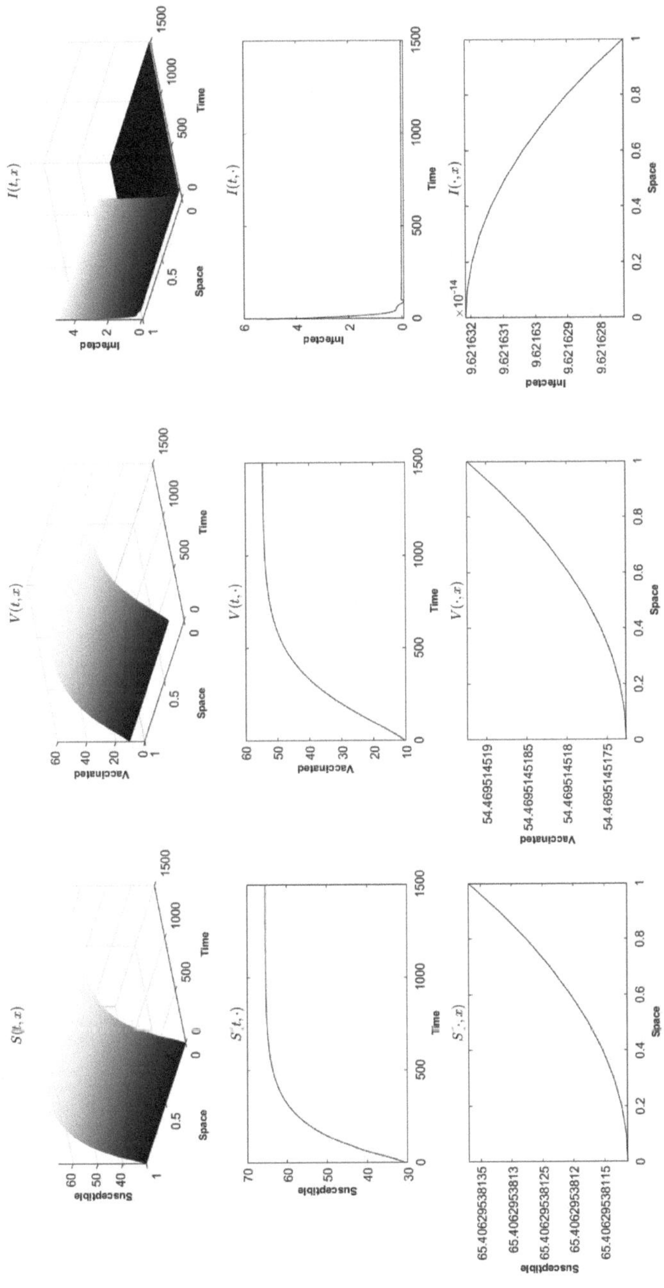

FIGURE 15.1 Numerical results of (1)–(2) when $\mathcal{R}_0 < 1$.

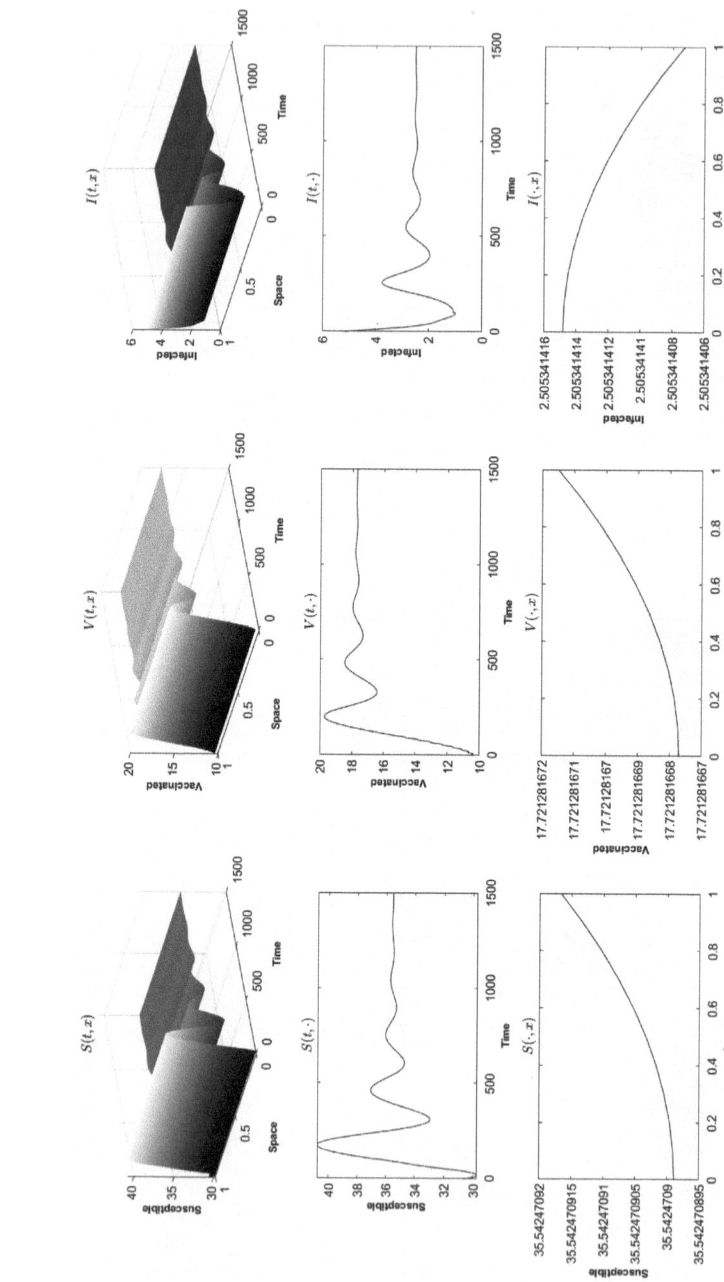

FIGURE 15.2 Numerical results of (1)–(2) when $\mathcal{R}_0 > 1$.

On the other hand, Figure 15.2, the case where $\mathcal{R}_0 = 2.1804 > 1$ is depicted, considers the parameters $\beta_1 = 0.002$ and $\beta_2 = 0.0016$. In this context, the endemic equilibrium is demonstrated to be asymptotically stable, and the system's solutions tend towards a positive balance.

15.7 CONCLUSION AND DISCUSSION

In conclusion, this chapter has investigated a reaction–diffusion SVIR infection model with distributed delay and nonlinear incidence rate. The well-posedness of the model was established, and through Lyapunov functionals, we demonstrated the global asymptotic stability of both the disease-free and endemic equilibrium states. The incorporation of delay in the SVIR model with reaction–diffusion yields several significant insights into the dynamics of infectious diseases within spatially distributed populations. The introduction of delay mechanisms captures the time lag between infection, onset of symptoms, and subsequent interventions. Moreover, it leads to the emergence of new dynamical behaviors, such as the existence of multiple equilibrium states or the occurrence of sustained oscillations. These phenomena arise due to the interplay between intrinsic disease dynamics, spatial diffusion, and the time delay in the system. Numerical simulations further illustrated these theoretical findings, showcasing the behavior of the system under different scenarios of the basic reproduction number $\mathcal{R}_0$. Specifically, when $\mathcal{R}_0 < 1$, the disease-free equilibrium was shown to be asymptotically stable, leading to the eventual eradication of the disease. Conversely, for $\mathcal{R}_0 > 1$, the endemic equilibrium was found to be asymptotically stable, resulting in a persistent state of infection within the population. These results contribute to our understanding of the dynamics of infectious diseases and have implications for public health interventions aimed at disease control and prevention. Further research could explore the effects of additional factors, such as spatial heterogeneity or seasonal variations, on the dynamics of epidemic models with distributed delay, providing valuable insights for disease management strategies in real-world scenarios.

15.7.1 Acknowledgments

The authors express their appreciation to the reviewers for their valuable comments. This work is carried out under the supervision of CNRST as part of the PASS program.

15.7.2 Funding

The authors assert that no funding was received for the preparation and execution of this manuscript.

15.7.3 Data Availability

All the information and data that were analyzed or generated to support the results of this work are provided within this manuscript.

15.7.4 Conflict of Interest

The authors declare that there are no problems or conflicts of interest between them that may affect the study presented in this paper.

REFERENCES

1. Ahmed, N., Elsonbaty, A., Raza, A., Rafiq, M., and Adel, W. Numerical simulation and stability analysis of a novel reaction–diffusion COVID-19 model. *Nonlinear Dynamics 106*, 2 (June 2021), 1293–1310.
2. Anderson, R. M., and May, R. M. Population biology of infectious diseases: Part I. *Nature 280*, 5721 (Aug. 1979), 361–367.
3. Annas, S., Isbar Pratama, M., Rifandi, M., Sanusi, W., and Side, S. Stability analysis and numerical simulation of SEIR model for pandemic COVID-19 spread in Indonesia. *Chaos, Solitons Fractals 139* (Oct. 2020), 110072.
4. Avila-Vales, E., and Pérez, A. G. C. Dynamics of a reaction–diffusion SIRS model with general incidence rate in a heterogeneous environment. *Zeitschrift für angewandte Mathematik und Physik 73*, 1 (Nov. 2021).
5. Chen, Y. C., Lu, P. E., Chang, C. S., and Liu, T. H. A Time-Dependent SIR Model for COVID-19 With Undetectable Infected Persons. *IEEE Transactions on Network Science and Engineering 7*, 4 (Oct. 2020), 3279–3294.
6. Cui, J., Sun, Y., and Zhu, H. The impact of media on the control of infectious diseases. *Journal of Dynamics and Differential Equations 20*, 1 (May 2007), 31–53.

7. Duan, X., Yuan, S., and Li, X. Global stability of an SVIR model with age of vaccination. *Applied Mathematics and Computation 226* (Jan. 2014), 528–540.
8. Groeger, J. Divergence theorems and the supersphere. *Journal of Geometry and Physics 77* (Mar. 2014), 13–29.
9. Guenther, R. B., and Lee, J. W. *Partial differential equations of mathematical physics and integral equations.* Courier Corporation, 1996.
10. Jamiluddin, M. S., Mohd, M. H., Ahmad, N. A., and Musa, K. I. Situational analysis for COVID-19: Estimating transmission dynamics in Malaysia using an SIR-type model with neural network approach. *Sains Malaysiana 50*, 8 (Aug. 2021), 2469–2478.
11. Kumar, A., and Nilam. Stability of a time delayed SIR epidemic model along with nonlinear incidence rate and holling type-II treatment rate. *International Journal of Computational Methods 15*, 06 (Sept. 2018), 1850055.
12. La Salle, J. P. *The stability of dynamical systems.* Society for Industrial and Applied Mathematics, Jan. 1976.
13. Luo, Y., Zhang, L., Zheng, T., and Teng, Z. Analysis of a diffusive virus infection model with humoral immunity, cell-to-cell transmission and nonlinear incidence. *Physica A: Statistical Mechanics and its Applications 535* (Dec. 2019), 122415.
14. Marinov, T. T., and Marinova, R. S. Adaptive SIR model with vaccination: Simultaneous identification of rates and functions illustrated with COVID-19. *Scientific Reports 12*, 1 (Sept. 2022).
15. Martin, R. H., and Smith, H. L. Abstract functional differential equations and reaction-diffusion systems. *Transactions of the American Mathematical Society 321*, 1 (Sept. 1990), 1.
16. Mohd, M. H. *Numerical bifurcation and stability analyses of partial differential equations with applications to competitive system in ecology.* Springer Singapore, 2019, pp. 117–132.
17. Pazy, A. *Semigroups of linear operators and applications to partial differential equations.* Springer New York, 1983.
18. Ren, X., Tian, Y., Liu, L., and Liu, X. A reaction–diffusion within-host HIV model with cell-to-cell transmission. *Journal of Mathematical Biology 76*, 7 (Jan. 2018), 1831–1872.

19. Salman, A. M., Ahmed, I., Mohd, M. H., Jamiluddin, M. S., and Dheyab, M. A. Scenario analysis of COVID-19 transmission dynamics in Malaysia with the possibility of reinfection and limited medical resources scenarios. *Computers in Biology and Medicine 133* (June 2021), 104372.
20. Sidi Ammi, M. R., Tahiri, M., and Torres, D. F. M. Global stability of a Caputo fractional SIRS model with general incidence rate. *Mathematics in Computer Science 15*, 1 (Mar. 2020), 91–105.
21. Sidi Ammi, M. R., Zinihi, A., Raezah, A. A., and Sabbar, Y. Optimal control of a spatiotemporal $\mathcal{SIR}$ model with reaction–diffusion involving p-laplacian operator. *Results in Physics 52* (Sept. 2023), 106895.
22. Smith, H. L. *Monotone dynamical systems: An introduction to the theory of competitive and cooperative systems.* American Mathematical Society, Mar. 1995.
23. Wang, J., Zhang, R., and Kuniya, T. A reaction–diffusion Susceptible–Vaccinated–Infected–Recovered model in a spatially heterogeneous environment with Dirichlet boundary condition. *Mathematics and Computers in Simulation 190* (Dec. 2021), 848–865.
24. Yang, B., Yu, Z., and Cai, Y. The impact of vaccination on the spread of COVID-19: Studying by a mathematical model. *Physica A: Statistical Mechanics and its Applications 590* (Mar. 2022), 126717.
25. Zhang, C., Gao, J., Sun, H., and Wang, J. Dynamics of a reaction–diffusion SVIR model in a spatial heterogeneous environment. *Physica A: Statistical Mechanics and its Applications 533* (Nov. 2019), 122049.
26. Zhou, Y., Xiao, D., and Li, Y. Bifurcations of an epidemic model with non-monotonic incidence rate of saturated mass action. *Chaos, Solitons Fractals 32*, 5 (June 2007), 1903–1915.
27. Zinihi, A., Sidi Ammi, M. R., and Ehrhardt, M. *Optimal control of a diffusive epidemiological model involving the caputo-fabrizio fractional time-derivative*, 2024. DOI: 10.48550/arXiv.2403.00364

CHAPTER 16

Gauss-Newton Methods for Convex Composite Optimization under Generalized Continuity Conditions

Ioannis K. Argyros, Santhosh George, and Michael Argyros

16.1 INTRODUCTION

A plethora of applications in programming, such as the penalization methods, nonlinear inclusions, and goal and minimax programming, can be reduced using mathematical modeling to solve the optimization problem

$$\min H(F(x)). \tag{1.1}$$

Here $H : \mathbb{R}^m \longrightarrow \mathbb{R}$ is convex, and $F : \mathbb{R}^k \longrightarrow \mathbb{R}^m$ is a differentiable mapping. The study of (1.1) is connected to the convex inclusion problem

$$F(x) \subset K = argumin\ H. \tag{1.2}$$

Notice that if $x^* \in \mathbb{R}^k$ satisfies (1.2), then x^* solves (1.1). However, if $x^* \in \mathbb{R}^k$ solves (1.1), it does not necessarily solve (1.2) [1, 5, 6, 8, 9, 10, 11, 12, 17, 18, 19, 20, 21, 22].

A semi-local analysis (SLA) is presented in [7] for the Gauss-Newton algorithm, which generates a sequence converging to x^* solving (1.1). The

DOI: 10.1201/9781003530602-16

convergence conditions involve majorant conditions used to control the derivative and the assumption that the initial point is quasi-regular for (1.2). Relevant work can be found in [13, 14, 15, 16], where the majorant function is replaced by Wang's condition [16].

The objective in this study is twofold: On one hand, we provide an SLV based on more general but tighter majorant functions than the ones considered in [7, 13]. On the other hand, we introduce hybrid Gauss-Newton algorithms in order to address some difficulties with the application of the Gauss-Newton algorithm (GNA), referring to our works in [2, 3, 4].

16.2 PRELIMINARIES

The symbols $S(x,\rho)$, $S[x,\rho]$ are used for the open and closed balls, respectively, centered at $x \in \mathbb{R}^k$ with radius $\rho > 0$. Let $V \subseteq \mathbb{R}^k$ be closed as well as convex. Then, the polar of V is

$$V^\circ = \left\{y \in \mathbb{R}^k : \langle y, v\rangle \leq 0, v \in V\right\}.$$

The distance from x to V is defined as $d(x, V) = inf\{\| x - v \| : v \in V\}$. The set $T(\mathbb{R}^k)$ stands for all subsets of $\mathbb{R}^k$, while $Ker(L)$ is the kernel of a linear operator L. If $P \subseteq \mathbb{R}^k$ is a vector subspace, then $dim(P)$ stands for its dimension. Furthermore, if $w \in \mathbb{R}^k$, then $w^\perp = \left\{u \in \mathbb{R}^k : \langle u, w\rangle = 0\right\}$. Finally, if $x \in \mathbb{R}^k$ and $E \in T(\mathbb{R}^k)$, then

$$y + E = \{y + e : e \in E\}.$$

Next, we define the GNA for solving (1.1): If $F \in C^1(\mathbb{R}^k, \mathbb{R}^m)$, $\delta \in (0, +\infty)$, and $x \in \mathbb{R}^k$, consider

$$D_\delta(x) = argmin\left\{H(F(x) + F'(x)d) : d \in \mathbb{R}^k, \|d\| \leq \delta\right\}. \tag{2.3}$$

The $D_\delta(x)$ is a solution set for

$$\min\left\{H(F(x) + F'(x)d) : d \in \mathbb{R}^k, \|d\| \leq \delta\right\}. \tag{2.4}$$

Take $x_0 \in \mathbb{R}^k$ as an initial point and $\lambda \in [1, +\infty)$ be fixed. Then, the GNA as considered in [7] is defined as follows:

Algorithm 1: Step 1. Select $\delta \in (0, +\infty]$, $\lambda \in [1, +\infty)$, and $x_0 \in \mathbb{R}^k$. Take $m = 0$. Step 2. Find $D_\delta(x)$. If $0 \in D_\delta(x)$, STOP. Otherwise go to Step 3. Select d_m satisfying

$$d_m \in D_\delta(x_m), \|d_m\| \leq \lambda d(0, D_\delta(x_m)),$$

then take

$$x_{m+1} = x_m + d_m,$$

$m = m + 1$ and go to STOP criterion.

16.3 ANALYSIS

The SLA depends on some conditions.

Definition 3.1 Let $x_0 \in \mathbb{R}^k, F \in C^1(\mathbb{R}^k, \mathbb{R}^m), \rho > 0$, and $M \in \mathcal{L}(\mathbb{R}^k, \mathbb{R}^m)$. We say that the operator F' satisfies the center-majorant condition on the ball $S(x_0, \rho)$ if there exists a continuous and nondecreasing function (CNDF) $h_0 : [0, \rho) \longrightarrow [0, +\infty), \alpha > 0$ with

$$\|F'(x) - M\| \leq \alpha h_0(\|x - x_0\|), \; x \in S(x_0, \rho).$$

Assume $\alpha h_0(t) - 1 = 0$ has solutions in $(0, \rho)$. Let ρ_0 be the smallest zero in $(0, \rho)$.

Definition 3.2 Let $x_0 \in \mathbb{R}^k, F \in C^1(\mathbb{R}^k, \mathbb{R}^m)$ and ρ_0 be as in Definition 3.1. The operator F' satisfies the restricted-majorant condition on the ball $S(x_0, \rho_0)$ if there exists CNDF $h : [0, \rho_0) \longrightarrow [0, +\infty), \alpha > 0$ with

$$\|F'(y) - F'(x)\| \leq \alpha h(\|y - x\|), \; x, y \in S(x_0, \rho_0).$$

Definition 3.3 Let $x_0 \in \mathbb{R}^k, F \in C^1(\mathbb{R}^k, \mathbb{R}^m)$ and ρ be as in Definition 3.1. The operator F' satisfies the majorant condition on the ball $S(x_0, \rho)$ if there exists CNDF $h_1 : [0, \rho) \longrightarrow [0, +\infty), \alpha > 0$ with

$$\|F'(y) - F'(x)\| \leq \alpha h_1(\|y - x\|), \; x, y \in S(x_0, \rho).$$

Remark 3.4 These definitions imply

$$h_0(t) \leq h_1(t) \tag{3.5}$$

and

$$h(t) \leq h_1(t) \tag{3.6}$$

for $t \in [0, \rho_0)$. *We shall assume that*

$$h_0(t) \leq h(t), t \in [0, \rho_0). \tag{3.7}$$

If not, the results hold with $\bar{h}$, *where* $\bar{h}$ *is the largest of the functions* h_0 *and* h *on the interval* $[0, \rho_0)$. *The function* h_1 *can specialize as follows:*

Case 1

$$h_1(\|v-u\|) = h_2'(\|v-u\| + \|u-x_0\|) - h_2'(\|u-x_0\|),$$

where $h_2 \colon [0, \rho) \longrightarrow [0, +\infty)$ *is a twice-continuously-differentiable function. Such a choice for a majorant function is considered in [7]. The function* h_2 *must satisfy some additional conditions.*

Case 2

$$h_1(\| v-u) = \int_{\|v-u\|}^{\|u-v\|+\|v-x_0} L(x)\,dx, \ \ u, v \in S(x_0, \rho),$$

where $L \colon [0, \rho) \longrightarrow [0, +\infty)$ *is a continuous function. This is Wang's condition employed in [13]. Other choices for* h_0, h *and* h_1 *are possible. A possible choice for the operator is* $M = I$, *the identity operator, or* $M = F'(x_0)$ *or* $M = F'(\tilde{x})$, *where* $\tilde{x} \in \mathbb{R}^k$ *is an auxiliary point. Such a choice is considered in [6]. In view of the above discussion, it is worth studying the semi-local convergence analysis for Algorithm 1 with the function* h *replacing* h_2 *or* h_3. *A crucial role is played by the concept of regularity. Let* K *be the set of all minimum points of the function* H. *If* $F \in C^1(\mathbb{R}^k, \mathbb{R}^m)$ *and* $u \in \mathbb{R}^k$, *consider the set* $D_k(u)$ *to be*

$$D_k(u) = \{d \in \mathbb{R}^k : F(u) + F'(u)d \in K\}.$$

There is a connection between $D_\delta(u)$ *and* $D_K(u)$.

Lemma 3.5 *If* $D_K(u) \neq \emptyset$ *and* $d(0, D_K(u)) \leq \delta$, *then*

$$D_\delta(u) = \{d \in \mathbb{R}^k : \| d \leq \delta, F(u) + F'(u)d \in K\} \subseteq D_K(u)$$

and

$$d(0, D_\delta(u)) = d(0, D_K(u)).$$

The proof is given in [7, Proposition 2.18].

The following definition has been used in the study of the Gauss-Newton method [7].

Definition 3.6 Let $F \in C^1\left(\mathbb{R}^k, \mathbb{R}^m\right)$ and let $H: \mathbb{R}^m \longrightarrow \mathbb{R}$ be as previously defined. We say $x_0 \in \mathbb{R}^k$ to be a quasi-regular for (1.2), i.e.

$$F(u) \in K = agrminH = \{x \in \mathbb{R}^m : H(x) \leq H(u), u \in \mathbb{R}^m\}.$$

If $\rho \in (0, +\infty)$ *exists as well as a function* $\lambda: [0, \rho) \longrightarrow (0, +\infty)$ *which is increasing and positive-valued so that*

$$D_K(u) \neq \emptyset, \ d(0, D_K(u)) \leq \lambda(\|u - x_0\|) \, d(F(u), K) \tag{3.8}$$

$u \in S(x_0, \rho)$, then ρ_{x_0} is the supremum of ρ for which (3.8) holds for $\lambda: [0, \rho) \longrightarrow [0, +\infty)$ which is increasing and positive-valued, i.e.

$$\rho_{x_0} = sup\{\rho : \text{there exists } \lambda : [0, \rho) \longrightarrow [0, +\infty) \text{ satisfying (3.8)}\}. \tag{3.9}$$

Let $\rho \in \left[0, \rho_{x_0}\right)$. The set $E(x_0, r)$ is defined as

$$E(x_0, r) = \{\lambda : \lambda : [0, \rho) \longrightarrow [0, +\infty) \text{ satisfying (3.8)}\}.$$

Let us also define

$$\lambda_{x_0} = inf\left\{\lambda(t) : \lambda \in E\left(x_0, \rho_{x_0}\right)\right\},$$

for all $t \in \left(0, \rho_{x_0}\right)$.

Definition 3.7 [7] We say $x_0 \in \mathbb{R}^k$ is regular for $F(x) \in K$ provided

$$Ker\left(F'(x_0)^T\right) \cap (K - F(x_0))^\circ = \{0\}.$$

Lemma 3.8 [7] Let $x_0 \in \mathbb{R}^k$ be regular for $F(u) \in K$. Then, there exist $\rho > 0$ and $\lambda > 0$ with

$$D_K(u) \neq \emptyset, d(0, \ D_K(u)) \leq \lambda d(F(u), K)$$

for all $u \in S(x_0, \rho)$.

Thus, we have $\rho_{x_0} \geq \rho$ and $\lambda_{x_0} \leq \lambda$.

Define the scalar sequence $\{s_n\}$ for $s_0 = 0, s_1 = \xi$ for some $\xi \in [0, \rho_0)$, and each $n = 0, 1, 2, ...$ by

$$s_{n+2} = s_{n+1} + \frac{\alpha \int_0^1 h((1-\theta)(s_{n+1} - s_n))\, d\theta\, (s_{n+1} - s_n)}{1 - \alpha h_0(s_{n+1})}. \tag{3.10}$$

The sequence $\{s_n\}$ is shown to be majorizing for $\{x_n\}$ in Theorem 3.9. But first let us develop a convergence condition for it. In condition (C), there exists $\rho_1 \in [\xi, \rho_0)$ such that for each $n = 0, 1, 2, ...$

$$\alpha h_0(s_n) < 1 \text{ and } s_n \leq \rho_1.$$

This condition and (3.10) imply

$$0 \leq s_n \leq s_{n+1} \leq \rho_1,$$

and there exists $s^* \in [\xi, \rho_1]$ such that $lim_{n \longrightarrow +\infty} s_n = s^*$. In the case when the function h_0 is strictly increasing, we can take $\rho_1 = h_0^{-1}\left(\frac{1}{\alpha}\right)$.

Theorem 3.9 *Let $F \in C^1(\mathbb{R}^k, \mathbb{R}^m)$. Suppose: $\rho > 0, x_) \in \mathbb{R}^k$, and h_0, h are as in Definition 3.1 and 3.2, respectively; the condition (C) holds.*

For $\lambda \in [1, +\infty), \delta \in (0, +\infty), H : \mathbb{R}^m \longrightarrow \mathbb{R}$ as before with minimizer set $K \neq \emptyset$, the initial point $x_0 \in \mathbb{R}^k$ is quasi-regular for $F(x) \in K$ with ρ_{x_0}, and λ_{x_0} as previously defined:

$$d(F(x_0), K) > 0, s^* \leq \rho_{x_0}, \tag{3.11}$$

$$\delta \geq \xi \geq \lambda \lambda_{x_0}(0)\, d(F(x_0), K) \tag{3.12}$$

and

$$\alpha \geq sup\left\{\frac{\lambda \lambda_{x_0}(t)}{\lambda \lambda_{x_0}(t)\, h_0(t) + 1} : \xi \leq t < s^*\right\}. \tag{3.13}$$

Then, the sequence $\{x_n\}$ produced by Algorithm 1 exists in $S(x_0, s^*)$

$$F(x_n) + F'(x_n)(x_{n+1} - x_n) \in K, \; n = 0, 1, 2, ... \tag{3.14}$$

Moreover, it converges to some $x^* \in S[x_0, s^*]$ satisfying $F(x^*) \in K$ so that

$$\|x_{n+1} - x_n\| \leq s_{n+1} - s_n$$

and

$$\|x^* - x_n\| \leq s^* - s_n.$$

Proof. Simply replace the function h_1 by (h_0, h) in Theorem 3.1 in [7]. □

Remark 3.10 Although the functions (h_0, h) are more general than h_2 or h_3 in view of (3.5) and (3.6), the majorizing sequences $\{s_n\}$ are tighter than the corresponding ones in [7] even if they are chosen as in Remark 3.4. Consequently, the new convergence criteria are weaker.

16.4 THE IMPLEMENTATION OF ALGORITHM 1

The computation of the iterates $\{x_n\}$ of Algorithm 1 requires expansion of the inversion $F'(x_n)^{-1}$. We developed hybrid methods in [2, 4] for solving nonlinear equations like $F(x) = 0$. The method is

$$F(x_n) + F_1(x_n)(x_{n+1} - x_n) = 0, \tag{4.15}$$

where for $\Delta = M^{-1}(M - F'(u))$,$B = B_j = I + \Delta + \dots\Delta^j$, where M is an invertible operator, j a natural number, and $F_1 = MB^{-1}$. Notice that $B_\infty M^{-1} = \lim_{j \longrightarrow +\infty} B_j M^{-1} = F'$, provided that this limit exists. One such condition can be $\|\Delta\| < 1$. We define hybrid methods to replace (4.15) and use the analogy of Algorithm 1 instead. Let us choose $M = F'(x_0)$ for simplicity. As in [2, 4], define the parameters

$$q = q(\|u - x_0\|) \geq \|\Delta\|,$$

$$\bar{b}_j = q\frac{1-q^j}{1-q}, \quad b_j = \frac{1}{1-\bar{b}_j}, \quad b_\infty = \frac{1-q}{1-2q}$$

and

$$\mu(\|u - x_0\|) = \mu_j(\|u - x_0\|) = b_j b_\infty \frac{q^{j+1}}{1-q}.$$

Define the sequence $\{w_n\}$ for $w_0 = 0, w_1 = \xi$,

$$w_2 = \frac{\alpha\int_0^1 h_0((1-\theta)w_1)\,d\theta w_1 + \mu(w_1)}{1-\alpha h_0(w_1)},$$

$$w_{n+2} = w_{n+1} + \frac{\alpha\int_0^1 h_0((1-\theta)(w_{n+1}-w_n))\,d\theta(w_{n+1}-w_n) + \mu(w_{n+1})}{1-\alpha h_0(w_{n+1})}. \tag{4.16}$$

The sequence $\{w_n\}$ is shown to be majorizing for the sequence $\{x_n\}$ generated as follows. First, as in (2.3),

$$D_{\delta}^{1}(u) = argmin\{H(F(u) + F_1(u)d) : d \in \mathbb{R}^k, \|d\| \leq \delta\}.$$

Algorithm 2: Step 1. Let $\delta \in (0, +\infty), \lambda \in [1, +\infty)$, and $x_0 \in \mathbb{R}^k$. Take $m = 0$. Step 2. Find $D_{\delta}^{1}(u)$. If $0 \in D_{\delta}^{1}(x_m)$, STOP. Otherwise, proceed to Step 3. Select d_m satisfying

$$d_m \in D_{\delta}(x_m), \ \|d_m\| \leq \lambda d(0, D_{\delta}(x_m)),$$

and take

$$x_{m+1} = x_m + d_m$$

$m = m + 1$ and go to STOP criterion.

Let us provide a stopping criterion for the sequence $\{x_n\}$. For condition (C1), suppose the equation $\alpha h_0(t) - 1 = 0$ has a smallest positive solution. Denote it by ρ_2.

There exists $\rho_3 \in [\xi, \rho_2)$ such that $\alpha h_0(w_n) < 1$ and $w_n \leq \rho_3$. It follows, as in condition (C), that

$$0 \leq w_n \leq w_{n+1} \leq \rho_3,$$

and there exists $w^* \in \xi, \rho_3]$ such that $\lim_{n \longrightarrow +\infty} w_n = w^*$. Then, we can show the analogy of Theorem 3.9 using Algorithm 1.

Theorem 4.1 *Suppose that conditions (3.11)–(3.13) of Theorem 3.9 hold with* $w^*, D_{\delta}^{1}(x)$ *replacing* $s^*, D_{\delta}(x)$ *,respectively and* $P \in \left[0, \frac{1}{2}\right)$. *Then, the sequence* $\{x_n\}$ *given by Algorithm 2 exists in* $S(x_0, w^*)$,

$$F(x_n) + F'(x_n)(x_{n+1} - x_n) \in K, \ n = 0, 1, 2, ... \tag{4.17}$$

Moreover, it converges to $x^* \in S[x_0, w^*]$ *so that* $F(x^*) \in K$. *Furthermore, the assertions hold*

$$\|x_{n+1} - x_n\| \leq w_{n+1} - w_n$$

and

$$\|x^* - x_n\| \leq w^* - w_n.$$

Proof. We need the estimates for $u \in S(x_0, w^*)$

$$\begin{aligned} \|B_j - I\| &\leq \|\Delta\| + \|\Delta\|^2 + \ldots + \|\Delta\|^j \\ &\leq q + q^2 + \ldots q^j \\ &= q\frac{1-q^i}{1-q} = \bar{b}_n < 1, \end{aligned}$$

(since $q \in \left[0, \frac{1}{2}\right)$). So, $B_j^{-1} \in \mathcal{L}(Y, X)$ and

$$\|B_j^{-1}\| \leq \frac{1}{1-\bar{b}_j} = b_j.$$

Then, we can write

$$\begin{aligned} F_1(u) &= F'(x_0)\, B_j^{-1}(u) \\ &= F'(x_0)\left(B_j^{-1}(u) - B_\infty^{-1}(u) + B_\infty^{-1}(u)\right) \\ &= F'(x_0)\, B_\infty^{-1}(u) + F'9x_0)(B_j^{-1}(u) - B_\infty^{-1}(u)) \\ &= F'(x_0)\, B_\infty^{-1}(u) + F'(x_0)\, B_j^{-1}(u)\left(B_\infty(u) - B_j(u)\right) B_\infty^{-1}(u). \end{aligned} \tag{4.18}$$

But we get

$$\begin{aligned} \|B_j^{-1}(u)\left(B_\infty(u) - B_j(u)\right) B_\infty^{-1}(u)\| &\leq b_j\left(q^{j+1} + \ldots\right) b_\infty = b_k b_\infty \frac{q^{j+1}}{1-q} \\ &= \mu_j\left(\|u - x_0\|\right). \end{aligned} \tag{4.19}$$

By switching F with F_1 in Theorem 3.9 and using (4.16), (4.18), and (4.19), we obtain

$$\|x_{n+1} - x_n\| \leq w_{n+1} - w_n \tag{4.20}$$

leading to the existence of $x^* \in S[x_0, w^*]$ solving $F(x^*) \in K$. It follows from (4.20) that for each $i = 0, 1, 2, \ldots$

$$\|x_{n+i} - x_n\| \leq w_{n+i} - w_n. \tag{4.21}$$

By letting $i \longrightarrow +\infty$ in (4.21) we complete the proof. □

Remark 4.2 If followed by the choice of P, (3.10) and (4.16), that as $j \longrightarrow +\infty$, the two sequences coincide.

16.5 CONCLUSION

In this chapter, a new SLV is presented for GNV to solve inclusion problems defined in Euclidean space. The analysis is based on generalized continuity assumptions used to control the derivative. Although the majorant functions are more general than the ones used in earlier studies, even if specialized to the earlier, they generate tighter majorizing sequences and weaker convergence criteria. The benefits are derived under the same computational effort because the new majorant functions can be specializations of earlier ones. We also developed hybrid Gauss-Newton methods to address the implementation problem of GNA.

REFERENCES

1. I. K. Argyros, *The theory and applications of iteration methods with applications*, Engineering Series, 2nd Edition, CRC Press, Taylor and Francis Publisher, Boca Raton, FL, 2022.
2. I. K. Argyros, S. George, S. Shakhno, S. Regmi, M. Havdiak, and M. I. Argyros, Asymptotically Newton-type methods without inverses for solving equations. *Mathematics,* 12 (2024), p. 1069. https:// doi.org/10.3390/math12071069.
3. I. K. Argyros and S. George, On a unified convergence analysis for Newton-type methods solving generalized equations with the Aubin property, *J. Complexity*, 81 (2024), p. 101817.
4. I. K. Argyros, S. George, S. Regmi, and C. I. Argyros, Hybrid Newton-like inverse free algorithms for solving nonlinear equations, *Algorithms*, 17 (2024), p. 154. https:// doi.org/10.3390/a17040154
5. J. V. Burke and R. A. Poliquin, Optimality conditions for non-finite valued convex composite function, *Math. Programming*, 57 (1992), pp. 103–120.
6. J.A. Ezquerro and M. A. Hernamdez-Veron, Domain of global convergence for Newton's method from auxiliary points, *Appl. Math. Lett.*, 85 (2018), pp. 48–56.
7. O. P. Ferreira, M. L. N. Goncalves, and P. R. Oliveira, Convergence of the gauss–newton method for Convex composite optimization under a majorant condition, *SIAM J. OPTIM.*, 23, 3, 1757–1783, https://doi.org/10.1137/11084160

8. O. P. Ferreira, M. L. N. Goncalves, and P. R. Oliveira, Local convergence analysis of the Gauss-Newton method under a majorant condition, *J. Complexity*, 27 (2011), pp. 111–125.
9. O. P. Ferreira, M. L. N. Gonc¸alves, and P. R. Oliveira, Local convergence analysis of inexact Gauss-Newton like methods under majorant condition, *J. Comput. Appl. Math.*, 236 (2012), pp. 2487–2498.
10. O. P. Ferreira and B. F. Svaiter, Kantorovich's theorem on Newton's method in Riemannian manifolds, *J. Complexity*, 18 (2002), pp. 304–329.
11. O. P. Ferreira and B. F. Svaiter, Kantorovich's majorants principle for Newton's method, *Comput. Optim. Appl.*, 42 (2009), pp. 213–229.
12. J.-B. Hiriart-Urruty and C. *Lemar´echal, Convex Analysis and Minimization Algorithms. I*, Grundlehren Math. Wiss. 305, Springer-Verlag, Berlin, 1993.
13. C. Li and K. F. Ng, Majorizing functions and convergence of the Gauss-Newton method for convex composite optimization, *SIAM J. Optim.*, 18 (2007) pp. 613–642.
14. C. Li and K. F. Ng, Convergence analysis of the Gauss-Newton method for convex inclusion and convex-composite optimization problems, *J. Math. Anal. Appl.*, 389 (2012), pp. 469–485.
15. C. Li and X. Wang, On convergence of the Gauss-Newton method for convex composite opti- mization, *Math. Program.*, 91 (2002), pp. 349–356.
16. C. Li and X. H. Wang, On convergence of the Gauss-Newton method for convex composite optimization, *Math. Program.*, 91 (2002), pp. 349–356
17. S. M. Robinson, Extension of Newton's method to nonlinear functions with values in a cone, *Numer. Math.*, 19 (1972), pp. 341–347.
18. R. T. Rockafellar, *Convex Analysis*, Princeton Math. Ser. 28, Princeton University Press, Princeton, NJ, 1970.
19. R. T. Rockafellar, *Monotone Processes of Convex and Concave Type*, Memoirs of the American Mathematical Society, 77, AMS, Providence, RI, 1967.
20. G. W. Stewart, On the continuity of the generalized inverse, *SIAM J. Appl. Math.*, 17 (1969), pp. 33–45.

21. P. Wedin, Perturbation theory for pseudo-inverse, *BIT*, 13 (1973), pp. 217–232.
22. R. S. Womersley, Local properties of algorithms for minimizing nonsmooth composite function, *Math. Programming*, 32 (1985), pp. 69–89

Index

Kolibri 360 GmbH
Ferdinand-Jühlke-Straße 7
99095 Erfurt, Deutschland
produktsicherheit@kolibri360.de